AF675432

AIRPORT, AIRCRAFT, AND AIRLINE SECURITY

Second Edition

AIRPORT, AIRCRAFT, AND AIRLINE SECURITY

Second Edition

Kenneth C. Moore

Butterworth-Heinemann

Boston London Oxford Singapore Sydney Toronto Wellington

Copyright © 1991 by Butterworth-Heinemann, a division of Reed Publishing (USA) Inc.
All rights reserved.

No part of this publication may be reproduced, stored in a retrieval system, or transmitted, in any form or by any means, electronic, mechanical, photocopying, recording, or otherwise, without the prior written permission of the publisher.

Recognizing the importance of preserving what has been written, it is the policy of Butterworth-Heinemann to have the books it publishes printed on acid-free paper, and we exert our best efforts to that end.

Library of Congress Cataloging-in-Publication Data
Moore, Kenneth C., 1934-
Airport, aircraft, and airline security / Kenneth C. Moore. -- 2nd ed.
p. cm.
Includes index.
ISBN 0-7506-9019-4
1. Airports--Security measures. 2. Hijacking of aircraft--Prevention. 3. Aeronautics, Commercial--Freight--Security measures. 4. Aeronautics, Commercial--Security measures.
I. Title.
HE9797.4.S4M66 1991
387.7'364--dc20 91-17003
CIP

British Library Cataloguing in Publication Data
Moore, Kenneth C.
Airport, aircraft, and airline security.
I. Title
387.74042

ISBN 0750690194

Butterworth-Heinemann
80 Montvale Avenue
Stoneham, MA 02180

10 9 8 7 6 5 4 3 2 1

Printed in the United States of America

To
Steve and Michelle
Mary Kay and Jim
and the precious gifts
they have given Betty and me

Morgan
Nicole
Jeffrey
Taylor
Meagan

and perhaps
other grandchildren
yet to come.

Praise God
from whom
all blessings flow.

CONTENTS

PREFACE

The first edition of this book was published in October 1976. Since then, the aviation industry has seen many changes. Some changes have been subtle. For example, the airlines have introduced new programs. Fifteen years ago, no one ever heard of frequent flyer programs or denied boarding compensation. In addition, roles have changed. While men have entered the domain of the flight attendant, women have entered the cockpit.

Other changes have been dramatic. In the past few years, the carriers who have disappeared from the scene are too numerous to mention; others are now barely hanging on. Of 150 airlines started after 1978, the first year of deregulation, 118 have failed or merged with other carriers.[1] Even the venerable Civil Aeronautics Board disappeared from the domestic scene with deregulation. After deregulation, carriers no longer filed for just one or two new destinations a year; their expansion was dramatic--too dramatic in some instances. Some carriers bit off more than they could chew.

New innovations in the security arena, such as computerized access control, video badging, fiber-optics, bar coding, and biometrics for personal identification and verification, are being applied to the aviation industry.

Profit from drugs has accelerated and exacerbated the problem of crime. School children in Seattle, Washington, and Oakland and Los Angeles, California were accustomed to "duck and cover" drills for earthquakes. Now, the same drills, called "gang-drop drills" or "bullet drills" are being used to protect school children from drive-by shootings.[2] An article in

---

1 "The Best (and Worst) Airlines," *Consumer Reports* (July 1991):465.

2 Seth Mydans, *New York Times*, seen in *The Maui News*, 16 June 1991.

the *New York Times* advised that manufacturers were putting bulletproof Kevlar™ material in many back-to-school items, including blazers, clipboards, and book bags, in an effort to protect New York school children from stray bullets.[3]

The threat to aviation security has changed dramatically. The previous emphasis on hijacking has been shifted to bombings and explosives. Perhaps most important, airline and airport security has assumed a new place of prominence. Airlines have taken out full-page advertisements in major newspapers to promote their security.[4] Some air carriers now market security more strongly than convenient schedules, reduced fares, elaborate meal service, advance check-in, preassigned seats, frequent flyer clubs, and all the other competitive frills.

It is indeed security, or the perceived lack of it, that has hurt air carriers, particularly transatlantic traffic. Because of the threat of terrorism from the Iraqis in early 1991, air travel declined and carriers laid off thousands of employees. First Lady Barbara Bush flew domestically in February 1991 to prove that no one should be afraid to fly. While I was Manager of Corporate Security at Apple Computer, I, rather than the Manager of Travel--was approached by a U.S. flag carrier to make a presentation on the security provided its flights. The marketing emphasis had shifted because it was the Security Managers who were advising their employees to fly or not to fly American flag carriers on transatlantic routes.

Security measures at airports have recently been at the highest level in the history of civil aviation. The challenge ahead for the aviation security professional demands the latest technology, the greatest of innovative thinking, and a high degree of cooperation between the private and public sector and among governments.

I want to extend my thanks to those who have contributed valuable assistance in taking this book from a raw manuscript to the finished product. The editorial staff of Butterworth-Heinemann; Gregory Franklin, Senior Editor, and Mary Cervantes, Production Editor, have been extremely helpful and patient with me. The editor's job is a difficult one. I count myself fortunate in having had good editors who not only made grammatical corrections, but challenged my thinking, wanted to know why some subjects were included and why others were not, suggested changes in format and organization, and in many ways made the book a better product than it ever would have been without their input. Walter H. Copley, PhD, Associate Professor, Criminal Justice and Criminology, Metropolitan State College, Denver, read the manuscript and made numerous helpful comments designed to make it a more helpful text and reference book. I am genuinely appreciative of his help.

Kenneth C. Moore
Dublin, California

---

3 *New York Times*, September 19, 1990.

4 *San Francisco Chronicle* (May 20, 1986):7
New York Times (May 16, 1990):A11

Section I

ATTACKS AGAINST AIR TRAFFIC

Chapter 1

THE HIJACKING THREAT

No one particularly noticed the man as he boarded Northwest Orient Flight 305 at Portland. He wore a dark suit and tie, carried a briefcase, and had a dark topcoat to protect him from the Oregon winter--he looked like any businessman on any flight. Besides, November 24, 1971, was the day before Thanksgiving, and travelers' minds were on holiday reunions, family, and friends. The unobtrusive man chose a rear seat on the right side of the Boeing 727. The flight's 36 passengers only partially filled the plane.

Just after 3:20 P.M., as the plane was taxiing for takeoff, the man, who had given the name of Dan Cooper, handed a note to the flight attendant, telling her, "Read it now." The note said, "I have a bomb in my briefcase."

The plane took off and headed north toward Seattle. Cooper announced that he would blow up the plane unless he was given $200,000 and four parachutes. He backed up his threat by allowing two of the flight attendants to look inside the briefcase, where they saw what appeared to be dynamite and wires.

At Seattle, Cooper received the money and parachutes. In turn, he allowed all the passengers and two of the flight attendants to deplane--37 of the lives at risk were now safe. The pilot took off again, following Cooper's instructions to fly to Reno at low altitude and minimum cruising speed with flaps down and ventral stairs extended. The crew was told to stay in the front cabin with the curtain closed. When the plane landed at Reno, Cooper was not aboard. The only indication as to when Cooper might have left the plane was a slight shift in control pressures felt when the plane was in the vicinity of Woodland, Washington.

EVOLUTION OF "SKYJACKING"

Cooper's feat was neither the first nor the last hijacking of an aircraft, but it is memorable because it introduced the parachuting hijacker and because "Dan Cooper" remains at large and unidentified. The earliest known act of air piracy occurred in 1930, when hijackers took over a Peruvian aircraft. In the late forties, numerous incidents were reported of individuals escaping or attempting to escape from behind the Iron Curtain by means of hijacked planes, trains, and other forms of transportation. In the years from 1930 to 1967, only 12 U.S. hijackings were attempted, seven of which were successful.

In 1968, however, a deluge of hijackings began. Many Americans had the frightening experience of being detoured to Havana's Jose Marti Airport before arriving at their original destinations. Of the 22 hijackings that occurred in the U.S. that year,, 19 were directed to fly to Cuba. The next year began with an even faster hijacking rate; March 5, 1969, marked the eighteenth Havana diversion of the new year. In 64 days, 1359 U.S. passengers and crew members had made the Cuban detour.

This was the peak, however. After 1969, trips to Cuba began to taper off, representing 55 and 52 percent of 1970 and 1971 hijackings respectively, and only 22 percent of the 1972 total.[1]

The decline of Cuba's popularity as a destination began after reports reached the United States that described the hijackers' lot in Cuba--an armed reception committee, months of interrogation and imprisonment, riots, suicides, guards emptying their weapons into occupied cells, injections, and a house in Havana where about one-third of the American hijackers were kept under close arrest. According to the reports, almost all of the hijackers had expressed extreme disillusionment with their chosen refuge.[2]

A stark demonstration of large-scale hijacking began on September 6, 1970, when terrorists hijacked a TWA 707 on a flight from Frankfurt to New York. On the same day, Palestinian guerrillas hijacked a Swissair DC-8 outbound from Zurich and a Pan Am 747 out of Amsterdam. Three days later a BOAC VC-10 was hijacked, completing the series. The hijackers directed the Pan Am plane to Beirut and then Cairo, where the plane was blown up the day after it was hijacked. The other three aircraft were forced to fly to Dawson Field, Jordan. Here the nearly 500 passengers and crew members suffered six days of terrible confinement in the plane before they were allowed to exit. The three planes were then blown up, causing an equipment loss of $40 million.

MOTIVATION OF HIJACKERS

---

[1] Federal Aviation Administration, Civil Aviation Service - Worldwide Reported Hijacking Attempts - Summarization, July 10, 1974.

[2] John P. Wallach, "The Nightmare World of the Cuba Skyjackers", San Francisco Examiner and Chronicle, September 17, 1972.

On April 21, 1980, a five-month-long boat lift known as the "freedom flotilla" began bringing Cubans from the harbor at Mariel to the United States. By the time it ended on Castro's orders, it had brought 125,000 refugees to the U.S. About 2000 of the "Marielitos" were in U.S. federal prisons because Castro had sent many of his prison inmates and mental hospital patients to the U.S. as a part of the boat lift. Some of these people had been separated from their families and did not want to leave Cuba.

The boat lift prompted a number of the "homesick" Cuban hijackings. Every year around Christmas time, warnings were sent to airports reminding security personnel that there would be a greater risk of hijackings during the holiday season. The Cubans seemed to favor a bottle of flammable liquid as a weapon and emphasis was placed on that factor during predeparture screening.

In July 1983, the Federal Aviation Administration (FAA) even broadcast radio and television commercials in English and Spanish in south Florida and New York City. The commercials warned would-be hijackers that they would face long stays in Fidel Castro's jails if they were to hijack an airplane. During May, June and the first three weeks of July of that year, eight U.S. flights had been hijacked to Cuba. The commercials were an unusual approach to what had become a serious problem.

Hijacking a plane appeals to extortionists who seek political and personal gain as a dramatic method of enforcing their demands. Cooper's escape sparked another wave of U.S. hijackings that resulted in extortion demands totaling $8,712,000, of which $5,355,000 was actually paid.[3] Temporarily, that is--all of the money has been recovered except for the $200,000 paid to the missing Cooper, and a portion of that may have been found. Other hijackers include fleeing felons, the mentally deranged, and the "homesick" Cubans, previously discussed.

GOVERNMENT RESPONSE

While the 500 persons and three aircraft were still hostages in Jordan, President Nixon proposed the placement of "specially-trained, armed United States Government personnel on flights of U.S. commercial airliners."[4] The White House statement also called for the use of electronic screening equipment and urged "the Departments of Transportation, Treasury, and Defense, the Central Intelligence Agency, the Federal Bureau of Investigation, the Office of Science and Technology and other agencies to accelerate their present efforts to develop security measures."[5] Responsibility for directing this program was given to the Secretary of Transportation.

---

[3] U. S. Congress, Senate Committee on Commerce, Aviation Subcommittee, Paul R. Ignatius, Executive Vice President, Air Transport Association of America, Testimony on S. 3815 and S. 3871, August 15, 1972.

[4] Statement by the President of the United States, issued by Office of the White House Press Secretary, September 11, 1970.

[5] Ibid.

Although the Federal Aviation Act of 1958[6] had been amended nine years earlier to include a new subsection dealing with and defining aircraft piracy, not until January 1969 was there enough public concern to move the government to further action. In that month, the President directed the FAA to research all aspects of aircraft hijacking and to create a defensive plan.

The public needed no directive; the airlines were flooded by suggestions from private individuals, prompting a government/industry agreement that all suggestions would be forwarded to the office of Air Transportation Security of the FAA, for consideration and reply. Many of the ideas reappeared with such frequency that form letters could be used for reply. One of the most frequent suggestions was that planes be equipped with a sleeping gas which would be pumped into the passenger cabin in the event of a hijacking. Actually, this suggestion had been considered, but medical advisors of the Joint Airline/Government Task Force had warned that a gas strong enough to disable hijackers could cause the death of passengers with certain chronic illnesses. Also, the air conditioning system in modern aircraft is so effective that it would be impossible to maintain the necessary concentration of the gas. Nor could the air conditioning be turned off; it is an essential part of the cabin pressurization system without which high-altitude passenger flights would be impossible.

Other suggestions were for arming flight attendants or training them in karate and judo, fingerprinting or photographing passengers as they board, issuing photo-identification credentials to frequent and known customers, and using armed guards aboard the aircraft (the Sky Marshal program concept). Photographing and fingerprinting were ruled out as being ineffective in deterring hijacking. After all, it was commonly known that less than 10 percent of hijackers remained unidentified--clearly, they did not fear identification.

Public Suggestions

Of the ideas most frequently mentioned, the suggestion that received the most consideration in official circles was that of the photo-identification credential. At least one carrier submitted such a proposal to the FAA, but there were so many inherent problems that the idea had to be dismissed. The principal drawback was the necessity of deciding who should receive such a card--and, worse, who should not. It was feared that in only a matter of days after such a system was initiated, charges of discrimination would be filed by someone who had been denied a card. The courts would, in all probability, compel the carriers to issue such cards more broadly and would demand that applicants be refused only for proven and related reasons such as openly advocating hijacking an aircraft. An official identification credential would create more problems than it solved.

Even if cards had been a feasible measure, grave questions arose as to their effectiveness. The October 1972 hijacking at Houston was led by an employee of the Department of Commerce. He earned a high salary, traveled extensively in his work, and had no police record. Certainly no carrier or

---

6 49 U.S.C. 1472

government agency would have refused his application for a photo-identification credential.

Another problem would have been deciding who would undertake the enormous task of issuing the cards. Most carriers believed that the cards should be issued by and on the responsibility of the government. The government wanted no part of such an administrative burden and felt the task should be undertaken by the carriers.

OTHER SUGGESTIONS

A few suggestions brought some humor into otherwise routine days. One was that a hypodermic needle filled with a poison or a drug that would immediately render the recipient unconscious should be installed in each airline seat. By pressing the proper seat button in the cockpit, the pilot could inject the hijacker. Such a system might well make passengers reluctant to fasten their seat belts snugly, let alone contemplate flying for hours in close proximity to a system that would not be fail-safe.

Another suggestion advanced regularly would have had passengers strip down to the buff in dressing rooms and don overalls for the flight. A rather more severe idea was to equip each airline seat with manacles and leg irons. They would automatically lock in place when passenger sat down and placed their arms, as instructed, on the armrests. No meals would be served, naturally, and passengers would need an armed escort to use the rest rooms.

The writers for Archie Bunker of the "All in the Family" television show probably topped them all. They had Archie suggest giving each passenger a gun as he boarded the flight.

An inaccessible, bulletproof pilot's compartment was a suggestion that got as far as a second round of proposed rule-making. It was suggested that there would be no voice communication from the cabin to the cockpit, nor could the door between the two be opened from the cabin side. In an emergency, a cabin attendant would push an alarm button, and the pilot would land at the nearest airport without further communication. The Air Transport Association (ATA) opposed the proposal, arguing that no pilot had ever refused to admit a hijacker to the cockpit in the face of threats against the life of a crew member or passenger. Although the Air Line Pilots Association (ALPA) came out in favor of the "inaccessible cockpit" idea, the FAA came to concur with the opinion that a Bulletproof cockpit would be vulnerable to hostage threats.

PSYCHOLOGICAL PROFILE OF THE HIJACKER

In response to the January 1969 Presidential directive to research all aspects of aircraft hijacking, the FAA Task Force on Deterrence of Air Piracy was established and promptly commenced its investigation. The study encompassed motivational and behavioral characteristics of hijackers, identification of the weapons used, and an analysis of the origin and intended destination of hijacked flights. The most significant achievement of the task force was the development of a psychological profile of the hijacker.

At the recommendation of the task force, the FAA set up a test program based on the profile. Passengers who appeared to match the profile were to be designated as "selectees," and special precautions were to be taken with them. This program also introduced the use of magnetometers (metal detectors), which were used in screening the selectees for weapons.

Eastern Airlines was selected for testing the proposed antihijacking program, probably because its East Coast flights had been particularly hard-hit by hijackers seeking to go to Cuba: 19 hijackings were attempted between September 1968 and March 1969. Following sampling tests at various locations, a relatively large field test was put into effect in October 1969, which involved one Easter Airlines gate in each of six cities. From October to February 1970, Eastern had no hijackings. In February, one hijacking did occur, but this was on a flight that had been enplaned through an unprotected gate. This hijacking proved to be significant to the test program, because it was later determined that the hijacker fit the profile.

This case and later ones confirmed the validity of the profile for identifying potential hijackers. Most air carriers voluntarily adopted the FAA's profile-based screening system. Between 80 and 90 percent of hijackers appeared to fit the profile. It was a good tool, and it was of material assistance, but hijackings did not cease, and the search for a solution continued.

SKY MARSHAL PROGRAM

In response to Nixon's call in September 1970, for armed government personnel on U.S. commercial flights, Secretary of the Treasury David M. Kennedy and Secretary of Transportation John A.Volpe announced that the Treasury Department would provide a permanent Sky Marshal force. The first 500 U.S. marshals were a temporary force recruited from the FAA, the Bureau of Customs, the Secret Service (Treasury), the FBI, and the CIA. In addition, 880 military personnel were added to the group, which was sent to Fort Dix (New Jersey) for training as Sky Marshals.[7] These 1300 people were later replaced by a civilian corps of 1500, recruited and trained by Treasury's Bureau of Customs, now the U. S. Customs Service. These Customs Security Officers (CSOs) were assigned to duty with the FAA under the direction of Lt. Gen. Benjamin O. Davis (USAF-Ret.), then Director of Civil Aviation Security for the Department of Transportation.[8] In addition to the Sky Marshal force, which was to be brought to its authorized strength of 1500 by the end of August 1971, a force of 230 U.S. marshals was authorized for assignment to airports.

While the Sky Marshal program enjoyed some success, it was clearly not the whole answer. Probably the most embarrassing indication of this was that a hijacking took place on a flight carrying both a Sky Marshal and an FBI

---

7 John Cashman, "Sky Marshals--How They Train and What They Do," Parade, November 19, 1970, p.7.

8 Department of the Treasury News Release, October 28, 1970.

agent. Statements made by General Davis fairly early in the program indicated that he, too, saw the advisability of shifting the primary anti-hijacking defense to the ground. Time and events proved him correct.

The Sky Marshal program also evoked criticism from the airlines, who considered the coverage too sparse and who were deeply concerned about the possibility and consequences of a midair shootout, particularly at the 35,000- to 40,000-foot altitudes at which the largest planes flew. Several carriers asked the government to eliminate the Sky Marshal program, claiming that it was ineffective and risky, to which the government finally responded with a countercharge that the airlines were not doing enough on their own initiative for fear of inconveniencing passengers.[9]

Although the Sky Marshal program appeared to have been phased out once 100 percent mandatory screening proved effective in preventing hijacking, in March 1976 the FAA advised that some of its security inspectors had been designated as "Federal Air Marshals, formerly Sky Marshals as part of the Federal Air Marshal (FAM) Program. Duties of these inspectors, who are deputized as Special Deputy U. S. Marshals, were said to be to provide protection against hijacking of specified aircraft. The reaction of some airlines to this indication of a reappearance of the Sky Marshal program was not enthusiastic.

There was a renewed emphasis on the Federal Air Marshal program following the hijacking of TWA Flight 847 on June 14, 1985. Three weeks later, the FAA imposed new regulations in the form of FAR 108.14 (14 CFR 108), requiring that scheduled carriers and public charter operators carry Federal Air Marshals on a priority basis, without charge, and that they provide the seating selected by the marshals, even though it might mean bumping full-fare passengers. The old regulations did not cover deadheading[10] air marshals who were repositioning themselves for another flight, but these new regulations took care of that error.

The new Public Law 99-83, the International Security and Development Cooperation Act of 1985, established an explicit statutory basis for the FAA's FAM program. This statute provided the Secretary of Transportation with the authority to authorize (with the approval of the Attorney General and the Secretary of State) civil aviation security Federal Air Marshals to carry firearms and to make arrests without warrant for any offense against the United States committed in their presence if they have reasonable grounds to believe that the person to be arrested has committed or is committing a felony.

The marshals have been used on certain high risk flights where good judgment dictated that there should be added coverage. On occasion, the selection of these flights was based on specific threats. During the last six months of 1989, Federal Air Marshals flew over 3,795,000 nautical miles on

---

9 Albert R. Karr, "Policing the Skies," The Wall Street Journal, Vol. LI, No. 238, September 21, 1971.

10 Traveling on a non-working status, but related to the job; i.e. traveling to a departure point to work a flight, or returning home from working a flight. The term is not used in relation to vacation or pleasure travel.

missions.[11] They are trained at the Federal Law Enforcement Training Center (FLETC), U.S. Department of the Treasury, in Marana, Arizona.

In July 1985, it was reported that a House report accompanying fiscal 1986 appropriations for the Treasury Department ordered the U.S. Customs Service to consider the establishment of its own air marshal program to assist in deterring the hijacking of U.S. aircraft. This was part of an overreaction to the TWA hijacking, but it also is an indication of some of the interdepartmental rivalry that has existed among various branches of the government. As Jimmy Durante used to say, "Everybody wants to get into the act." I'm sure the FAA was not pleased with the thought of Customs starting a rival air marshal program. Fortunately, it never came to be.

AIRPORT AND AIRLINE SECURITY PROPOSALS

On September 28, 1971, the FAA issued notices of proposed rule-making for the landmark revisions of airport and airline security measures. One notice proposed the issuance of Part 107 of the Federal Aviation Regulations (FARs) to give to airport operators the responsibility for providing protection against unauthorized access to air operations areas.

The second notice proposed a new Section 538 to FAR Part 121, which would require each scheduled carrier to develop and implement a security program designed to prevent or deter the carriage aboard aircraft of sabotage devices or weapons and to prevent or deter unauthorized access to aircraft. It also attempted to ensure that passenger baggage would be checked in by a representative of the carrier and that neither cargo nor checked baggage would be loaded aboard an aircraft unless it had been cleared in accordance with the carrier's security procedures. Certain notifications and security inspections were to be required in the event of bomb threats or suspected acts of aircraft piracy.

These proposals indicated a change in antihijack theory away from the Sky Marshal Program. They stated it specifically: "To be effective, security measures must be employed and concentrated on the ground."[12] Ninety days were given for comment.

Not surprisingly, the proposals met with resistance. Carriers and operators termed the proposals vague. The airport operators took the position that the required security was the job of either the airlines or the government. In any event, they said, they were financially unable to bear the costs of implementing the plan. Most U.S. airports are owned by a city or owned jointly by two states or cities, and they are operated by appointed commissions or authorities subject to relatively inflexible local budgets and operating revenues.

Controversy also arose over the delegation of approval power over the programs to be developed under the new rules. Federal Aviation

---

[11] Semiannual Report to Congress on the Effectiveness of the Civil Aviation Security Program (July 1, 1989--December 31, 1989), Federal Aviation Administration, February, 1991.

[12] Federal Register, 36:19172-19174, Docket 11431, 11432, September 30, 1971.

Administration Regional Directors, rather than a central authority, were to approve both airport and carrier security plans. Inconsistencies developed between the programs of the various airlines because of the FAA's decentralization of the approval authority and acceptance of the airlines' individual approaches. Somc carrier programs were developed in great detail, while others were drafted in broader, more general terms.

The differences between airport security programs were even greater. Carriers complained that the FAA appeared to be too lenient toward the airports, approving limitations in airport security programs not allowed carriers. They also saw an imbalance in the enforcement of penalties for airport security violations compared with the penalties assessed against air carriers. Between April 1972, when the first violation was charged, and June 1974, eighty-eight civil penalties were levied against air carriers, and one was charged against an airport, according to a Security Enforcement Summary published by FAA. Carriers also complained that enforcement differed in severity from region to region.

Approval of security plans for each carrier was later made the responsibility of an FAA employee designated the Principal Security Agent (PSA), now known as the Principal Security Inspector (PSI). The PSI coordinates security matters and acts as FAA liaison contact for the carrier. Thus, each airline has its own Principal Security Inspector who reports to the regional FAA authority. Some people in the industry suggested that the PSIs report to a single, central authority in Washington D.C., rather than to a regional office, to provide for greater consistency. The adoption of the Air Carrier Standard Security Program (ACSSP) has met that need.

NEW CARRIER RULES

Part 121.538, the rule affecting security by air carriers, was issued on January 31, 1972,[13] and required each certificate holder to adopt and implement a screening system that would detect weapons and explosives in carry-on baggage or on the person of passengers. Because of the emergency situation, the amendment was to be effective on three day's notice.

Thirty days later,[14] the order was amended to require each carrier to submit its screening program to the FAA Administrator no later than June 5, 1972. The carrier's security program was to be designed to:

1. Prevent or deter unauthorized access to its aircraft.

2. Ensure that baggage would be checked in by a responsible agent or representative of the certificate holder.

---

13 Ibid., 37:2500, February 2, 1972.

14 Ibid., 37:4904, March 7, 1972.

3. Prevent cargo and checked baggage from being loaded aboard its aircraft unless handled in accordance with the certificate holder's security procedures.[15]

Several other provisions incorporated in the amendment reflected comments received on September's Notice of Proposed Rulemaking (NPRM).

Part 121.538 was amended again on March 9, 1972,[16] in response to a presidential statement ordering a renewed emphasis on air security. The screening requirement was to be put into immediate effect, and the 60-day deadline for submission of new carrier programs to the FAA became May 8, 1972.

On September 11, 1981, the new Part 108 of the FARs became effective. Part 108 was a rewrite of §121.538 of the FARs. This meant there was now a Part 107 for the airport operator and a Part 108 for the air carrier. These sections are once again undergoing extensive rewrite, partially as a result of the *Report of the President's Commission on Aviation Security and Terrorism*, issued May 15, 1990.

NEW AIRPORT RULES

Also in May 1972, new airport security requirements, proposed the previous September, were issued. The new Part 107 required the airport operator to "immediately adopt and put into use facilities and procedures designed to prevent or deter persons and vehicles from unauthorized access to air operations areas."[17] Operators were given 90 days in which to prepare and submit to the FAA a master security plan for their airport. The plan had to show the means and procedures used, or intended to be used, in:

1. Improving or establishing protection against unauthorized access to air operations areas.

2. Identifying persons authorized access to air operations areas through "suitable identification."

3. Identifying vehicles operating in air operations areas.

The airport security plan did not generally apply to "Access from a nonair operations area to an air operations area where both areas are adjacent and exclusively occupied or controlled by a certificate holder that is required to have a security program under Section 121.538 of this chapter."[18] The time allowance for implementation of airport programs given in the original notice of proposed rule-making had been five years, but this did not appear in the final version of Part 107.

---

[15] Ibid., p. 4905

[16] Ibid., 37:5254, March 11, 1972.

[17] Ibid., 37:5689-5691, March 18, 1972.

[18] Ibid.

The next emergency order was prompted by two hijackings of Pacific Southwest Airline (PSA) flights occurring on successive days--July 5 and 6, 1972. (A California man achieved the unenviable distinction of being hijacked twice in a single trip.) The emergency order, which was to go into effect in 30 days, was limited to high-density "shuttle" flights and thus primarily affected PSA, Southwest, and the shuttle operations of Eastern Airlines.[19] All carry-on items were to be searched. Passengers were to be required to show two forms of personal identification before boarding, and those who could not were to be screened by a metal detector.

The affected carriers protested that they were being made to bear the expense of extensive screening procedures that the larger carriers did not have, and they argued that the rule was discriminatory. In fact, some of the larger carriers were already exceeding the current FAA requirements by giving selected "high-risk" flights 100 percent screening.

THE LAST STRAW

Resistance to mandatory 100 percent screening was broken by two hijackings that introduced a new type of hijacker: the "fleeing felon." On October 29, 1972, the Eastern Airlines agent on duty at Houston had boarded all the passengers on a flight and proceeded down the jetway to "button up"[20] the aircraft. When he turned to receive four last-minute arrivals who followed him into the jetway, one of them shot him without warning. The men boarded the aircraft, firing a second shot at a ramp serviceman who was trying to tell the flight crew that they had started their engines while fueling was in progress. The passenger agent was killed, and the ramp serviceman was wounded. Three of the four hijackers were subsequently identified as bank robbery and murder suspects.

The second case followed shortly. On November 10, three alleged rapists, one of whom was also an escaped prisoner, took over a Southern Airways DC-9 out of Birmingham and demanded $10 million. They were given $2 million, but the nightmare had only begun. For 29 hours, they held 31 passengers and crew, eventually shooting and wounding the copilot. They directed the plane back and forth across the country, at one point threatening to force the pilot to crash the plane into the Oak Ridge, Tennessee nuclear facility. The flight finally terminated safely in Havana on its second attempt.[21]

THE END BEGINS

These events prompted the government to issue two emergency regulations on December 5, 1972. One gave carriers 30 days in which to institute a 100

---

19 Aviation Daily, 202:42, July 11, 1972, (Washington, D.C.: Ziff-Davis Co.).

20 Close the door prior to departure.

21 Frank Brennan, "Anti-hijacking: Who Pays Bill?" Chicago Daily News, February 23, 1973.

percent search of all passengers and carry-on items, and the other gave airport operators 60 days in which to station at least one law enforcement officer at each passenger checkpoint during boarding or reboarding. The heavy financial burden of the new requirements was to be borne by the carriers and airport operators.

Their response to these orders was immediate, vocal, and negative. The carriers were presented with an insoluble problem in logistics; they were ordered to institute 100 percent electronic screening within 30 days, but with what? At most airports, carriers had been sharing a single magnetometer. This had been sufficient because they were required to screen only selectees . Now every passenger would be a selectee. There were not enough magnetometers in existence to meet the new requirement; the FAA itself had 1090 walk-through metal detectors on order for installation at 87 key airports, and only 400 had been delivered so far.[22] Carriers had just 30 days in which to recruit, hire, and train the number of additional employees, or to retain the services of sufficient contract agency personnel needed to conduct the 100 percent carry-on baggage hand search.

The airports' dilemma was equally acute. Funds were not budgeted for full-time police coverage, and the airports and their city or joint authority would be hard-pressed to find the necessary additional funds on such short notice. Also, in most cases the order would require the recruitment and training of additional law enforcement officers.

Nevertheless, on January 5, 1973, 100 percent screening of air passengers and their carry-on baggage was instituted. The result was unforgettable, and only the virtually infinite patience of the travelers made it possible for the planes to get off the ground that first day. Their tolerance revealed that the American traveling public was ready, willing, and eager to be protected; they had had enough of hijacking.

I recall standing at the boarding gate at O'Hare International Airport in Chicago on that first day, observing the screening of a United Airlines flight to San Francisco, which I was about ready to board. I noted Melvin Belli, the noted attorney and "King of Torts," standing in line to be screened. I hoped our people would be "careful" with him, not wanting to incur a lawsuit. Naturally, Murphy's law immediately went into effect. The security personnel went over everything he was carrying with a fine-toothed comb, even to the extent of leafing page by page through a *Time* magazine he was carrying. To cap it off, they broke the zipper on his carry-on bag. The incident was also witnessed by a reporter from the *Wall Street Journal.* I went ahead and boarded the flight, but because I was then Manager of Security for United, I sought out Belli and apologized. He was most gracious; he told me he was the General Counsel for Japan Air Lines and was very understanding of the need for screening.

When the remarks of an FAA official four months later were misinterpreted to imply that the FAA was contemplating the relaxation of the stringent security measures instituted in January, and the public newspaper editors

---

22 Aviation Daily, 205:143, January 26, 1973.

protested strongly. Soon thereafter, the FAA announced it would continue the 100 percent screening indefinitely.

PROBLEMS OF IMPLEMENTATION

Despite the demonstrated and continuing support of the public, initial opposition to the new emergency orders was strong. The new regulations were attacked in the Senate for not providing federal funding to cover the costs. Civil Aeronautics Board Chairman Secor Browne criticized the program on the ground that airport security should be a federal, rather than a local, responsibility.[23] The Airport Operators Council International (AOCI) sought and received a ten-day restraining order. It contended that hijacking was a chronic problem, not an emergency, and hence the FAA, by not allowing time for comment before issuing the rules, was acting in violation of the Federal Administrative Procedures Act.[24]

There was ample evidence, however, that the real issue was whether the required law enforcement was to be federal or local. In arguments before the U.S. District Court, the general counsel for the Department of Transportation (DOT) stated that AOCI's "real argument is who should pay for this; the federal establishment or the airport operators?"[25] The administration clearly preferred to keep security a local responsibility; as DOT Under-Secretary Egil Krogh testified before a congressional committee, "We see no rationale for distinguishing the airport from the bus depot or the train station in the provision of police protection."[26]

The air carriers also were faced with problems resulting from this federal-local controversy. Some airports threatened to shut down operations on the grounds that they had no funds for the additional officers. At other airports, police refused to make the "pat down" inspections and body searches required for passengers who could not be cleared by the carrier's "no hands" screening. For fear of legal repercussions, a number of airlines originally prohibited their employees or contract guards from performing these personal searches. Some carriers faced a choice between allowing their employees to conduct body searches, or instructing their people to advise those passengers who could not be cleared by walk-through or hand-held magnetometers that they would not be permitted to board.

The restraining order obtained by the AOCI was vacated on February 12, 1973. The judge said that the safety of the traveling public would be jeopardized by a failure to implement the regulations and that the FAA Administrator was within his rights in issuing the emergency rules.[27] A last-minute appeal to a higher court brought some amelioration. A three-judge panel of the U.S. Court of Appeals ruled that the FAA could proceed with

---

23 Ibid., 205:114, January 14, 1973.

24 Ibid., 205:214, February 8, 1973.

25 Ibid., 205:226, February 12, 1973.

26 Ibid., 205:319, February 28, 1973.

27 Ibid., 205:233, February 13, 1973.

implementing the requirement of a law enforcement officer at airport boarding gates, but penalties were not to be imposed without a hearing if the airport operator cited was making a reasonable effort to comply with the requirement.[28] The decision also held that the FAA could station federal law enforcement officers at any airport unable to provide local officers and would be reimbursed by the airport authority for those services.

TRAINING LAW ENFORCEMENT OFFICERS

While the new rules were being implemented, the FAA established a training program for local law enforcement officers. Classes in airport security techniques were held, and the first class graduated from the Transportation Safety Institute on January 26, 1973.

The Transportation Safety Institute continues to provide training for law enforcement officers supporting the airport security programs. Currently funded by the FAA, this training has been expanded to include the Federal Aviation Administration's inspectors and special agents, as well as aviation and airport security officials from all over the world. In addition to the basic civil aviation security course, training now includes seminars and classes on such security-related subjects as hijacking seminars, explosives detection K9 teams, aviation explosives security, investigations, and civil aviation security inspection and reports.

As of the end of Fiscal Year 1988, over 15,700 students had participated in this training at the Oklahoma City facility, or at seminars conducted throughout the United States. This figure includes 555 international students from 110 different countries.

Although some courses, such as the investigations course and the inspection and reports course, are restricted to FAA personnel, other courses are available to the aviation/airport security industry. Interested applicants who are either law enforcement officials responsible for airport security or aviation/airport security officials can contact the nearest office of the FAA's Civil Aviation Security. Foreign nationals should apply through their respective governments to the embassy of the United States.

FUNDING

Faced with increased security costs estimated at $56 million,[29] carriers filed in 1973 for special tariff permission to increase one-way fares by $1 and round trip fares by $2, to cover the new costs. The Civil Aeronautics Board denied these requests and refused to waive the 30-day statutory notice required for tariff permissions. January 19, 1973, was set as the date by which carriers were to submit fully documented data on their security costs.

---

[28] Ibid., 205:267, February 10, 1973.

[29] Ibid., 205:319, February 28, 1973.

As a result of the hearing, the CAB concluded that the best approach would be to establish a standard charge for all carriers rather than a different charge for each. The CAB set 34¢ as the average industry cost, based on its analysis of the data presented by the carriers, and directed that the charge be assessed on each flight coupon.[30] Carriers resubmitted their tariff proposals at the CAB figure, and the surcharge became effective on April 1, 1973.[31]

This did not end the financial problems. The 34¢ surcharge was to reimburse the carriers for their costs for predeparture screening and was not intended to cover the cost of the law enforcement officers required of the airport operator. The indignant airport operators sought to recover the cost of these officers from the carriers, without result. At the request of the Department of Transportation, the CAB resolved the conflict by approving an additional 25¢ surcharge per coupon with which the carriers were to reimburse the airports for the cost of the law enforcement officers.[32] This additional surcharge became effective on May 12, 1973,[33] and brought the total surcharge to 59¢.

Sixteen months later, the CAB canceled the 59¢ combined surcharge and replaced it with a 34¢ maximum for virtually all airlines, with minor exceptions involving intra-Hawaiian and intra-Alaskan flights. The change stemmed from a conflict between airport operators and carriers involving the 25¢ surcharge. The AOCI charged that the airlines had realized as much as $30 million during 1973 from the difference between actual carrier reimbursement to airports and the amount collected by the carriers through the 25¢ surcharge. In addition, the AOCI complained that the airlines refused to reimburse the airports for their costs they incurred between the date the emergency order went into effect (January 4, 1973) and the date the airport reimbursement surcharge became effective (May 12, 1973).[34]

On August 19, 1975, CAB Order 75-8-103 disallowed all security surcharges, effective February 1, 1976. The thrust of this order was that a surcharge could be justified only to recoup "unexpected, temporary or rapidly changing costs." Because security measures did not meet this test, they were abolished. In the same order, the CAB permitted a 62¢ fare increase per coupon within the U.S., recognizing that the security program should be a component of the passenger fare, but not specifically identified as such.

There have been attempts by airlines, airports, and the government to recoup various security-related costs by appending additional charges to the fares. Several airports experimented with "head taxes," which were similar to departure taxes at U.S. and foreign airports. Public reaction was so negative that these plans were abandoned.

In July 1985, the U.S. Customs Service proposed a $5 entry fee for travelers entering the U.S. The Air Transport Association strongly protested the fee, claiming that the current $3 departure tax, the 8 percent tax on

---

30 Ibid., 206:82, March 15, 1973

31 Ibid., 206:90, March 16, 1973.

32 Ibid., 206:145, March 27, 1973.

33 Ibid., 207:26, May 4, 1973.

34 Ibid., 209:141, September 27, 1973.

domestic airline tickets, and the 5 percent charge on air freight waybills had created a $3 billion surplus in the aviation trust fund. Nonetheless, the House Ways and Means Committee, and eventually Congress, approved the fees, including a $25 charge to owners of general aviation aircraft. The fees went into effect on July 7, 1986. Transiting passengers and airline crews were exempted.

Pan Am, TWA, American and People Express began charging a $5 per passenger surcharge in May and June of 1986 for flights to Europe. The fees were intended to pay for increased security screening costs on overseas flights.

Under legislation passed by the House on August 2, 1990, airline passengers would be forced to pay up to $12 in airport fees for each round-trip ticket. Major airport authorities would be allowed to charge up to $3 for each passenger using the facility. If a traveler made a stopover at a second fee-charging airport on the same trip, another $3 could be charged, but the maximum fee would be $12 for a round trip. Only the nation's 71 largest airports would be authorized to impose the tax. Income from the tax is projected at $1.4 billion. The money would be used for airport expansion, runways, taxiways, new gates, and noise reduction. The measure had not yet passed in the Senate, however, at the time of this writing. Opponents noted that the existing Federal Aviation Administration's Airport and Airway Trust Fund, also known as theAviation Trust Fund, supported largely by an 8 percent ticket tax, already had a surplus of about $7.8 billion and argued that passengers shouldn't be hit with a new tax while there was still a surplus. When the law was passed by Congress, it raised the 8 percent tax to 10 percent, the additional 2 percent designated for the Treasurey's General Fund rather than for aviation.

The passenger facility charges (PFCs) are a part of a bill authorizing $18.3 billion in FAA programs in 1991 and 1992. The bill authorizes $1.8 billion in fiscal year 1991 and $1.9 billion in fiscal 1992 for the Airport Improvement Program (AIP) and $2.5 billion in 1991 and $2 billion in 1992 for facilities and equipment. It also involves a formula designed to reduce the Airport and Airway Trust Fund surplus from $7.6 billion to $1.1 billion by 1995. It authorizes $438.6 million per year from 1992 to 1998 for essential air service to small communities to come from the trust fund. The bill contains an option for airports to impose a PFC of $1 to $3 to finance capital development projects. Large and medium hub airports imposing a PFC would forgo half of their AIP funds, while small airports would be able to retain all their AIP monies.

GROWING PAINS

These problems, and others that are be mentioned later, were only a part of the great volume of problems, protests, inconsistencies, and difficulties that inevitably arose from this unprecedented change in the federal regulation of an essential U.S. industry. The succeeding years have seen a sometimes slow and not always steady progress toward the effective incorporation of the new regulations into the primary operations of both the airports and the carriers.

The magnitude of the challenge is evident in the statistics. In the *Semiannual Report to Congress on the Effectiveness of the Civil Aviation Security Program* for the period from July 1, 1989, through December 1, 1989, the FAA reported that over 535 million persons were processed through screening checkpoints at U.S. airports. That's over a billion passengers a year and almost 3 million a day.

Much thought and effort has gone into making the administration of the regulations more effective and efficient. Although it once appeared that no general uniformity of airport security could be achieved, in 1974 carriers came to believe it was possible that a single, general security program could be developed that might be acceptable to all airlines.

AIR CARRIER STANDARD SECURITY PROGRAM (ACSSP)

Early in 1975, a committee was appointed from members of the Security Committee of the Air Transport Association to work out a proposal for a single standard security program. The draft program was accepted by the ATA in July 1975 and submitted to the FAA for comment. After an exchange of written comments and two meetings between the FAA and the ATA Security Committee was reached on the Standard Security Program (SSP), later to be known as the Air Carrier Standard Security Program (ACSSP). By January 1976, all but a few U.S. carriers were operating under the uniform security program. Today, their participation is mandatory.

Chapter 2

TERRORISM

Terrorism! The word defines itself. The act of placing someone in extreme fear. It brings to mind the song title "Far Away Places with Strange Sounding Names,": Entebbe, Mogadishu, Munich, Beirut, Dublin, Athens, Rome, Karachi, Lockerbie, even the *Achille Lauro*. Each incident seems to hold the world's attention while it is occurring, a drama in real life played out on the TV screen. Viewers have come to know flight crews and passengers by name. And indeed, that is part of it. It has been said that terrorism is theater. It is a small group of actors and "all the world's a stage," as Shakespeare said. The *sine qua non* of terrorism is the media. Without the media, the terrorists would be deprived of their audience, and the show, most probably, would not go on. Yet the media cannot be expected to ignore terrorist captives wherever they are.

The airplane provides a capsule container of ready made hostages, all organized, sitting in rows, strapped into their seats, and totally defenseless. They represent a good cross section of the population, including women, children, old people, and citizens of many countries; they will surely command the attention of the world. The ticket office of a national flag carrier, the embassy, and the diplomatic office also represents an opportunity to make a forceful statement against a particular country without going there in person.

It is often said that one man's terrorist is another man's freedom fighter. However, there is a difference between the two. Terrorism is the deliberate and systematic murder, maiming and menacing of the innocent to inspire fear for political ends. The late Senator Henry "Scoop" Jackson said, "Freedom fighters or revolutionaries don't blow up buses containing non-combatants; terrorist murders do. Freedom fighters don't set out to capture and slaughter school children; terrorist murders do. Freedom fighters don't assassinate innocent businessmen, or hijack and hold hostage innocent men, women and children; terrorist murders do. It is a disgrace that democracies would allow the treasured word "freedom" to be associate with acts of terrorists." Guerilla fighters or freedom fighters work toward a political goal of overthrowing a government through an irregular type of warfare aimed at interdiction of supplies, destruction of personnel and property, and general harassment. They do not particularly seek to bring attention to their cause. They are more likely to hide out in the hills than command the attention of TV cameras. Terrorists, on the other hand, attempt to focus attention on their particular issue or grievance.

The airline industry was concerned for a number of years with hijacking. Now the threat has changed, but the target is the same. Although the threat of hijacking will continue to exist, the current threat is that of explosives. On April 14, 1986, the United States launched an attack against Libya. That attack followed an explosion on April 3 aboard TWA Flight 840, traveling from Rome to Athens, which ripped a hole in the side of the 747. Four passengers, all of them American, were killed. Two days later, a bomb ripped a West Berlin discotheque frequented by American troops. It killed U.S. Army Sergeant Kenneth Ford and a Turkish woman and injured 230 people, 79 of them American. Both incidents were positively tied to Libyan terrorism. While this retaliatory attack by the U.S. was designed to let Muammar Khadafi know that the U.S. had reached the limit of what it would tolerate, there were those whose views were different. On April 15, 1987, former U.S. Attorney General Ramsey Clark said he intended to file claims totaling millions of dollars in compensation for the Libyan victims of the "terrorist bombing" of Libyan targets by U.S. warplanes. We also observed Ramsey Clark in Iraq, protesting the bombing of Baghdad. Obviously, Mr. Clark had a different view of who the terrorists were.

Terrorism has been around for a long time, and as the saying goes, it's probably going to get worse before it gets better. Yet, people still hold conferences and meetings to define terrorism. The *Report of the Task Force on Disorders and Terrorism*, published by the National Advisory Committee on Criminal Justice Standards and Goals, defined "terrorism" as a tactic or technique by means of which a violent act or the threat thereof is used for the prime purpose of creating overwhelming fear for coercive purposes. The report included a definition of "quasi-terrorism" as those activities incidental to the commission of crimes of violence that are similar in form and method to true terrorism but which lack its essential ingredient of inducing terror in the victim. As I write these words, there is an incident of quasi-terrorism involving hostages being held in two federal prisons by prisoners who don't want to be sent back to Cuba. The hostages are being held as bargaining pawns, but there is no attempt to induce a high degree of fear. Nevertheless, the means of combatting such an act are essentially the same as those of combatting true terrorism.

The CIA has defined "terrorism" as government-or-institution supported acts of violence against foreign governments, citizens, or institutions for the purpose of drawing attention to a cause or situation. The definition of "terrorism" given by the U.S. Department of State and the one commonly used by the U.S. government is, "premeditated, politically motivated violence perpetrated against noncombatant targets by subnational groups or clandestine agents, usually intended to influence an audience." International terrorism is terrorism involving the citizens or territory of more than one country.

The United States has been faced with many different forms of terrorism. The National Abortion Federation (NAB) reports that there were five incidents involving bombing and arson at abortion facilities in the U.S. and nine bomb threats during the first nine weeks of 1987. We have seen terrorism from the Ku Klux Klan, the Order, the White Patriot party, the Aryan Nation and other right-wing, neo-Nazi, and white supremacist groups. And of course, there are left-wing radicals also.

Seemingly, everyone with a cause, including animal-rights activists, has taken to terrorist acts to further the cause. On June 3, 1990, Secretary of Health and Human Services Louis Sullivan said, "I am saddened and a bit angry that we have to put up with major disruptions to science by so-called animal activists who are, in fact, nothing more than animal-rights terrorists."[1] Just a week or so later, in London, a bomb attached to the underside of a veterinarian's car exploded as he drove down the street. A 13-month-old toddler nearby was struck by shrapnel.[2] This was the second bomb directed at animal researchers in a week.

Even the person without a cause, such as the individual who poisons food or drug products on market shelves for extortion purposes, is a terrorist. Incidents of product tampering increased from 127 in 1985 to over 1600 in 1986.

In Colombia, the drug cartels have been waging a war of terrorism. During one week in May 1990, the toll was 503 people murdered and 18 kidnapped, bringing the year's total to more than 5000 killings and 342 abductions in less than five months. Among those most vulnerable were candidates for the presidential election, which occurred on May 27. Police officers were also vulnerable. Pablo Escobar, Medellin cartel leader, offered a $4300 bounty for each police agent killed. During the first half of 1990, 110 law officers were slain in Medellin, 34 in May alone.

WHO ARE THE TERRORISTS?

Internationally, terrorists are generally anarchists, separatists, or insurgents. Domestically, they are usually left-wing radicals or right-wing racist or survivalist groups, separatists, or émigré groups. Separatist groups include the Irish Republican Army, the Argentine Montoneros, the Puerto Rico independence groups (FALN), the anti-Castro Cubans, the Basques in Spain, the Croatians in Yugoslavia, and the Armenians in Turkey.

The groups have changed somewhat. The Red Brigades, the Baader-Meinhof and the Action Directe groups that flourished in the seventies were an outgrowth of the student generation of the sixties, nourished on neo-

---

[1] San Jose Mercury, June 8, 1990: 10A

[2] San Jose Mercury, June 12, 1990.

an outgrowth of the student generation of the sixties, nourished on neo-Marxist libertarian ideas. They discovered they could not change the world by peaceful means and resorted to violence.

The terrorist of the eighties and the nineties is different. Numerous groups have ties with nationalist or communalist movements in the Middle East, Africa, and Asia. Shi'ites in Lebanon, Iraq, and the Gulf; Kurds in Iran, Iraq, and Turkey; Armenians still seeking revenge on Turkey for the massacres of 1915; Sikhs in India; Tamils and Muslims in Sri Lanka; Muslims in the Philippines; blacks in South Africa; Eritreans and Tigrayans in Ethiopia; and many others are all defending their "cause" and resisting "oppression." In some cases, the cause is a good one, and that makes it all the more difficult to criticize the terrorism involved.

The Department of State in 1988 classified as lethal, the Abu Nidal Organization, pro-Iranian militant Shia, the Red Brigades, the Red Army Faction, and the November 17 Group of Greece, all of whom specifically target U.S. military and diplomatic officials.

Dr. Thomas Strentz, an FBI Special Agent attached to the Behavioral Science Instruction and Research Unit of the FBI Academy, has written several articles on the profile of the terrorist. He points out that the terrorist of the eighties is more likely to be poorly educated, unskilled, unemployable and illiterate than the terrorists of the sixties and seventies, who were generally college educated, sophisticated, and well trained[3]. The modern terrorist may be trained for a specific exercise, but does not have the general training that was prevalent in the past. The Defense Intelligence Agency describes a typical Hizballah terrorist as 18 to 24 years old, poor, from rural Lebanon, a resident of Beirut, religiously motivated, an idealist, recruited by a relative, and unfamiliar with the outside world.

Strentz points out several instances of lack of education, sophistication, and training in the modern terrorist. For example, the individuals who hijacked Pan Am Flight 73 out of Karachi left most of their gear on the ramp and had difficulty finding the cockpit. Indeed, they gave the impression they might never have been on a plane before. Those who hijacked TWA Flight 847 had planned to hijack an El Al flight, but did not know that El Al had security procedures exceeding those of other air carriers, a fact known to most of the general public. The modern terrorists are more likely to come from street gangs and refugee camps than from colleges and universities.

I have talked with Captain John Testrake, who was in command of TWA Flight 847. The first group who hijacked the aircraft were little more than 17-year-old boys who were unnecessarily brutal to the passengers. They are a part of the new group of terrorists who, despite the myths, are not as willing to die for their country as one might think. They may be prepared to kill, but not to die. Dr. Strentz refers to them as "unsophisticated, untrained, and minimally dedicated." They are a far cry from the September 1977 hijacking out of Bombay where the Japanese Red Army so effectively manipulated their passengers that each hostage, at the request of the JRA, completed a critique of their experience and turned it in as they left the air-

---

3 Thomas Strentz, Ph.D, "A Terrorist Psychosocial Profile," *FBI Law Enforcement Bulletin* (April, 1988): 13.

craft. The critiques included a review of the strengths and weaknesses of the operation, as well as recommendations on how to improve on the next hijacking, according to Dr. Strentz.

In a thought-provoking article, William T. Corbett categorizes terrorists as fundamentalists, anarchists, and cause terrorists[4]. He makes an excellent case for this categorization, arguing that their motivations are different, their targets are different, and the methods of dealing with them are different. Fundamentalists are those associated with the Islamic Jihad (Holy War). Their targets are Western influences, which they are trying to eradicate from Moslem areas. They generally don't conduct suicide attacks. The anarchists' goals are to destroy existing social, economic, and political orders. Examples of anarchists are West Germany's Red Army Faction, Italy's Red Brigades, France's Action Directe and similar groups. Their primary targets are senior officials and activities identified with existing social, political, military, and economic systems. The cause terrorists are those who feel disenfranchised and who strike from a sense of outrage over perceived injustices. They strike indiscriminately, often involving the innocent. Corbett says cause terrorists are responsible for 97 percent of Americans killed and maimed in Europe and the Mediterranean during the past five years.

He believes the U.S. policy against terrorism should be two-pronged: first, establish strategic goals and programs to neutralize and defeat terrorist movements that threaten U.S. interests, and second, develop antiterrorist security measures to defend U.S. personnel and activities. Corbett also makes a good case for listening to the legitimate causes of some groups identified with terrorism. He states, "Unjust causes fail from lack of popular support. Legitimate causes generate the support of many, active dissent of some and violence of a few."[5] He also effectively expresses arguments that support his conviction that military retaliation does more harm than good.

In his book, *Triumph Over Terror on Flight 847,*[6] Captain John Testrake reflects that in the Middle East, as everywhere else, there are two sides to every story. He states that while we need to be firm in our measures dealing with violence against innocent citizens, we also need to be fair in our dealings with all the people in the Middle East. He believes that fairly addressing legitimate grievances of all concerned will go a long way toward "defueling the engine of hate that is directed against us."

The following list of the major terrorist organizations now in existence was set out in the State Department's report on *Patterns of Global Terrorism, 1989.* [7] One can get lost in the "alphabet soup" of these organizations if you don't have a guidebook.

ANO	Abu Nidal Organization
Al-Fatah	

---

[4] William T., Corbett, "A Time to Redefine Policy," *Security Management* (June, 1990).

[5] Ibid.

[6] Captain John Testrake, , *Triumph Over Terror on Flight 847* (Old Tappan, NJ: Fleming H. Revell,) 1987. 251

[7] U.S. Department of State, *Patterns of Global Terrorism: 1989,* (April 1990): 55.

Hawari Group	
ASALA	Armenian Secret Army for the Liberation of Armenia
DFLP	Democratic Front for the Liberation of Palestine
15 May Organization	
Force 17	
Hizballah (Islamic Jihad)	
JCAG	Justice Commandos of the Armenian Genocide
LARF	Lebanese Armed Revolutionary Faction
PLF	Palestine Liberation Front
PIJ	Palestinian Islamic Jihad
PLO	Palestinian Liberation Organization
PFLP	Popular Front for the Liberation of Palestine
PFLP-GC	Popular Front for the Liberation of Palestine -- General Command
PFLP-SC	Popular Front for the Liberation of Palestine -- Special Command
PSF	Popular Struggle Front
FAL	Armed Liberation Forces Zarate Willka
PRTC	Central American Revolutionary Worker's Party
CERF	Clara Elizabeth Ramirez Front
FMLN	Farabundo Marti National Liberation Front
MJL	Lautaro Youth Movement
FPMR	Manuel Rodriguez Patriotic Front
FPM	Morazanist Patriotic Front
MIR	Movement of the Revolutionary Left
M-19	Movement of 19 April
ELN	National Liberation Army
EPL	People's Liberation Army
FRP-LZ	Popular Revolutionary Forces -- Lorenzo Zelaya
FARC	Revolutionary Armed Forces of Colombia
SL	Sendero Luminoso
MRTA	Tupac Amaru Revolutionary Movement
AD	Action Directe
ETA/M	Basque Fatherland and Liberty
CCC	Communist Combatant Cells
PKK	Kurdish Worker's Party
GRAPO	October 1st Antifascist Resistance Group
FP-25	Popular Forces of 25 April
PIRA	Provisional Irish Republican Army
RAF	Red Army Faction

ERCA	Red Army for the Liberation of Catalonia
BR	Red Brigade
BR-PCC	Communist Combatant Party
BR-UCC	Union of Combatant Communists
17 November	Revolutionary Organization 17 November
ELA	Revolutionary People's Struggle
TL	Terra Lliure
Chukaku-Ha	
JRA	Japanese Red Army
NPA	New People's Army
Sikh Groups	
ANC	African National Congress
RENAMO	National Resistance Movement

THE INNOCENT VICTIM

To the innocent victim, the justness of the terrorist's cause is of little comfort. Omega 7, an anti-Castro group, had a grudge against TWA because the airline had supplied aircraft for charter flights to Cuba. That prompted Omega 7 to put a bomb in a suitcase in 1979 and check it through from Kennedy Airport to Los Angeles. Fortunately, the bomb exploded before it was loaded on the plane, and the 155 people on that flight were spared their lives. This action was taken by a group that claims to love democracy and to hate communism.

Indeed, the innocent victim is a characteristic of terrorism. The victim provides an easy target and attracts great attention. Terrorist acts against innocent victims can create havoc among the population and have great economic impact. Witness the decline in tourism in Europe and the Middle East in 1986 following the outbreak of terrorism in those areas.

Terrorism has always existed in some form and probably will never disappear. One form of terrorism did not start with the innocent victim, however. It first began with the Russian revolution, where the anarchists assassinated government officials. It was soon discovered that terrorists could have as much or more impact by attacking defenseless victims. Sometimes even a minor incident can have a far-reaching effect.

On March 26, 1977, a bomb exploded in the florist shop of the Gondo Airport in Las Palmas, Canary Islands. The airport was temporarily closed, and flights were diverted to nearby airports, including Santa Cruz, Tenerife. Two of the Boeing 747s rerouted to Tenerife were a Pan Am and a KLM aircraft that collided on the runway, killing 580 persons in the world's worst airline disaster. The bombing was intended to further the cause of the Canary Island Independence Movement.

At about the same time, another obscure group, the South Moluccan Independence Group, had hijacked trains, occupied schools taking teachers and children hostage and taken over a government building and held its occupants as hostages. All were innocent victims of a cause no one but the perpetrators cared about. One can only feel anger at the tragedies caused by groups such as this.

Senator Daniel Patrick Moynihan has said,

> "Terrorism denies the distinction between state and society, public and private, government and individual, the distinction that lies at the heart of humane belief. For the terrorist, as for the totalitarian state, there are no innocent bystanders, no private citizens. Terrorism denies that there is any private sphere, that individuals have any rights or any autonomy separate from or beyond politics. There are thus no standards according to which the individual citizen, or the threatened society, can attempt to come to terms with the totalitarian terrorist. There is no way to satisfy his demands."[8]

INTERNATIONAL COOPERATION AMONG TERRORIST GROUPS

No longer are these local movements remaining confined to their own countries. They have become international in scope. Thus Croatian terrorists hijacked a plane out of New York in September 1976, causing it to fly to London and Paris to drop leaflets espousing their cause. There has been cooperation between various international terrorist groups over the past few years. Armenian terrorist groups have claimed close military cooperation with the Red Brigades, the Japanese Red Army, African Liberation Movements, and the Palestinians.

Alliances have been established and have developed between some of the older terrorist groups, some of whom are now out of business, such as the Baader-Meinhof Gang, with the Popular Front for the Liberation of Palestine, with the Japanese Red Army, and among guerrilla groups in South America. These alliances are evidenced by such joint operations as the June 27, 1976, Air France hijacking out of Athens, in which a West German, a Turkish woman and Arabs participated; the explosion at Ben Gurion Airport at Tel Aviv on May 25, 1976, in which a German carried a suitcase of explosives; the training of Dutch nationals by Arab terrorists in 1976; and the December 21, 1975, raid on the Vienna OPEC meeting, in which Arabs and West Germans participated.

After the attacks at the Rome and Vienna airports on December 27, 1985, it was believed that the Rome gunmen's Kalashnikov rifles and hand grenades, some of Chinese manufacture, had come from a weapons depot the Palestinians once shared in Italy with the Red Brigades. The Libyan news agency described the airport attacks as "acts of heroism." The toll at Rome was 15 dead and 74 wounded; at Vienna, 3 were dead and 47 were wounded.

In 1986, members of the El Rukns, a Chicago Islamic group described by an Assistant Cook County State's Attorney as "the closest thing to the Mafia that we have in Chicago, other than the Mafia," approached Khadafi offering to burn or blow up U.S. Government buildings and commit other terrorist acts for a price of $2 million. The plot was uncovered by federal agents who collected over 3500 hours of taped conversations, initially as a

---

[8] *Time*, April 14, 1986:50.

On April 14, 1988, five people were killed and about 17 injured when a car bomb went off outside an American servicemen's club in Naples. Junzo Okudaira, whom the Italians suspect of having planted the Naples car bomb, is a known member of the Japanese Red Army. He had been released from a Japanese prison in 1977 in exchange for 156 people in a Japanese airliner that had been hijacked to Bangladesh.

This incident illustrates the problem that results when terrorists are imprisoned. Imprisonment usually prompts another terrorist action designed to free the the terrorists. At one time, Moshe Dayan, former Minister of Defense and Foreign Minister of Israel, suggested that all terrorists should be executed to prevent this.

Indeed, the hijacking of Kuwait Airways Flight 422 on April 5, 1988, was committed for the purpose of attempting to free 17 terrorists jailed by Kuwait since 1983 for the bombings of the U.S. and French embassies in that country. Additionally, it is suspected that Hassan Izz-al-Din, one of four men indicted by a federal grand jury for the slaying of Robert Stethem during the seizure of TWA Flight 847 in 1985, was one of the hijackers of the Kuwaiti plane. Unfortunately, Algeria allowed the hijackers of Flight 422, who had murdered two hostages, to escape during the night, an act deplored by the United States. One of the spoils of war that accrued to Saddam Hussein when he invaded Kuwait in August 1990, were these 17 prisoners who had been held in the Kuwait City jail. These prisoners are now free.

It was revealed in June 1990 that the East German secret police, the Stasi, had run a type of halfway house for West German terrorists since 1980. Following the tearing down of the Berlin Wall, the East Germans admitted that these terrorists were given new identity papers, apartments, and jobs. In the first few days of June, 1990, seven people, believed to be members of West Germany's Red Army Faction (RAF) were arrested. The most notable arrest was that of Susanne Albrecht, who is accused of taking part in the 1977 killings of Jurgen Ponto, chairman of the Dresdner Bank, and industrialist Hanns-Martin Schleyer.

SURROGATE WARFARE

International terrorism is overwhelmingly an extension of warfare sustained and supported by the states built on the foundations of Marxism and radical Islam. The Soviet Union, several of its East European satellites, Cuba, North Korea, and Middle Eastern states such as Libya, Iran, Syria, and South Yemen have given terrorists weapons, training, and money. They have also provided sanctuary, safe passage, and safe houses--often their embassies. Terrorism is part of the broader trend of waging war by proxy.

The terrorist is now the spearhead of the theory and practice of surrogate warfare. Nations who had grave differences with one another used to resolve those differences by a declaration of war. Troops were drafted, factories converted to defense industries, battle lines drawn, and the nations went to war. A much more subtle, but less costly method of warfare, "war by proxy," has evolved through terrorism, the costs being paid by a sponsoring nation. The U.S. air carriers, in pleading for more government help in by proxy," has evolved through terrorism, the costs being paid by a sponsoring nation. The U.S. air carriers, in pleading for more government help in combating terrorism, have claimed correctly that the attack on Pan Am 103 was, in reality, an attack on the United States. The airlines are simply vulnerable surrogate targets.

Libya under Colonel Khadafi became the prime example of surrogate warfare. Libya would not have fared well with an open declaration of war against the United States and its allies. But it could finance terrorism around the world, from Ireland to Japan. Libya turned into the principal refuge for hijackers and their hostages. Libyan diplomatic pouches were used to smuggle terrorist arms into every trouble spot in the world, and Libyan embassies and consulates served as arms distribution centers. Most recently, Iran and Syria have reportedly assumed that role. The U.S. military action against Libya in March 1986 seems to have been highly effective in cooling down Colonel Khadafi. Many believe that the closing of Libyan diplomatic missions following the U.S. retaliatory attack on Libya was the most significant development leading to the decrease in terrorism in Europe.

The United States has maintained its formal designation of six countries as state supporters of terrorism--Cuba, Iran, Iraq, Libya, North Korea, and Syria. This list is maintained pursuant to Section 6 (j) of the Export Administration Act of 1979, which imposes certain trade restrictions on countries determined by the Secretary of State to have repeatedly provided support for acts of international terrorism.

The number of terrorist incidents that could be attributed to state sponsors declined in 1989. According to the U.S. Department of State, evidence indicates that 58 incidents involved state sponsors in 1989, a drop of 67 percent from 1988 when 176 such attacks were noted. In 1990 the figure dropped to 54.[9] The decrease was partly due to the ability of a number of states that have aided terrorist groups to effectively mask their involvement. The greatest portion of the drop resulted from Kabul's apparent curtailment of its bombing campaign in Pakistan following the removal of Soviet military forces. Iran's involvement in terrorism was not detected as frequently in 1989, but the State Department suspects an upturn in its support during the second half of the year reflects a return to a greater pace of operations. Libya and Syria were not directly tied to any attacks in 1989, but they continue to provide various forms of support for several terrorist groups. Libya's involvement in terrorism during 1990 went beyond support for the May 30 attack on Israel. Tripoli continued to shelter and aid the Abu Nidal organization, to fund other radical Palestinian groups such as the Popular Front for the Liberation of Palestine--General Command, and to support terrorist groups elsewhere in Africa, Latin America, and Asia. It should be noted that the investigations into the terrorist bombings of Pan Am Flight 103 in December 1988 and of UTA Flight 772 in September 1989 continue and could uncover involvement of state sponsors.[10]

---

[9] Patterns of Global Terrorism: 1990, U.S. Department of State, April, 1991: 32.
[10] Ibid.

the strongman of the 22-nation Arab bloc. Hussein has in the past pledged to "scorch half of Israel" with poison gas. His use of chemical weapons is in violation of the 1925 Geneva accords. A few weeks after becoming president of Iraq in 1979, he had 21 senior members of the ruling Baath Party executed for treason, including Adnan Hussein al-Hamdani, one of his closest friends. He once shot a colleague point-blank at a Cabinet meeting and has obliterated entire villages of ethnic Kurds. He committed his first murder at age 14, attempted his first assassination at age 22, and has been described in news reports as an Adolph Hitler, Khadafi, or Ceausescu.

The fear was that Hussein, a Sunni Muslim ruling a nation that is largely Shiite, would attempt an invasion of Saudi Arabia and capture 35% of the world's oil reserves. Obviously, the world could not stand by and let that happen, nor could it sanction the brutal take over of Kuwait. Arab states agreed to send Arab troops to join forces of other nations in defending Saudi Arabia. Only Libya and the Palestine Liberation Organization openly backed Iraq in the crucial decision by the Arab League.

In January, 1991, after repeated attempts at negotiation, coalition forces began the bombing of Iraq. Hussein launched Scud missiles at Israel, attempting to draw that nation into the conflict and cause defections among the Arab coalition. He also called for a holy war, including terrorist acts around the world. International air travel fell off dramatically, causing air carriers to cut back flights and lay off over 40,000 employees in January and February of 1991. Even the threat of terrorism has dramatic results. But terrorism was more than a threat. Hussein pursued a "scorched earth" policy of ecoterrorism and set over 550 oil wells on fire. He released into the Persian Gulf ten times more oil that was released in the Alaska disaster. The ground war in Iraq began on Saturday, February 23, 1991 and ended four days later. It is already being referred to as the Four-Day War or 100 Hour war. The humiliating defeat of the Iraqis could spark terrorist acts for years to come by those who supported Hussein.

COMBATING TERRORISM

The U.S. government's official policy toward terrorism is:

1. The government will make no concessions of any kind to terrorists.

2. State sponsors of terrorism will be made to pay a price for their actions through military retaliation or more peaceful means such as political, diplomatic, and economic actions, public diplomacy, and sanctions.

3. The government will follow a program of action based on practical measures to bring terrorists to justice, to disrupt their operations, and to destroy their networks.

4. The Department of State's Anti-Terrorism Training Assistance Program (ATA) provides training in antiterrorism techniques to law enforcement officials around the world.

4. The Department of State's Anti-Terrorism Training Assistance Program (ATA) provides training in antiterrorism techniques to law enforcement officials around the world.

When a terrorist incident occurs, things can be done to manage the incident. The action taken will depend on a large number of variables.

First, you need to decide if you are willing to negotiate with the terrorists at all. Most governments have policies either for or against negotiating with terrorists. Even though a government may have a policy against negotiation, it doesn't necessarily follow that they are unwilling to talk to them. And quite often, negotiations and agreements may go on behind the scenes in spite of stated policies.

Negotiation techniques--particularly the field of hostage negotiation--require special training and cannot be dealt with in this book because of the confidentiality of those techniques. Such negotiations are generally conducted by skilled and trained members of special military or law enforcement teams. The techniques, along with those of threat analysis, have been developed by psychologists, psychiatrists and students of human behavior to such a degree that the skills of hostage negotiation can be taught to those who have a need for such training.

One option sometimes exercised in the event of a hijacking is to storm the plane or location by force. This was done most notably in 1976 at Entebbe, Uganda, and in 1977 in Mogadishu, Somalia. There have been some successes and some failures. It has been proposed often that a multinational SWAT[11] team be formed. It would operate out of the Middle East or Europe and be made up of all nations committed to interdiction, so that there would be no need to wait for the Delta Force to arrive from the United States and no country need feel under political pressure not to intervene. The Holloway Commission, appointed by President Reagan to explore new ways of combating terrorism, has recommended "forward deployment" of some special forces and equipment to overseas locations.

The U.S. has a number of units trained for possible military intervention. The best known is Delta Force, the Army's elite antiterrorist unit, based in Fort Bragg, North Carolina The Navy has SEALS (Sea, Air and Land forces), the Army helicopter unit is Task Force 168 and the Air Force has First Special Operations Wing (1st SOW). The West German counterpart of Delta Force is G.S.G. 9, a highly trained antiterrorist unit, and the French have the GIGN. The FBI has the highly professional Hostage Rescue Team (HRT) based at the FBI Academy at Quantico, Virginia.

The one failure of the Delta Force was the attempt to rescue the Iranian hostages in 1980 in an operation called Desert One. The Delta Force was to be flown to a desert rendezvous and then to the outskirts of Tehran. Three of the helicopters broke down, and one crashed into a C-130. Part of the failure of this mission has been blamed on interservice rivalry and poor preparation. As a result, the Joint Special Operations Command was formed to prevent such interservice problems. The U.S. Government increased the budget of

---

11 Special Weapons and Tactics. A team specially trained to handle crisis situations.

Special Forces from $441 million in 1982 to $1.2 billion in 1986, and the number of troops grew from 11,000 to nearly 15,000.

A 1977 study completed by the Rand Corporation showed that 79 percent of all hostage deaths in terrorist situations occur during the rescue attempt. These included all terrorist situations and not just hijacking or aviation related incidents. Those statistics aren't that shocking or surprising. We must assume that the bulk of the other 21% occurred at the hands of the terrorists. With the high loss of life from planes being blown out of the sky, those percentages have most certainly changed.

It is a widely held belief that a country's national sovereignty cannot be violated by sending in an outside rescue team. It can be argued, however, that the government on whose soil the incident takes place has a responsibility to do all it can to secure the release of the hostages. If it does not fulfill that obligation, as has often happened in the past, then the country whose citizens are being held hostage, has a right to act when the host country refuses to do so. Benjamin Netanyahu, Israel's Ambassador to the United Nations, suggests the host country can either do it themselves or let someone else do it.[12]

Military intervention is a reasonable and legitimate course of action in many cases. The primary objections to such action are the threat of reprisals and the fear that innocent people will be injured or killed. Buckling under to the threat of reprisals is nothing more than the old protection racket. Concern for the safety of innocent victims is certainly legitimate, but to yield to that concern each and every time conveys only one message to terrorists: All they have to do is to take some hostages, and the world will meet their demands. To resist terrorism at the cost of innocent lives is a difficult but often essential course of action. The failure to act decisively, in order to guarantee the safety of hostages, condemns many future victims to death.

The *Report of the President's Commission on Aviation Security and Terrorism* (May, 1990) recommended, "The United States should pursue a more vigorous counterterrorism policy, particularly with respect to nations sponsoring terrorists." Current U.S. policy is to work with allies to identify, track, and apprehend, prosecute, and punish terrorists. The Commission found unacceptable the idea of holding ourselves in all cases to a criminal standard of proof before we act. The final recommendation was this: "The United States should ensure that all government resources are prepared for active measures--preemptive or retaliatory, direct or covert--against a series of targets in countries well-known to have engaged in state sponsored terrorism."

This controversial recommendation has been interpreted by some to mean assassination if necessary. In his article, William T. Corbett suggests that the "selective execution of individual terrorists guilty of crimes against humanity" might be a better national policy than actions such as air strikes that kill innocent people.[13]

---

12 Benjamin Netanyahu, "Terrorism, How the West Can Win." Time (April 14, 1986):48.

13 William T. Corbett, "A Time to Redefine Policy," Security Management (June 1990).

An argument can be made that one act of terrorism begets another and that to make a preemptive or retaliatory strike simply prompts a "get even" strike in return. There was, for example, speculation that the bombing of Pan Am Flight 103, presumably by the Popular Front for the Liberation of Palestine--General Command (PFLP-GC) was to avenge the Iranian airliner shot down by the U.S. Navy Cruiser Vincennes on July 3, 1988, in which 290 people were killed. The thought was put forth that Ahmed Jabril, who heads the PFLP--GC from Syria, was hired by the Iranians for the task. Evidence has now been put forth which shows that the bombing was carried out by Libyan intelligence in revenge for the 1986 U.S. bombing of Tripoli.[14]

Following the 1986 bombing of a Berlin disco frequented by U.S. military personnel, the U.S. bombed Libya, narrowly missing Colonel Muammar Khadafi, but killing his adopted daughter. Subsequently, three hostages in Lebanon and a U.S. diplomat in Sudan were murdered.

Similarly, on September 25, 1987, PLO terrorists killed three Israelis on a yacht in Cyprus. The Israelis then bombed PLO headquarters in Tunisia on October 1, nearly killing Arafat. The seizure of the *Achille Lauro* a week later was probably too soon to be considered retaliation, but on December 27, Abu Nidal's men attacked El Al ticket counters in Rome and Vienna airports, killing 19 and wounding more than 100 people.

FINANCES

In his book, *The Financing of Terror*,[15] James Adams argues that terrorism has gone capitalist. He believes that bombs, gunfire, and arrests won't do the job, and that terrorists must be struck at their economic roots. The PLO has investments valued at $5 billion in world financial centers and in the Near East. The total income of the various PLO factions runs to $1.25 billion. It is the richest terrorist organization in the world, with a bigger budget than some Third World countries. It extracts a 5 percent income tax from every Palestinian.

The Irish Republican Army now gets only $200,000 a year from donors in the U.S., a small proportion of its total income of $7.5 million. The rest comes from activities favored by organized crime, including protection rackets, frauds, smuggling, kidnap for ransom, and control of most Belfast taxicabs, according to Adams.

Kidnaping, in this and other parts of the world, is one of the best cash sources for terrorists. In 1988, kidnapings accounted for approximately 5 percent of terrorist incidents. It is also currently a big "fund-raising" business in Rio de Janeiro. From January through the middle of July 1990, there were about 30 kidnapings in Rio. That is more than the total for all of 1989 and three times as many as in 1987. One twist in these kidnapings is that by choosing mostly middle-class businessmen and demanding small ransoms,

---

[14] San Jose Mercury News, June 24, 1991.

[15] James Adams, *The Financing of Terror: Behind the PLO*, Simon and Schuster, (1986).

sometimes as low as $2,000, the kidnapers have been able to negotiate settlements more quickly and cleanly. So far, all the ransom demands have been met, earning more than $16 million for the kidnapers.

The financing of terrorism is not the problem we thought it was, however. In May 1986, South African commandos who took part in a raid at Harare, Zimbabwe, traveled there in three Hertz cars rented in Bulawayo. The raiders paid a deposit of $1240 and complied with normal requirements for renting a car.

WHERE ARE WE NOW?

According to the report of the United States Department of State entitled, *Patterns of Global Terrorism: 1989*, published in April, 1990, the level of international terrorism worldwide in 1989 declined sharply from that of 1988, dropping by almost 38 percent from 856 incidents to 528. The Middle East continued to experience the largest number of incidents of international terrorism, incurring 193 attacks -- 37 percent of the total worldwide. The proportion of international terrorism connected with the Middle East increases to 45 percent, however, when Middle East spillover attacks into other regions are added. With 131 attacks, 25 percent of the total, Latin America ranked second. Western Europe was third with 96 incidents.

United States interests continued to be the most frequent targets of international terrorists. In 1988, the attacks directed against the United States numbered 193, compared with 165 in 1989, a decline of 15 percent. Despite this decline in international attacks, the State Department warns that terrorists retain the potential for resuming a greater level of violence, particularly against the United States.

The State Department attributes the decline to several factors. The Afghan government curtailed its terrorist campaign in Pakistan after Soviet troops were withdrawn. Yasser Arafat's renunciation of terrorism resulted in a sharp decline in operations by groups affiliated with the PLO. Dissension within the Abu Nidal Organization (ANO) -- previously one of the most active and deadly terrorist groups -- and its focus on Lebanese militia matters decreased the group's operations. A number of states involved in terrorism, including Libya and Syria, remained wary of getting caught sponsoring terrorists and reduced their support. Iran was a notable exception to the trend.

Brian Jenkins, formerly of the Rand Corporation, has forecast an increase of terrorism, and he suggests that large-scale incidents will become more common.[16] He feels that as prime targets such as government buildings and airline terminals are protected, terrorists will become less discriminate in their attacks, creating greater public outrage.

A listing of Significant Incidents of terrorism are included as Appendix A of this book.

---

[16] Brian Jenkins *Terrorism Violence Insurgency Report*, The Rand Corporation (December, 1985).

Chapter 3

TERRORISM COUNTERACTION

From January 5, 1973, when the mandatory screening of passengers and carry-on items went into effect, through 1989, over 11 billion persons and 12 billion carry-on items have been screened. Over 44,000 firearms have been detected resulting in over 20,000 related arrests.

During the 16 years from 1973 through 1989, there were 120 hijacking incidents involving U.S. carriers (an average of 6.6 per year). This compares favorably with 133 hijackings (an average of 27 per year) recorded in the 5 years (1968 to 1972) immediately preceding the implementation of the mandatory screening procedures. Since the addition of new security procedures following the hijacking of TWA Flight 847 in June 1985, the number of hijackings has been reduced to an average of 4 per year. In 1988 and 1989, U.S. airline hijackings were at their lowest levels since 1973, with 2 and 3 incidents a year, respectively.

For many years there have been international differences of opinion on the question of the relative morality of air piracy as an instrument for political change. A parallel has been drawn to the civil disobedience demonstrations of the fifties in the U.S. and elsewhere. However, nonviolent disobedience to laws believed to be unjust is not comparable to air piracy as a force for change. The first is passive, and the perpetrator alone risks injury; the other is active and invariably carries a high risk of death and injury for uninvolved persons. Indeed, as we have seen, some activists have quite deliberately killed innocent persons, appearing to believe that this offers force to their argument.

While the "relative morality" debate may seem hardly debatable, another related question is quite familiar to Americans. In the eyes of some United States citizens, commandeering an airplane to escape to the West from behind the Iron Curtain is seen as justified, admirable, and even praiseworthy, while the hijacking of an airplane to Cuba by persons seeking to flee from the United States is seen as wrong and even despicable. Clearly, general agreement is needed if air piracy is to be abated everywhere.

The International Civil Aviation Organization (ICAO), formed in Chicago in 1944, took no action on air piracy until June 1970, and even omitted the problem in its 1963 Convention on Offenses and Certain Other Acts Committed Onboard Aircraft. The 1970 ICAO action was a resolution condemning all acts of violence directed against personnel, airports, and facilities, and it called for strengthening existing extradition arrangements.

The ICAO has no monopoly on procrastination, of course. Three months after commercial aircraft hijacking came to a halt in the U.S., the United States and Cuba signed a pact agreeing that both nations would either extradite or punish with "the most severe penalty" any aircraft hijacker landing in one of the countries from the other.

NEW STRENGTH IN U.S. LAW

On August 5, 1974, Public Law 93-366 was signed into law. Title I is known as the Anti-Hijacking Act of 1974, and Title II, the Air Transportation Security Act of 1974. The law implements in the United States the Hague Convention for the Suppression of Unlawful Seizure of Aircraft, and significantly changes the Federal Aviation Act of 1958.

On an international basis, the law gives the President statutory authority to suspend "without notice or hearing and for as long as he determines necessary" the right of any U.S. or foreign air carrier to engage in air transportation between the United States and any nation that permits its territory to be used as a base of operations or sanctuary for terrorist organizations that engage in aircraft piracy. A "secondary boycott" provision gives the President the right to suspend foreign air commerce between the United States and any foreign nation that maintains air service between itself and an offending foreign nation.

The Secretary of Transportation, with the approval of the Secretary of State, may withhold, revoke, or impose conditions on the operating authority of an airline which does not maintain security standards equal to or above the minimum standards set by the Convention on International Civil Aviation.

Subsequently, the FAA issued FAR 129, entitled "Foreign Air Carrier Security Program," which was published in the *Federal Register* on July 11, 1975. It adds a new section, FAR 129.25, which states that, after October 9, 1975, foreign air carriers coming into or leaving the U.S. shall screen passengers and their carry-on baggage. It also contains provisions for searching aircraft that have been identified with bomb threats, and prohibits the carriage of weapons in the cabin. While many foreign carriers complied with this requirement, some question existed as to whether the section had enforcement provisions. It appeared to only require screening of passengers and property carried in the cabin "in accordance with that air carrier's security program." On March 13, 1989, the FAA issued its final rule requiring foreign air carriers to submit their security programs (in writing and in English) to the FAA for acceptance by the Administrator.

Among other provisions, Public Law 93-366 expanded the legal definition of "in flight" from the previous wording ("from the time power is applied for take-off until the landing run ends,") to read "from the moment when all external doors are closed following embarkation until the moment when one such door is opened for disembarkation." This change was necessary to meet the problem presented by the suspect apprehended in the aircraft before the plane has started its takeoff roll.

A similar necessity dictated the amendment of the wording of the regulation concerning the carriage of weapons on board aircraft. The new wording prohibits attempting to carry a weapon on board an aircraft, either on one's person or in one's property, if the weapon would be accessible to that person in flight. Previously, there was no penalty for carrying a weapon that was detected at the screening checkpoint.

A major jurisdictional change gave the FAA Administrator "exclusive responsibility for the direction of any law enforcement activity affecting the safety of persons aboard aircraft in flight," a provision apparently inserted in response to criticism of FBI actions in some hijackings. In the case of the TWA 707 hijacked from California to Rome on October 31, 1969, the flight captain complained that the FBI agents "needlessly endangered the lives of the entire TWA crew as well as themselves" by crowding around the plane as it was being refueled in New York. Substantial public criticism was also expressed when FBI agents shot out the tires of a Southern Airways DC-9 hijacked 24 hours before. The crew had had no relief, 31 passengers were still on board, and the plane was on the ground in Orlando, Florida, refueling for a flight to Havana with the pilot held at gunpoint. Against all probability, the pilot managed to get the plane airborne and to land it safely at Havana.

Another provision of the new law imposes the death penalty or life imprisonment if the commission of a hijacking results in the death of another person, subject to specific provisions of the law.

One section appears to be an omnibus clause that gives the Administrator the power to "prescribe such reasonable rules and regulations requiring such practices, methods, and procedures, or governing the design, materials, and construction of aircraft, as he may deem necessary to protect persons and property aboard aircraft...against acts of criminal violence and aircraft piracy."

Significantly, the law requires the FAA Administrator to ensure courteous and efficient treatment of passengers by air carriers during the screening process. This seems reasonable; if passengers and visitors are not treated courteously during the screening process, long-range public tolerance of the program will be adversely affected. Both the government and the carriers must be aware of this fact. The government must avoid imposing unnecessary rules and practices that would encumber and delay the screening process, while the carriers must continuously train and prompt their screening personnel in courtesy and friendliness. Historically, the airlines have been most conscious of the role played by the courtesy and friendliness of their "public contact" people, but the training of screening personnel has sometimes been neglected in this regard. Nowhere is a consistent courtesy more needed; a careless or abrupt attitude in an already tense situation can nullify the efforts of all other personnel to make the flight a pleasant one. The passenger who boards with "feathers ruffled" presents the flight attendant with a real job.

INTERNATIONAL LAW

One nation cannot do the job alone. Efforts to achieve multilateral action through international law have met with some success, and the United States is a signatory to the three agreements developed so far. The first, the *Convention on Offenses and Certain Other Acts Committed on Board Aircraft* (the Tokyo Convention of September 1963), the so-called "Crimes Aboard Aircraft Convention," contained a commitment by the party nations to take custody of the perpetrators of criminal acts aboard aircraft and expedite the continued journey of the aircraft, its crew and its passengers. Some 123 nations are parties to this treaty. In the U.S., it has been implemented by PL 91-449.

The second treaty, the *Convention for Suppression of Unlawful Seizure of Aircraft* (the Hague Convention of December 1970), commonly referred to as the "Hijacking Convention" provides that party nations, some 128 in number, will apprehend and prosecute or extradite hijackers and ensure that hijacking is subject to severe penalties. The Hague Convention has been implemented in the U.S. as PL 93-366.

The third treaty, the *Convention for the Suppression of Unlawful Acts Against the Safety of Civil Aviation* (The Montreal Convention of September 1971), usually cited as the "Sabotage Convention," calls for apprehension and prosecution or extradition of aircraft saboteurs, and it also contemplates severe penalties for perpetrators of these heinous crimes. It also calls on the 128 party nations to take all practicable measures to prevent sabotage and to forewarn all affected nations of such offenses when it has reason to believe they will be committed.

Taken collectively, these three treaties would appear to ensure the peaceful, orderly, and expeditious conduct of international airline operations and swift and appropriate justice for those who unlawfully interfere with such operations. Unfortunately, that has not been the case, despite the fact that Israel, Lebanon, Libya, and Syria are signatories, along with the United States, of all three treaties.

By 1973, many nations concluded that the three treaties did not suffice because some parties were refusing to extradite, imposing slight penalties for heinous crimes in deference to their perceived political nature, or otherwise violating the spirit and letter of their treaty obligations by offering safe havens to hijackers and terrorists in general.

Consequently, there was a ground swell of support for the development of a sanctions convention, that would entail a collective suspension of air service to any nation that did not live up to its commitment under the three treaties. However, shortly before the Rome conference was convened, yet another attack on civil aircraft generated a very heated protest from much of the civil aviation community. The Rome diplomatic conference degenerated to exhaustive, unproductive polemics and produced total disillusionment and abandonment of the initiative. Even an alternative suggestion by the international airline community, through the International Air Transport Association (IATA), that the Chicago Convention be amended by incorporating the operative provisions of Tokyo, Hague, and Montreal, with the potential for expulsion from ICAO for violations of the amended Chicago Convention, gained no support.

The need for some enforcement strength survived, however. Thus, at the July 1978 economic summit in Bonn, after an impromptu mention of the problem by the Japanese Prime Minister, Prime Minister Pierre Trudeau of Canada pressed for a joint declaration on the subject by the seven powers. The so-called Bonn Declaration, the only affirmative agreement reached at that summit, asserted that Canada, the Federal Republic of Germany, France, Italy, Japan, the United Kingdom, and the United States were intensifying their joint efforts to combat international terrorism by agreeing to cease all flights to any country that refused to extradite or prosecute hijackers or did not return a hijacked aircraft and to halt all incoming flights from any country by the flag airline of the offending nation. The President's Commission on Aviation Security and Terrorism has recommended that the signatories to the Bonn Declaration adopt an annex declaring that members would also halt air services in cases of unpunished attacks at airports and airline ticket offices.

The spontaneous joint declaration, lacking the benefits of advance preparation, triggered a series of meetings of experts from the seven countries seeking agreement on how to implement the understanding in an appropriate situation. In the course of that immediate post-Bonn flurry, some twenty other nations formally registered their concurrence with the declaration, and another fifty nations informally conveyed concurrence.

However, partly due to inherent legal problems repeatedly cited by the French, the euphoria subsided. Indeed, except for some collective saber-rattling at South Africa for offering refuge to the mercenaries who failed in an attempt to overthrow the government of the Seychelles and a temporary collective suspension of service to Afghanistan for its failure to prosecute or extradite hijackers of a Pakistan International Airways flight, the Bonn Declaration has never been used.

The airline industry was gratified by the renewed effort to address international terrorism at the Tokyo Summit and most particularly by the intention expressed by the seven chiefs of state to make the Bonn Declaration more effective in dealing with all forms of terrorism against civil aviation.

The Tokyo Summit Statement on International Terrorism, issued May 5, 1986, committed the head of the seven major democracies to "make maximum efforts to fight against the scourge." Specifically, they agreed to the following measures:

- Refusal to export arms to states which sponsor or support terrorism.
- Strict limits on the size of the diplomatic and consular missions and other official bodies abroad of states which engage in such activities, control of travel of members of such missions and bodies, and,

where appropriate, radical reductions in, or even the closure of, such missions and bodies.

- Denial of entry to all persons, including diplomatic personnel, who have been expelled or excluded from one of our states on suspicion of involvement in international terrorism or who have been convicted of such a terrorist offense.

- Improved extradition procedures within due process of domestic law for bringing to trial those who have perpetrated such acts of terrorism.

- Stricter immigration and visa requirements and procedures in respect of nationals of states which sponsor or support terrorism.

- The closest possible bilateral and multilateral cooperation between police and security organizations and other relevant authorities in the fight against terrorism.

James E. Landry, Senior Vice President and General Counsel of the Air Transport Association, has suggested the following as necessary follow-up action:[1]

- Expand the scope of the Bonn Declaration to cover acts other than hijackings.

- Closely monitor incidents to assure proper and timely actions by governments.

- Refine and expand appropriate sanctions with improved criteria and mechanism for application.

- Collaborate, particularly with like-minded nations, in such international fora as ICAO, drawing upon their expertise to improve and extend countermeasures against terrorism and those who sponsor or support it.

- Strengthen and accelerate ICAO's existing process for the promulgation of international security standards and recommended practices (SARPs)Security standards and recommended practices (SARPs).

- Authorize and provide resources for ICAO to evaluate the extent and methods of states in applying the SARPs and their effectiveness in combating terrorist and other crimes against aviation.

- Convene nations again for the adoption of a "Sanctions" convention to supplement the coverage of the Tokyo, Hague, and Montreal Conventions by providing for identification and sanction of states that fail to adhere to the obligations of those conventions, who fail

[1] Remarks of James E. Landry before the National Forum Foundation's Conference on "Terrorism and Transportation: Threats and Responses" (May 12, 1986).

> to effectively apply the ICAO SARPs, or who otherwise support or condone terrorist or other criminal acts against civil aviation.

Additionally, it should be pointed out that the airline industry is doing its part. In May 1990, the Air Transport Association and the Air Line Pilots' Association (ALPA) offered a $2 million reward for information about terrorist activities, past or planned, against any ATA member airline worldwide. Added to the $2 million reward offered by the U.S. Department of State, it amounts to a rather healthy sum. The reward was first proposed by Captain Bruce Smith, a commercial airline pilot who lost his wife in the bombing of Pan Am 103. Rewards of more than $500,000 have been paid under this program.[2]

WHAT IS AN ATTEMPT?

The FAA keeps statistics on hijackings or related crimes that may have been prevented by airline and airport security measures. Obviously, not all of the 40,200 firearms detected during the 15 years of the screening program, would have been used to hijack aircraft. Some situations, however, reasonably give rise to a belief that a crime may have been prevented. It is one thing if a weapon is discovered in a purse or briefcase of someone who carries it for self-protection. It is another thing entirely if an individual has a pistol taped to the center of his back or if the pistol is disassembled and hidden in a portable radio. By the end of 1987, the FAA concluded that in 117 such incidents, a hijacking was contemplated. Other statistics furnished by the FAA stated that for the 10 years from January 1978 through December 1988, there were 93 actual hijackings of U.S. air carrier aircraft and 46 prevented hijackings.

Whether hijack attempts or incidents, these cases also offer a serious indication that the point of confrontation is moving from the better protected aircraft to the more open terminal. The killing of 25 persons and wounding 76 in a pointless machine-gunning of tourists within Tel Aviv's Lod Airport Terminal on May 31,1972, the killing of 32 persons in the Rome-Athens attack in December 1973, and the Rome-Vienna killings of 18 people in December 1985, awakened much of the world to the need for increased airport security. In each of these instances, passenger terminals were sprayed with automatic weapons fire. It is sobering to realize that passenger and carry-on baggage inspections would not have averted any of these attacks.

FOREIGN AIRPORT SECURITY

A number of the procedures that passengers on international flights must pass through have a secondary security benefit. In routine processing, the passenger is required to check in at the ticket counter and check their baggage. They

---

[2] *Patterns of Global Terrorism: 1990,* U.S. Department of State, April, 1991: iii.

then go through passport control, where they are further identified. They return to their baggage, which is now on a belt at customs control, to be inspected by customs. Following this, the passengers go through a carry-on search before they are allowed onto the airport bus or into a jetway. There someone takes the boarding passes and in many European airports, there is a guard at the foot of the boarding stairs to make certain only authorized persons get on the plane.

During recent years, I've traveled in the Far East and have seen very good security at a number of airports, particularly Narita. Passengers are searched at Narita before they even get onto airport property. One need only look at the armored personnel carriers at various checkpoints leading to the terminal to know that the authorities mean business. And they are still doing a pat-down search on a random basis of all passengers. Other airports, such as Bangkok, Singapore, Taipei and airports in Australia and New Zealand, were observed and are doing an excellent job of screening.

Athens came under severe criticism for the effectiveness of its screening, particularly after the June 1985 hijacking of TWA Flight 847, but it appears to have upgraded the quality of its security. A travel advisory was published but was lifted on July 22, 1985. Shortly thereafter, an Egyptair jet was hijacked from Athens and flown to Malta, an indication that adequate security was still lacking in Athens. In October 1986, a European bank agreed to lend Greece's socialist government $5 million for security improvements and modernizations at Athens Hellenikon International Airport. The government had already renovated perimeter fences around the airport and increased the security staff. Armored cars were stationed at airport exits and security police with automatic weapons guarded departure lounges and aircraft ramp areas.

There were such severe problems with the screening in Manila that sanctions were imposed by the U.S. government and warnings were posted indicating that it was an unsafe airport, in order to force the Philippine government into acting. The problem was not that screening wasn't being done, but allegations were made that it was extremely easy to bribe your way past screening, as was done in at least one case by gun runners. Some U.S. carriers set up their own screening in addition to that provided by the airport, because they had so little faith in the airport screening.

In January 1991, notices were observed in U.S. airports stating, "Passengers are advised that the U.S. Secretary of Transportation has determined that Jorge Chavez International Airport, Lima, Peru, does not maintain and administer effective aviation security measures." On January 25, 1991, a car bomb exploded in the parking lot of Lima's airport, killing one person and injuring at least four. Minutes later, leftist guerrillas fired on the U.S. Embassy. A caller claimed responsibility for the bomb in the name of the Shining Path "in defense of Arabs everywhere."

Public Law 99-83, the Foreign Airport Security Act which was contained in the International Security and Development Cooperation Act of 1985, was enacted on August 8, 1985. Title V, Part B of the Act amends section 1115 of the Federal Aviation Act of 1958 and directs the Secretary of Transportation to assess the effectiveness of security measures at those foreign airports being served by U.S. air carriers, those foreign airports from which foreign air carriers serve the United States, those foreign airports that pose a high risk of introducing danger to international travel, and other airports as the Secretary may deem appropriate. The act requires that specific action be taken regarding airports that do not maintain and administer effective security measures.

The FAA has been delegated the responsibility for the implementation of certain legislative requirements in Public Law 99-83. These include the assessment of security measures at foreign airports and consultation with the Secretary of State concerning threats to U.S. citizens traveling abroad.

Nearly 250 foreign airports currently meet the assessment requirements of Public Law 99-83. This number fluctuates as changes in air carrier service take place at these airports. During the last half of 1989, 100 assessments were conducted at 96 airports in 47 foreign countries.

Assessments consist of an in-depth analysis of the security measures at the airports visited, using a standard that is based, at a minimum, on the Standards and Appropriate Recommended Practices (SARPs) contained in Annex 17 to the Convention on International Civil Aviation. If the FAA develops information indicating that an airport does not maintain and administer effective security measures, these assessments are reported to the Secretary of Transportation.

Public Law 99-83 provides for notification to the foreign country involved when a determination is made by the Secretary of Transportation that a foreign airport does not maintain and administer effective security measures. Notifications include recommended steps to remedy the problem. The law also specifies when and how the public is to be notified of that determination. Public notice occurs when the foreign government fails to bring security measures up to the standard within 90 days of being notified of the Secretary of Transportation's determination.

If the Secretary of Transportation at any time determines, after consultation with the Secretary of State, that a condition exists which threatens the safety or security of passengers, aircraft, or crew traveling to or from a specified airport, the Secretary of Transportation must immediately initiate the public notification procedures and, in addition, apprise the Secretary of State, who must issue a travel advisory. Under these circumstances, the Secretary of Transportation is also required to consider whether the public interest necessitates the immediate suspension of service between the United States and the specified airport.

On December 22, 1988, the General Accounting Office (GAO) recommended that the FAA perform additional analyses during foreign airport assessments, including a review of the host country's security system testing procedures.

In their report, the Presidents Commission on Aviation Security and Terrorism pointed out some of the difficulties in securing foreign airports . The report deals with government relationships rather than specific airports, referring to differing procedures in such countries as West Germany, the United Kingdom, France and Scandinavian countries. The International Civil Aviation Organization, a U.N. body, recommends standard practices for aviation security, but these standards are at a basic level and not stringent enough for high-threat international airports. Additionally, the ICAO has no enforcement mechanism. Several state sponsors of terrorism are ICAO members and the ICAO must provide them with a copy of their security manual--a case of letting the fox in the hen house if there ever was one.

Although the Foreign Airport Security Act tasks the FAA to evaluate the security of foreign airports, it only has jurisdiction over foreign carriers on the last leg of their flight to the U.S. There are no surprise inspections of foreign carriers; FAA inspectors are required to make appointments and to give advance notification of any inspections, and they are not allowed to roam over the airport unaccompanied. In some cases, they are not permitted in certain areas, such as the air operations area. When the FAA does make recommendations, the foreign carriers are free to accept or reject them.

Not only is the FAA handicapped with regard to foreign carriers, but U.S. air carriers operating out of foreign airports are equally handicapped. In many cases, FAA mandates that the U.S. carrier take certain action and have been blocked by foreign governments. A *Wall Street Journal* article[3] dealing with the difficulty of trying to improve security abroad cited some examples. For Instance, when the FAA ordered U.S. carriers to X-ray or search every bag checked in Western Europe and the Middle East, British authorities initially blocked the move at London's Heathrow airport. Britain argued that it would create long lines that would jam the terminal. However, in the end, the British relented.

In Amsterdam, however, the Dutch refused to permit X-raying of every bag, arguing that concentrating on a few high-risk passengers was more effective. Timothy R. Thornton, a former Northwest Airlines Vice President, told the President's Commission, "We were in the middle. Sometimes we were in violation of American laws; sometimes we were in violation of the foreign law....For six months, it was whose jail did we want to go to?"[4]

Similarly, the security inspections used by TWA at Brussels failed to pass the FAA inspection in 1989. TWA wanted to hire their own screeners, but the Belgian government insisted that the airline use a local airport-security company instead. Pan Am and American Airlines have the same problem in Brussels, and a similar situation exists in Oslo. Sometimes the road blocks erected by foreign governments are politically motivated, sometimes they are the result of parochial concerns; and sometimes they are put up in the name of sovereignty. Often, however, they are honest differences over security techniques.

This also creates a competitive disadvantage for U.S. carriers operating out of foreign airports. The hand search of baggage and extensive questioning of certain passengers create lines, space problems, time delays, and frustration. Another disadvantage is that they expose passengers to an added risk of terrorist actions by concentrating the passengers in the terminal for longer periods of time. Passenger who choose to fly an American flag carrier and undergo this added security "hassle" may feel better about the safety of their flight, but the frequent flyers may take a non-U.S. carrier and avoid the

[3] *Wall Street Journal,* July 11, 1990.

[4] *Report of the President's Commission on Aviation Security and Terrorism*, May 15, 1990: 32.

extended screening process. A similar situation might exist in the U.S. between two major carriers if one required screening and one did not.

Additionally, many, if not most, of the foreign flag carriers are financed by their governments, while the U.S. carriers are privately owned. This puts the U.S. carriers at a further financial disadvantage. The President's Commission found that 45 percent of U.S. international passengers fly on foreign carriers, many on the belief that they are "safer." Actually, security measures performed by U.S. carriers are more extensive than that of many foreign carriers. The Commission pointed out these problems with a strong recommendation that the U.S. Department of State assume from the carriers the role of negotiating aviation security measures with foreign governments and remove the carrier from the middle of any conflict between the FAA and the foreign government.

The U.S. air carriers feel there is a need to establish and maintain a "level playing field" by requiring foreign air carriers to operate under the same security rules as U.S. air carriers. In 1985 and again on December 29, 1988, the FAA mandated extraordinary security requirements for U.S. airlines at some 45 high-threat airports. But the same heightened security has not been required of foreign airlines that serve the U.S. from the same high-risk locations. The Air Transport Association notes that of the 12 explosions aboard commercial aircraft from 1984 to 1988, nine involved foreign air carriers.

Commenting on British Airways' doubling of passenger revenues, *Business Week* [5] said "Improved service accounts for part of the gain, but so does the Pan Am crash at Lockerbie, Scotland. which fed fears of flying U.S. carriers... BA's record still isn't perfect: Capacity crowds on transatlantic flights since the Lockerbie disaster have overstretched staff"

On March 14, 1989, in testimony before the Senate Committee on Appropriations, Secretary of Transportation Samuel K. Skinner announced the decision to amend FAR 129.25 to require foreign carriers to submit their security plans to the FAA for approval. The FAA uses the standards and recommended practices contained in ICAO Annex 17 as the yardstick against which those plans are measured.

Poorer nations that cannot afford the expense of airport security and currently have little or no security present an additional problem. As one official of a Third World country pointed out, "We cannot afford even an extra X-ray machine for the hospital, let alone any new machine for airport security."[6] It may be that as security is beefed up at European and Far East airports, terrorists may choose to strike through weaknesses in airports in Central and West Africa and other low-income countries.

---

[5] *Business Week*, March 12, 1990.

[6] Paul Wilkinson, *The Lessons of Lockerbie*. Research Institute for the Study of Conflict and Terrorism , London (December 1989):7

WHAT IS REASONABLE SECURITY FOR A U.S. AIRPORT?

Just how tight should security be at U.S. airports? It can be tighter. Should it be? One hears warnings that although we have been spared serious terrorist incidents at airports in the United States, that one of these days, it's going to happen. Therefore we should prepare for such an eventuality as if our airports were in the Middle East.

For a while, it seemed to be fashionable among the media for a crusading newspaper or TV reporter to attempt to penetrate screening with a toy pistol. We would then watch ourselves on the 11:00 news. Even "60 Minutes" got into the act. Let's face it. Our defenses are not impenetrable, and they are not designed to be. They are designed to prevent and deter hijackings. They have served that purpose. Granted, some weapons do get through, but a great number are detected.

On January 4, 1989, a Dutch TV station reporter smuggled a bomb onto a London-bound flight, and a French television worker was arrested and charged with checking fake bombs in New York on three different air carriers bound for London. Schiphol Airport, Amsterdam, was also infiltrated by a TV reporter carrying an attaché case holding a fake bomb. The event was filmed by the station. The three fake bombs checked in New York by the French TV worker were left for shipment in the cargo areas of TWA, Pan Am, and Air France. One of the fake explosives was noticed by TWA personnel, which resulted in the revelation of the hoax. The producer and reporter for the station were arrested and charged by the United States Attorney with "willfully and maliciously" conspiring to violate the Federal Air Safety Act, a crime that carries a maximum five-year sentence upon conviction.

On July 1, 1990, a man whose daughter had been killed on Pan Am 103 said he had smuggled a fake bomb past security officials at Heathrow several weeks earlier and checked it aboard a flight to the United States. In this case, instead of being arrested, the individual was invited to meet with Department of Transportation officials and give them details on how he had accomplished his feat. Although I deeply sympathize with this gentleman, I do not agree with the action of attempting to smuggle fake bombs aboard an aircraft.

It is an axiom of security that the protection should be commensurate with the threat. During a trip to Jordan and Israel, I crossed the Allenby Bridge over the Jordan River on the road from Amman to Jericho and Jerusalem. My passport was checked about five times in a distance of about 100 yards. We were under the surveillance of soldiers with machine guns trained on us from bunkers on both sides of the Jordan River. We were required to leave the bus and undergo a thorough search that took about two hours for 20 of us. We were told we were fortunate to get through so quickly because it was not a busy day. The search included a pat down in a booth, and we were required to take all items out of our pockets. One member of the party had to undergo a strip search. The soldiers then emptied our suitcases onto a table item by item, X-raying individual items if there was any question. Then, the empty suitcase was X-rayed. Finally, we were allowed to repack our bags and board the bus again. We tolerated the search for a number of reasons. The automatic weapons were actually reason enough. We also understood that the country of Israel was in a real state of war and border forays and commando raids were the order of the day. We had been told in advance what to expect and were prepared for it. In essence, we felt the search was justified, and there really wasn't anything we could do about it.

In the end, the security measures employed were great because the threat was great. I suspect we would hear screams of public outrage if we were to institute similar measures today at San Francisco International Airport. Why did we tolerate in Israel something we would object to in the United States? In the Middle East, we felt the threat was real, and thus, the search was justified. In the same way, the American people seem to accept present screening procedures and feel they are justified in light of the present threat. They will accept and even demand adequate security to cope with the reality of the threat. They will not accept security that appears to go beyond that threat.

What factors are involved in assessing of just how much security is needed in a given situation?

First, you have to assess your *vulnerability*. How easy a target are you? Is the property exposed, unprotected, accessible, or unguarded? Then, you need to evaluate the *criticality* of what you are protecting. What will be the impact if a serious loss or event occurs? Would it interrupt current operations? Would it have long-range effects? Third, you need to determine, as best you can, the *probability* of the serious loss or event occurring. If the protection is not afforded, is it highly probable that the loss will occur, or are you protecting against a relatively remote possibility? Fourth, the *cost of protection* must be weighed. The burden of risk must be compared to the cost of the control. As a general rule, the cost of the control should not ordinarily exceed the value of that which it is designed to protect, although there are obvious exceptions to this general rule.

As they relate to airports in the United States, what is the vulnerability, criticality, probability and cost of protection? In November 1978, a man rammed his car through a gate at the Dane County Regional Airport in Madison, Wisconsin. He drove up to a North Central Airlines aircraft, boarded it, and announced that he had a bomb in a white plastic bag and that he could detonate it by puncturing the bag with a small knife he was carrying. He was captured and underwent psychiatric treatment. What do we do to protect gates at our airports against such attacks--not just at Madison, Wisconsin but all over the U.S.? At the railroad depot and airport fuel farm at Narita Airport outside Tokyo, it has been recommended that measures be taken to prevent the breaching of security fences by heavy-duty vehicles. Such measures include the use of low, reinforced concrete walls, tank trap posts and the construction of six-foot deep ditches outside the security fence.

If it is appropriate for Tokyo, why not San Francisco or Los Angeles? We are certainly vulnerable. The gates we have now wouldn't stand up against such an attack. However, the probability of the loss occurring, which we know to be slight, and the cost of protection, which is great, generally rule out the construction of tank traps at San Francisco. But at Narita, it is prudent security, because the airport has already sustained such attacks. I want to stress that San Francisco cannot be said to have poor security because it doesn't have tank traps. It has good security, which means that the protection is adequate in view of the vulnerability, criticality, probability, and the cost.

Therefore, after the United States experiences its first serious air transportation terrorist incident, then and only then, will it raise its level of security to a level adequate to protect against terrorism. There are those who would say that Pan Am 103 was that incident, and to a degree that is true. But psychologically, that tragedy was an "overseas" incident. If that plane had been headed from New York to San Francisco, rather than London to New York, the impact on the average citizen would have been considerably different. As a former head of law enforcement for one of the country's largest airports testified before the President's Commission, "FAA should move in the direction of closing the gaps now, not wait until we have a significant domestic problem."[7]

Several incidents in recent history have prompted the demand for increased security at airports. It needs to be recognized that when a highly publicized incident occurs, government officials feel a need to take immediate action so that they will appear to be taking the proper corrective action. On occasions, this stance can be more over reactive than appropriately reactive.

Following the June 14, 1985, hijacking of TWA Flight 847, there was immediate pressure for the institution of new security measures. The Secretary of Transportation mandated several new security measures that created concern among airline officials as to the meaning and substance of some of the procedures. Airline officials expressed the thought that the new changes "for the most part disregard the input of the airlines."[8] There were indications that the FAA officials had wanted to consult with the airlines before mandating the procedures but had been overruled by the Secretary of Transportation. Additionally, the new procedures imposed by Transportation Secretary Elizabeth Dole were anticipated to cost the airlines millions of dollars. An eight-fold increase in crew training alone was expected to cost carriers at least $30 million.[9] One source called the measures "a political play...a reaction designed for public consumption and political benefit."[10]

Among the procedures put into effect was a prohibition of curbside baggage check-in for flights to, from, or outside the U.S. by U.S. carriers, restrictions relating to handling of freight and mail, expanded use of Federal Air Marshals, and the establishment of a Ground Security Coordinator and an Inflight Security Coordinator for each flight, who was the pilot in command of each flight. Forty hours of training and eight hours of annual recurrent training were required for the Ground Security Coordinator. The final rule on Security Coordination and Training became effective on July 11, 1985, as FAR §108.27.

Another reaction occurred following the downing of PSA Flight 1771 by a fired employee who had not been screened, either because he was recognized as an employee or because he had kept his I.D. badge. It was de-

---

[7] *Report of the President's Commission on Aviation Security and Terrorism*, May 15, 1990: 57.

[8] *Aviation Daily*, July 10, 1985: 50

[9] *Aviation Daily*, 280:1, July 1, 1985.

[10] *Aviation Daily*, June 27, 1985: 322

cided following this incident that all airline employees who passed through a security checkpoint had to be screened and were no longer exempt. The only weakness in that procedure is that most airline employees have numerous ways to get to the air operations area other than through the screening checkpoint. The new procedure did not cure the problem of an airline employee hijacking an aircraft, but it did make it more difficult for people working at the airport to travel from one part of the airport to another. And, of course, it was often necessary to set up a separate checkpoint for employees only--an added cost.

As an outgrowth of this incident, however, the FAA, on January 4, 1989, mandated the use of computerized access card control or alternative methods providing equal security for 270 airports that handle larger aircraft carrying more than 60 passengers. With computerized access control, it is possible to zone an access point by time or location. That is, an individual card can be programmed to allow an employee into a door just during certain hours or to provide access to some doors, but not others. If cards are lost or employees are terminated, they can be instantly programmed out of the computer. The cost to carriers for this program is estimated at $10 million over the next ten years .

EL AL SECURITY

The argument is frequently made that if U.S. carriers would just implement the methods used by El Al, it would solve all the security problems and make flights much safer. If all carriers were to introduce the military-style security used by El Al, air traffic in the U.S. would come to a halt. El Al has no short-haul flights, only long international flights. They carry about 150,000 people annually to 23 destinations.[11] They have a fleet of 20 aircraft as contrasted with the 300 to 500 aircraft of several major U.S. carriers, with hundreds of flights departing out of major airports each day. If you have one flight a day out of an airport, certain extra precautions can be taken. If you have over 200 flights a day out of that airport, the expense of those extra precautions would put the carrier out of business and delay flights to an intolerable degree.

[11] Noel C. Koch, "Airlines on the Front Lines," *Wall Street Journal*, September 12, 1986.

Chapter 4

PAN AM 103

On December 21, 1988, Pan Am Flight 103, which had originated in Frankfurt, pushed back from the gate at Heathrow Airport, London, at 6:07 P.M. local time with 243 passengers and a crew of 16. The plane also carried about 20 tons of cargo, including 43 bags of U.S. Military mail. The flight took off at 6:25 P.M., climbed out enroute to New York City, and leveled off at 31,000 feet. At 7:17 P.M., the aircraft exploded over Lockerbie, Scotland and disappeared from radar screens. The passengers and crew were killed, along with 11 people on the ground. Although in several other recent incidents aircraft were destroyed with explosives and a heavy loss of life resulted, this flight extracted a heavy toll of American lives. But why should books be written about this one disaster?[1] Why have studies been made of it, and why is a chapter in this book being devoted to it?

One thing, totally unique in the annals of terrorism, has caused that flight number and the name "Lockerbie" to become etched into our memory. Few, for example, remember UTA Flight 772 which exploded over Niger on September 19, 1989, killing all 171 people aboard. It is not the number of people killed or even the fact that more Americans were killed on Pan Am 103. I believe that the totally unique factor on Pan Am 103 is that the families of the victims banded together and have pursued the facts and issues surrounding this disaster with the tenacity of a bulldog that just won't let go.

At least two groups were formed: the Victims of Pan Am 103 and the Families of Pan Am 103/Lockerbie. They were instrumental in getting President Bush to issue Executive Order 12686 on August 4, 1989, establishing the President's Commission on Aviation Security and Terrorism. The

---

1 Steven Emerson and Brian Duffy, *The Fall of Pan Am 103*: G.P. Putnam's Sons (1990).

thorough report of that committee was issued on May 15, 1990, and is essential reading for anyone seriously interested in aviation security.

A number of issues were raised in that report. It is not necessary to mention or go into detail again on all of those points. Several of them will be discussed briefly because many have not yet been resolved and because they are pertinent to the subject matter of this book. A number of these issues are discussed in other chapters where they are pertinent.

RESPONSIBILITY

Although no one has stepped forward to claim responsibility for the downing of Pan Am 103, it was generally thought to be the work of the Popular Front for the Liberation of Palestine--General Command (PFLP-GC), particularly because of the radio-cassette bombs discovered in the hands of the PFLP-GC prior to the bombing.

Ahmed Jabril is the leader of the Damascus-based PFLP-GC. This organization is pro-Syrian, anti-Arafat, and anti-P.L.O. It has strong ties to Syria, although Libya has also long supported it. They have claimed responsibility for the 1970 explosion that downed a Swissair flight shortly after takeoff from Zurich on its way to Tel Aviv, killing 47 people, and for a 1972 blast aboard an El Al airliner that landed without casualties.

Authorities initially believed that Iran commissioned the attack in revenge for the accidental downing of an Iranian passenger jet in 1988 by a U.S. Navy warship in the Persian Gulf. They also felt that Syrian officials were aware of the activities of the PFLP-GC, but took no action to stop them. Articles published in October 1990 suggest that Jabril, hired by Iran, paid agents of Libyan leader Khadafi to carry out the bombing after discovering in October 1988 that Western intelligence agents had penetrated his own operations.

The latest evidence indicates the regime of Muammar Khadafi is responsible for the bombing.[2] This refuted the conclusions which had been held by investigators for almost two years. The key link was an obscure case involving the arrest of two men--Mohammed Marzouk, alias Mohammed Naydi, and Mansour Omran Saber--Libyan intelligence agents, at Dakar, Senegal airport in February 1988. The men had in their possession 20 pounds of Semtex plastic and TNT explosives, weapons and triggering devices. The triggering devices matched that microchip fragment from the Pan Am bomb. The same triggering device had been present in the 1984 bombing of a UTA airliner in Chad and a 1986 attempt to blow up the U.S. Embassy in Togo. In the latter case, nine people with two suitcases full of plastic explosives had been arrested.

One theory to surface, largely discounted now, addressed the possibility that terrorists infiltrated a Drug Enforcement Administration (DEA) undercover operation, resulting in Nazir Khalid Jafaar, suspected of being a drug courier or "mule," unwittingly bringing the bomb on board. It was alleged that arrangements were in place to put suitcases of heroin on planes

---

[2] San Jose Mercury News, June 24, 1991.

without the usual security checks, under an arrangement between the DEA and the German government. The thought was that the terrorists might have discovered this and substituted a suitcase with a bomb in it for a suitcase of heroin. Most sources indicated that this theory had little credibility.

THE EXPLOSIVE

A double-detonator device was used, the first trigger was activated by barometric pressure, which in turn activated a timing device. It is believed to have been designed to thwart the airline practice of putting cargo through a pressure chamber to activate barometric triggers. The explosive used was Semtex, a Czech-made explosive consisting of pentaerythrite tetranitrate (PETN) combined with mineral oil as a plasticizer. It is more powerful than TNT, has enormous destructive power and can easily be shaped, molded, or disguised as almost anything.

DISSEMINATION OF THREAT WARNINGS

The bomb was enclosed in a Toshiba radio-cassette player. Two such bombs had previously been found. In October 1988 a Toshiba radio-cassette player rigged as a bomb was found during a series of raids on residences where members of the Popular Front for the Liberation of Palestine--General Command had been observed. Another such device, a Toshiba Boombeat Model 453 radio-cassette player, was discovered several days later in an automobile of a PFLP-GC member. It had been fully rigged as a bomb with a barometric triggering device. A warning bulletin describing the device was sent by the FAA on November 18, 1988.

One of the lessons learned from this incident is the need for a formalized dissemination procedure for the handling of such bulletins, both within the government and in the airline industry. One airline official first saw the bulletin on his desk when he returned from a three-week vacation on November 28. Procedures have been changed so that carriers now provide written acknowledgment to the FAA of the receipt of such bulletins and the action they are taking. The red tape has been eliminated to provide for fast and direct dissemination of such information.

THE HELSINKI THREAT

Much has been said and written about a telephone threat received from an anonymous caller on December 5, 1988, at the American Embassy in Helsinki. The caller said that a Finnish woman would carry a bomb aboard a Pan Am flight from Frankfurt to the U.S. sometime within the next two weeks. The U.S. State Department was notified, and on December 7, the FAA disseminated a security bulletin to all of its U.S. regions. On December 9, the State Department forwarded the warning to all European diplomatic posts. By December 10, the Finnish police had determined that the threat was not credible, and they informed senior U.S. government officials who did not pass this information on to the FAA.

PUBLIC NOTIFICATION OF THREATS

The threat information was posted at the U.S. Embassy in Moscow, but was not given to the passengers of Flight 103. One of the more important issues to evolve from this incident is the question of who should be advised of such threats.

It can be argued that these threats come in such volume that it would shut down most carriers if all threats were passed on to the public. Anyone who had a grudge against an airline, including a drunk who might have been thrown off a flight or someone with a disputed baggage claim, could effectively shut an airline down by making anonymous phone threats. The other side of the coin is that this threat was passed on to embassy people who had the option to fly another carrier, but not to the general public. It has been compared to insider trading on the stock market.

The recommendation of the President's Commission on Aviation Security and Terrorism, published on May 15, 1990, was in favor of public notification of threats to civil aviation. In two instances subsequent to the Lockerbie disaster, Delta and Northwest had made announcements to the media regarding threats that had been received to two specific flights.

At the press conference at the National Press Club on May 15, 1990, following the release of the report of the Commission, Bert Ammerman of the Victims of Pan Am 103, advised that his committee only asked that the intended passengers on any flight that has received a threat be so advised, so that they can make an informed decision as to whether to take that flight. They were not urging a broad notification to the general public. The U.S. government position has been that threats of this type are most effectively handled by security professionals who are in a position to implement countermeasures. In a press conference on April 5, 1990, Admiral James Busey, FAA Administrator, praised the two airlines for their actions, but reaffirmed an overall policy of nondisclosure of threat information. It is the position of the FAA that if a specific, credible threat cannot be countered, the flight should be canceled. The question of whether passengers in the boarding area should be advised of a threat is still under consideration, but it is the position of the Department of Transportation that they probably should not be.

Section 109 of the Aviation Security Act of 1990 directs that Title III of the Federal Aviation Act of 1958[3] is amended to provided that "the President shall develop guidelines for ensuring notification to the public of threats to civil aviation in appropriate cases."[4]

---

[3] 49 U.S.C. (App. 1342-1358).

[4] Aviation Security Improvement Act of 1990, Public Law 101-604, Section 321(c)(1).

PASSENGER/BAGGAGE RECONCILIATION

The Commission's report states, "The undisputed facts before the Commission show that passenger/baggage reconciliation is a bedrock component of any heightened security system"[5]

At the time of this incident, the Air Carrier Standard Security Program (ACSSP) required the carrier to conduct a positive passenger/checked baggage match resulting in physical inspection or noncarriage of all unaccompanied bags. This provision came into play primarily in the area of interline bags. Pan Am was X-raying *all* interline[6] bags rather than identifying and physically searching *unaccompanied* interline bags, and the airline claims this was being done in compliance with an approval from the FAA, although that has been disputed by the FAA.

Investigation has disclosed the presence of an extra bag when the flight left Frankfurt. One passenger who was registered on the computer for Flight 103 did not arrive at the gate. However, the plane departed without removing the passenger's baggage.[7] This was apparently due to the failure to reconcile the number of interline bags loaded into the belly of any plane leaving Frankfurt with the number of bags previously checked by the interline passengers who actually boarded the plane. The ACSSP now requires both a positive match *and* an X-ray or a hand search of all checked baggage in specific designated countries.

OTHER ISSUES

Liability on international carriers is limited to $75,000 per passenger under the Warsaw Convention of 1929. It was not until November 1965 that the limit was raised from the original amount of $8,300 per passenger to $75,000. By contrast, domestic airlines have paid families of recent victims of U.S. air crashes an average of more than $500,000. The family of one business executive killed in a 1985 crash in Dallas won a record $7.9 million verdict. Thus, it is possible for a situation to occur wherein two passengers sitting next to each other on a flight from Los Angeles are killed in an accident. The ticket of one passenger lists the destination as New York, and his seatmate is going to London via New York. The family of the passenger headed for London can collect no more than $75,000, while that of the passenger headed for New York has no limit on what it can collect. A pending revision known as "Montreal Protocol 3," would establish absolute liability in

---

5 *Report of the President's Commission on Aviation Security and Terrorism*, May 15, 1990 (Washington, D.C.,U.S. GPO, 1990), 22.

6 Involving more than one airline or air carrier. Activities shared between two or more airlines. In this case, it refers to bags being transferred to Pan Am from another carrier. Differentiated from "online," which means it relates only to one airline; i.e. it doesn't go "offline," to another carrier.

7 *Report of the President's Commission on Aviation Security and Terrorism*, May 15, 1990 (Washington, D.C.,U.S. GPO, 1990), 13.

cases of death or injury occurring on international flights, and it would increase the current liability limit to $130,000.

RECOMMENDATIONS OF THE PRESIDENT'S COMMISSION

The *Report of the President's Commission on Aviation Security and Terrorism* contained more than 60 recommendations. The major recommendations are:

- The United States should pursue a more vigorous counterterrorism policy, particularly with respect to nations sponsoring terrorists.

- Congress should enact legislation to create a position of Assistant Secretary of Transportation for Security and Intelligence, an appointment with tenure to establish a measure of independence.

- The FAA security division should be elevated within the agency to a position that reports directly to the Administrator. (Coast Guard Vice Admiral Clyde Robbins has been appointed as the first Director of DOT's newly formed Office of Intelligence and Security)

- Through existing FAA resources, the federal government should manage security at domestic airports through a system of federal security managers.

- The State Department should conduct negotiations with foreign governments to permit U.S. carriers operating there to carry out FAA required screening and other security procedures. Airlines cannot be expected to conduct international negotiations in order to comply with regulations of their own government.

- The FAA should launch a top priority research and development program to produce new techniques and equipment that will detect small amounts of plastic explosives, operationally at airports. The program to require U.S. carriers to purchase and deploy the existing TNA (thermal neutron activation) machine should be deferred. However, the Commission expects the FAA to continue aggressively its new emphasis on upgrading the aviation security system's human and technical capabilities.

- Public notification of threats to civil aviation should be made under certain circumstances. As a rule, however, such notification must be universal to avoid any appearance of favored treatment of certain individuals or groups.

- Victims of terrorist actions aimed at the United States Government should qualify for special financial compensation as victims of acts of aggression against their country.

- The State Department must take major steps to ensure that the families of victims receive prompt, humane, and courteous treatment and service in overseas disasters.

The Commission's report dealt with numerous other issues. The members of the commission believe there should be a coordinated and compassionate response to the families of victims of terrorism by both the air carrier and by the State Department. They suggest improvements in the field of intelligence gathering and dissemination, although they state the current system is working reasonably well.

AVIATION SECURITY IMPROVEMENT ACT OF 1990

Congress reacted swiftly to the report of the President's Commission on Aviation Security and Terrorism. Just six months after the report was issued, the President signed into law the Aviation Security Improvement Act of 1990.[8] Title I of that law deals with aviation security and Title II with the United States response to terrorism affecting Americans abroad. Most of the recommendations of the President's Commission were implemented in this new law.

The findings of Congress that were remedied by this law are

1. The safety and security of passengers of United States air carriers against terrorist threats should be given the highest priority by the United States Government.

2. The report of the President's Commission on Aviation Security and Terrorism, dated May 15, 1990, found that current aviation security systems are inadequate to provide such protection.

3. The United States Government should immediately take steps to ensure fuller compliance with existing laws and regulations relating to aviation security.

4. The United States Government should work through the International Civil Aviation Organization and directly with foreign governments to enhance aviation security of foreign carriers and at foreign airports.

5. The United States Government should ensure that enhanced security measures are fully implemented by both United States and foreign air carriers.

6. All nations belonging to the Summit Seven[9] should promptly amend the Bonn Declaration to extend sanctions for all terrorist acts, including attacks against airports and air carrier ticket offices.

---

[8] Public Law 101-605 (HR 5732), signed into law on November 16, 1990.

[9] The 17th Economic Summit of Industrialized Nations, held July 9-11, 1990 in Houston, Texas.

7. The United States Government, in bilateral negotiations with foreign governments, should emphasize upgrading international aviation security objectives.

8. The United States Government should have in place a mechanism by which the Government notifies the public, on a case-by-case basis and through the application of a uniform national standard, of certain credible threats to civil aviation security.

9. The United States Government has a special obligation to United States victims of acts of terrorism directed against this Nation and should provide prompt assistance to the families of such victims and assure that fair and prompt compensation is provided to such victims and their families.

10. The United States should work with other nations to treat as outlaws state sponsors of terrorism, isolating such sponsors politically, economically, and militarily.

11. The United States must develop a clear understanding that state-sponsored terrorism threatens United States values and interest and that active measures are needed to counter more effectively the terrorist threat.

12. The United States must have the national will to take every feasible action to prevent, counter, and respond to terrorist activities.

The legislation establishes a number of new offices and positions. It creates a Director of Intelligence and Security within the Office of the Secretary of Transportation and the position of Assistant Administrator for Security who will report directly to the Administrator. It also establishes Federal Security Managers (FSMs) at category X airports and Foreign Security Liaison Officers (FSLOs) for designated airports outside the United States.

The law calls for an annual report to Congress on the status of aviation security, background investigations on individuals who will have access to aircraft or secured areas, the establishment of employment standards for those who contract with air carriers to work in airport security, the establishment of standards for the education and training of air carrier and airport security personnel.

The FAA and the FBI are directed to jointly conduct periodic assessment of threats to domestic airport security by conducting surveys of individual airports. Section 107 of the act mandates a program to accelerate research, development, and implementation of technologies and procedures to counteract terrorist acts against civil aviation. Public notification of threats is dealt with in Section 109 as previously discussed. The heads of agencies in the intelligence community[10] are directed to promulgate policies and proce-

---

[10] Section 111, Aviation Security Improvement Act of 1990. The intelligence community is specifically defined as including the Central Intelligence Agency, the Department of Defense, the Defense Intelligence Agency, the National Security Agency, the Department of the Army, the Department of the Navy, the Department of the Air Force, the Department of State, the

dures to insure that intelligence reports concerning international terrorism are made available to other members of the intelligence community, the Department of Transportation, and the Federal Aviation Administration. The question of screening of mail and cargo is also dealt with in this legislation. A study is mandated to consider the extent to which it is practicable to require for mail and cargo the same screening procedures as are required to for checked baggage.

Title II deals with terrorism affecting Americans abroad. Section 201 states the following policy:

> United States Policy: It is the policy of the United States -
>
> 1. to seek bilateral agreements to achieve United States aviation security objectives with foreign governments;
>
> 2. to continue to press vigorously for security improvements through the Foreign Airport Security Act and the foreign airport assessment program; and
>
> 3. to continue to work through the International Civil Aviation Organization to improve aviation security internationally.

The act places upon the Department of State the responsibility for negotiating requisite aviation security agreements with foreign governments. It requires that U.S. air carriers provide to the Department of State within one hour the passenger manifest for any flight involved in an aviation disaster. It allocates $5 million a year in passport fees to be available for the acquisition and production of machine-readable United States passports and visas and compatible reading equipment.

Many of the provisions of Title II relate to the handling of disasters by the Department of State. It provides for disaster training for State Department personnel, for designation of a specific individual and alternate as liaison for the family of each citizen involved in a disaster, and establishes a toll-free communications system for the families. Section 211 directs the President to submit to Congress recommendations on whether or not legislation should be enacted to authorize the United States to provide monetary and tax relief as compensation to United States citizens who are victims of terrorism. An Overseas Security Electronic Bulletin Board, updated daily and accessible to the general public, is established by the act.

This act contains a number of other provisions of importance to aviation security. The FAA and the Department of State have been busy implementing the provisions of the act within the time frames specified, which in most cases are relatively short.

Department of the Treasury, the Department of Energy, the Federal Bureau of Investigation, and the Drug Enforcement Administration.

Section II

AIRCRAFT SECURITY

Chapter 5

PREDEPARTURE SCREENING

As plans were formulated for instituting the 100 percent screening requirement, three general concepts emerged for physical arrangement of facilities: the sterile concourse, the sterile boarding area, and the departure gate screening.

STERILE CONCOURSE

The sterile concourse proved to be the most attractive plan, and this method was used in virtually every location in which it could be used. The sterile concourse represents an exceptional cost saving for both the carrier and the airport. It is an area to which access is controlled by the inspection of persons and property in accordance with an approved security program. Instead of having to bear the cost of a sufficient workforce to search passengers at each gate and having to post a law enforcement officer at each gate, a central screening point established at the throat of a concourse serviced numerous gates. This approach also proved of great value in making possible the later use of X-ray for screening carry-on items, a measure which had initially been ruled out as prohibitively costly in view of the numbers that would be required to cover each boarding gate. Once a single unit was able to service five to ten gates, X-ray screening became practical.

The same economies applied to the utilization of walk-through and hand-held metal detectors. The greatest saving, however, was probably the economy of personnel.

Another advantage of the sterile concourse approach, one that is discussed more fully under "Airport Security," is that it greatly increases the distance between the screening checkpoint and the boarding gate. Thus, it helps to deter attempts to storm an aircraft and gives law enforcement authorities and carrier personnel the advantages of time and distance with which to cope with the potential hijacker.

On a sterile concourse, predeparture screening of passengers is accomplished gradually as passengers arrive at the terminal, rather than delayed until the aircraft is ready to depart. It has resulted in less crowded concourses and gate areas, particularly at those terminals where visitors are discouraged or restricted from accompanying or meeting passengers within the sterile area. Some carriers were able to reduce the number of passenger agents on duty due to the reduction in the number of people in the concourse and a more steady flow of passengers arriving at the boarding gate.

The main disadvantage of the sterile concourse is the floor space needed to screen passengers at the entrance to the concourse. Although there was some early resistance from airport managers, the advantage of the sterile concourse quickly became so evident that the first makeshift screening points soon were replaced by more permanent and better organized arrangements. The strongest objectors were concessionaires and vendors whose shops and services were located on the sterile side of the checkpoint; they feared that the central screening of passengers and visitors would reduce their business. In most cases, this did not prove true because, as a general rule, visitors and passengers are still allowed access to sterile areas, provided they are screened.

To maintain the integrity of the sterile concourse, it is essential that each day, before screening is resumed in concourses not secured around the clock, a thorough search be made for any weapons or dangerous objects that may have been secreted during closed hours. Because a large part of the task consists of looking under the seats in boarding lounges, a carrier might wish to consider obtaining one of the mirrors attached to a pole, in much the same fashion as a dental mirror. I had suggested the construction of this type of device in the first edition of this book in 1976. Now they are commercially available from a number of sources. Larger units mounted on wheels assist security guards in checking under vehicles at embassies and other sensitive locations.

The perimeter of the sterile concourse must be secured by physical barriers such that persons who have been screened cannot receive items passed to them by persons outside the sterile area. Any passenger who leaves the sterile area for any reason must of course, be screened again before being allowed to re-enter.

In the first few weeks of experimentation with the sterile concourse, it appeared as if some serious legal questions could arise. Both airports and carriers were holding to the line that any visitor wanting to pass the checkpoint must be screened. Some visitors were neither seeing passengers off nor meeting arrivals; they merely wanted to have a drink at a bar or lunch at a restaurant located beyond the checkpoint. There were instances of these visitors refusing to be screened. In these cases, the airlines stood fast in their refusal to admit unscreened visitors. The Federal Aviation Regulations were eventually changed to require that anyone entering a sterile area be screened. Most concourses are open to the public, but some airports have restricted some or all of their concourses to ticketed passengers. It is also frequently a part of an airport's contingency plan in the event of a threat to impose such a restriction. In some cases, conflict has existed between the carrier and the airport over this issue; the airport has sought to impose the restriction, and the carrier did not want it. Both felt they were responsible for the security of the concourse.

STERILE BOARDING AREA

The second physical arrangement proposed for predeparture screening was the sterile boarding or sterile holding area. This seemed to be the next best alternative where a sterile concourse was not practical. It was achieved by creating a sterile area at the flight check-in point, usually by securing the boarding lounge from the concourse or other adjacent terminal areas so that passengers who had been screened would be isolated from physical contact with persons outside the sterile boarding area.

Enclosing the boarding area challenged the ingenuity of both the carriers and the airports in making these areas as practical and attractive as possible. At some locations, the enclosure was made of windowless walls of regular stud and sheet-rock construction. Other places utilized expanded metal mesh to create a barrier that would not obstruct the circulation of air, a method that avoided revamping heating, ventilation, and air conditioning ducts and outlets in some cases.

As with the sterile concourse, visitors could legitimately enter the boarding area if they were screened, but both carriers and airports were more likely to restrict visitors from these smaller sterile boarding areas because of the limitations of space.

The primary advantage offered by the sterile holding area is that screening is accomplished as passengers arrive at the boarding gate, rather than after an aircraft is available for loading.

DEPARTURE GATE SCREENING

The least desirable of the three alternatives is the departure gate screening. Personnel must be available for screening at each individual gate, and similarly, the city or other operating authority must provide a law enforcement officer at each gate; the personnel costs are high. In addition, passenger screening cannot begin until the aircraft is available for loading. This type of screening plan is most likely found at small airports where flights are few and screening does not begin until about 20 minutes before departure time.

Variations on the departure gate plan can be of some help in speeding up the screening process. If there is a location to which access can be controlled so that safe temporary storage can be provided, carry-on items may be inspected in advance, leaving little more than the electronic screening of passengers to be accomplished once the plane is ready for boarding. The storage area need be no more than the jetway or an open space behind the baggage inspection table. Passengers pick up their prescreened carry-on items after they pass through the walk-through weapons detector on their way to board the airplane.

An exception to this, of course, is women's purses; these are the only carry-on items that should not be prescreened. Carriers receive the most complaints about the loss of money or other valuable items during searches of purses. Of course, the same reason requires that carry-on items must always be searched in the presence of the passenger or visitor, but it is only in the departure gate plan that passengers may be separated from their carry-on items. Purses generally contain cash, credit cards, and checkbooks, as well as such small valuables as the passenger wishes to keep with her at all times. When the aircraft is ready to board, the purses are inspected in the presence of their owners before they pass through the weapon detector.

The same precaution applies, of course, to any carry-on luggage identified by a passenger as holding valuables. Always let the passenger retain possession of any bag so identified. Additionally, where the contents are of obviously high value, such as jewelry, the search should be done where other passengers will not see the valuables.

SCREENING CHECKPOINT ENHANCEMENTS

At several locations, one carrier has experimented with enhancements to improve the level of security. In addition to the metal detectors and X-ray machines at the throat of the sterile concourse, a control booth has been constructed. The booth is behind the checkpoint and is raised above the floor level to give the operator in the booth greater visibility. The operator is a contract screening employee, generally a supervisor. This individual has a duplicate set of monitors so that he or she can see the X-ray image on each X-ray unit on the floor and can communicate with X-ray operators at each machine by means of a two-way communications system. Each of the X-ray operators wears a headband microphone and earpiece to allow conversation with the supervisor in a hands-free mode.

Essentially, the control booth provides a second set of eyes looking at the X-ray images. The supervisor has the capability of videotaping the activity on any X-ray machine or of taping the view of any of the cameras that are focused on the general checkpoint area. The supervisor can even view screening at other gates and other X-ray machines that are not at the sterile concourse and are out of view by means of closed circuit television (CCTV). In addition to videotaping the activity, if a weapon is suspected a contact print can be made of any video screen and kept for future study or review.

Additionally, two or more large-screen color monitors are suspended from the ceiling and visible to the general public. These monitors show the view from the cameras monitoring the checkpoint itself, rather than the X-ray images. These monitors act as a deterrent and also permit law enforcement officers to observe what is occurring before they enter the area.

THE LAW ENFORCEMENT OFFICER

Part 107.4 of the Federal Aviation Regulations originally required the presence of at least one law enforcement officer

> At the point of, and prior to and throughout, the final passenger screening process prior to boarding, for each flight conducted by a certificate holder...continuously until all doors on the aircraft being boarded are closed and the aircraft has taxied away from the boarding area; and...in the event that the aircraft returns to the boarding area before take-off.

The FAA originally interpreted this to mean that the law enforcement officer should position himself at the predeparture screening point, wherever that might be--at the boarding gate, at the entry into the boarding area, or at the entrance to a sterile concourse. As the rule stated, the presence of the officer was required during the entire process, but the responsibility for seeing that the officer was present rested upon the airport operator, not the carrier. Nothing in the regulation prevented the carrier from processing passengers in the absence of a law enforcement officer.

Misinterpretation of this requirement occasionally led to carrier personnel believing that they must suspend screening operations because the law enforcement officer was late or did not arrive at all. Such was not the case. In the absence of a law enforcement officer, however, any passenger who could not be cleared of unaccounted-for metal could not be boarded unless an air carrier representative conducted the required pat-down search.

Another problem associated with this regulation was that some airports and communities had built sizable airport police departments and presented the bill for the entire development to the airlines. This was partially ironed out through a joint effort in which the Airport Operators Council International was included in the negotiations and discussions. A proposed statement of general principles and suggested guidelines was developed to determine which of these costs were included under the FARs. In it, carriers indicated that they would regard as legitimate expenses such costs as vehicles, radios, and weapons.

One of the good things about the Air Carrier Standard Security Program (ACSSP) is that it has been a constantly changing document. When new concepts arose that met the test of better security, they were implemented. There have been no sacred cows and the FAA has proved itself cooperative in those areas where no degradation of security would result.

Two arguments began to take shape for a less rigid requirement for a law enforcement officer at the checkpoint. One was the extreme cost, and the other was the belief that officers would be better able to make a planned response to a threat if they were not stationed right at the checkpoint. They could deal with the incident rather than become a part of it. The following two instances seemed to back up the logic of this approach.

Samuel Joseph Byck proved on February 22, 1974, that an officer stationed at a checkpoint could be a sitting duck for a gunman. Byck approached a gate screening position at Baltimore where Delta Flight 523 was being processed. Without provocation, Byck drew his weapon and murdered the law enforcement officer at the screening point. He thereafter boarded the aircraft, killed the copilot, and wounded the captain. He stated it was his intention to crash the airliner into the White House. The aircraft never got off the ground. Byck was shot by police officers and ultimately committed suicide on the airplane.

In a similar incident on October 20, 1977, a lone male hijacker approached a screening checkpoint at Grand Island, Nebraska, took a shotgun from his carry-on luggage at the screening point, and neutralized and disarmed the law enforcement officer. He thereafter boarded Frontier Airlines Flight 101. After the flight landed in Kansas City for refueling, the hijacker released some of his passengers, and the flight continued to Atlanta, where the hijacker committed suicide with his own gun.

Financially, it was estimated that the carriers would save more than $45 million, per year if the requirement of a law enforcement officer at each screening point was eliminated. Annual costs to the nation's airlines were running about $100 million. At O'Hare Airport in Chicago, the costs were approximately $2 million a year for 180 officers.

Flexible Law Enforcement Response Program

The concept that evolved was one of a flexible response or a "flex response," as it became known. An officer was not required to be at the screening checkpoint at all times but had to be able to arrive at the checkpoint within a given response time. It became necessary to emphasize that the FAA was not requiring the airport to "guarantee" these response times. It was recognized that emergencies might occur that would preclude the officer from arriving on time in certain situations, and that philosophy was acceptable. This resolved the complaint from some airport police departments that felt they would have to have back-up officers available in order to meet the "guaranteed" response time. Carriers, of course, viewed back-up officers as an additional to them, so it was necessary to recognize that the response time was one that might be anticipated under normal conditions, but that was not necessarily guaranteed.

As it developed, airports were divided into categories based on size and passenger boardings. Smaller airports where hijackings might be less likely to occur were allowed greater response times than major airports where an officer had to be in the immediate vicinity of the gate in order to make the required response time. At Washington National Airport, the first in the U.S. to implement the flex response mode on September 11, 1981, the number of law enforcement officers was reduced from 51 to 17.

To make certain that the public did not view the primary purpose of the program as strictly to reduce costs, it was first named the Strengthened Screening Point Security (SSPS) program. And, indeed, it probably did strengthen security at the checkpoint. The assignment of fully trained law enforcement officers at a checkpoint for hours at a stretch certainly did nothing to sharpen their skills and most probably resulted in apathy and boredom. It was believed if they patrolled in the terminal instead, they would be more alert and have a better perspective of what was taking place at the airport.

There is no need to delineate the response times in this book. That information is available from the FAA and is known to those who have a need to know. It is also a part of the Air Carrier Standard Security Program.

The categories of airports are based to some degree on the number of passengers boarded each year. They are as follows:

Category X	Can be any number boarded. Airports in this category are given special security attention because of a perceived threat.
Category I	More than 2 million passengers boarded.
Category II	500,000 to 2 million passengers boarded.
Category III	Less than 500,000.
Category IV	Any number.
Category V	No screening.

The FARs were amended so that §107.15, "Law Enforcement Support," which became effective April 1, 1981, now states,

> (a) Each airport operator shall provide law enforcement officers in the number and in a manner adequate to support--
>
> (1) Its security program; and
>
> (2) Each passenger screening system required by Part 108 or §129.25 of this chapter.

Checkpoint Security Supervisor

As a part of the flex response concept, a new position was created at the checkpoint: the Checkpoint Security Supervisor (CSS). The CSS was to take over many of the functions of the law enforcement officer and was to have a more extensive background than the average contract security officer. The CSS was required to be a high school graduate and have three years of law enforcement, industrial security, military police, or supervisory experience. At some airports, this individual was known as the Screener-in-Charge.

While it is the responsibility of the airport to provide the law enforcement officer, it is the carrier's responsibility to hire and train the CSS. This person is responsible for verifying the credentials of law enforcement officers who want to board with a weapon, for summoning law enforcement officers or station management when required, for conducting pat-downs, and for handling false statements, communications equipment, and other supervisory duties.

What the CSS was to wear was the subject of numerous discussions. Some thought he or she should have a uniform similar to that of a police officer and was to wear a hat and carry a nightstick and handcuffs. The police thought differently. It was finally decided that the CSS should wear a blazer. Then the discussions really became heated. The color arrived at was burnt orange. This went well at most airports, but carriers whose customer service representatives wore orange coats were not amused. Thus, the color became a local airport option, with a strong recommendation for orange. The color had to be uniform throughout the entire airport. Several airports decided on Kelly green, even though the employees of at least one car rental agency at the airport wore the same color. There was virtually no color that someone else wasn't already wearing.

A number of situations require the presence and the intervention of a law enforcement officer. Under the Federal Aviation Act, it is illegal to attempt to board an aircraft while having concealed on or about one's person a deadly or dangerous weapon (with certain exceptions that are dealt with later). Federal Aviation Regulations require that passengers or visitors discovered to be carrying an undeclared weapon, whether in carry-on luggage or on their person, must be referred to a law enforcement officer. In the absence of a law enforcement officer at the checkpoint, the passenger with an undeclared weapon should be detained until an officer can arrive and interview the passenger. If the weapon is observed on X-ray, the article of baggage should remain in the X-ray machine until the law enforcement officer arrives so that the weapon is inaccessible to the passenger.

In the case of armed law enforcement officers who declare their weapon, the matter can be handled by the CSS. It has been found advisable to handle this discreetly. Officers may be following another passenger, and narcotics officers have been inadvertently exposed in the notification process. Several of the larger airports have drug task forces with officers working at the airport around the clock.

It must be remembered that the final decision--and the legal responsibility--for boarding a passenger rests with the airline, not the law enforcement officer. Of course, if the passenger is arrested, the carrier has no further concern as to whether to board him or her,, but situations can arise in which the law enforcement officer may clear a passenger for boarding while the carrier representatives may have decided they will not board the passenger for one of the reasons available to them under their own regulations. Drunks are not boarded, and in fact the airline can be fined by the FAA if it does board an intoxicated passenger. Airline regulations, not FAA rules, provide for refusing to board any passenger who is filthy, has an offensive odor, or otherwise might prove repugnant to other passengers. Also, the pilot has the ultimate authority for the safety of the flight.[1] If the pilot believes, for example, that a passenger is under the influence of drugs, he or she may require that the passenger deplane and can refuse to carry that person on the flight, although such cases are rare.

Section 107.17 of the FARs now defines a law enforcement officer as follows:

> (a) No airport operator may use, or arrange for response by, any person as a required law enforcement officer unless, while on duty on the airport, the officer--
>
> (1) Has the arrest authority described in paragraph (b) of this section;
>
> (2) Is readily identifiable by uniform and displays or carries a badge or other indicia of authority;
>
> (3) Is armed with a firearm and authorized to use it; and
>
> (4) Has completed a training program that meets the requirements in paragraph (c) of this section.
>
> (b) The law enforcement officer must, while on duty on the airport, have the authority to arrest, with or without a warrant, for the following violations of the criminal laws of the State and local jurisdictions in which the airport is located:
>
> (1) A crime committed in the officer's presence.
>
> (2) A felony, when the officer has reason to believe that the suspect has committed it.

A security guard who has not been given the necessary powers of arrest does not and cannot fulfill the requirement of FAR Part 107.17. In some areas, this power of arrest (which must be granted by the appropriate agency) is strictly limited to the airport property and only covers arrests for certain offenses. Where this is in effect, there is a law enforcement presence at the airport, but it is not at the screening checkpoint. As a practical matter, if a guard feels there is a need to arrest someone, the guard will, in all probability, attempt to get a police officer there to make the arrest, but he or she does have that power if necessary.

The motivation for some municipalities to go to a private security agency is obvious; it's simply cheaper to hire a guard than to hire a police officer. Other cities have created an airport police force, with officers being specifically trained only for airport duties and paid at a rate less than that of a regular police officer. In some airports, there are airport police officers and in addition, a small contingent of regular police officers from the downtown department.

---

[1] Federal Aviation Regulation (FAR) 91.3a states, "The pilot in command of an aircraft is directly responsible for and is the final authority as to, the operation of that aircraft."

Much has been said and written about the duties of the law enforcement officer at the checkpoint. As originally conceived, passengers who could not be cleared of unaccounted-for metal were to be referred to the law enforcement officer for a pat-down body search. Passengers found to have weapons in their possession were also to be referred to the law enforcement officer.

The latter rule is still in effect, but the FAA holds that it is the responsibility of the carrier to perform the pat-down search and not the responsibility of the law enforcement officer, and the industry's Standard Security Program states that this search must be performed by the carrier or its contract agency. In support of its ruling, the FAA cites Section 315(a) of Title 2 of the Air Transportation Security Act of 1974, which requires that "all passengers and all property intended to be carried in the aircraft cabin in air transportation or interstate air transportation be screened by weapon-detecting procedures or facilities employed or operated by employees or agents of the air carrier."

The carriers still feel that airlines are not in the law enforcement business, and there is considerable reluctance to perform this function. The carriers generally believe that this clause can be interpreted to mean that the legal responsibility for a pat-down search is the carriers', but that they can have it carried out by a law enforcement officer as long as they retain the ultimate responsibility. As a practical matter, with the flexible response program, law enforcement officers now rarely do the pat-down searches. They are generally done by the Checkpoint Security Supervisor.

With the passage of time and the accumulation of experience, most other problems associated with the presence of a law enforcement officer at the screening point have been worked out. Generally, excellent cooperation now exists between police departments, airport managers, and air carriers.

SCREENING PASSENGERS

Whenever possible, passengers who alarm the metal detector should be taken aside for further screening, rather than allow their examination to slow the screening process for other passengers. These examinations may require additional personnel as well as a back-up metal detector, but in high-volume traffic, this practice is virtually a necessity.

First, the passenger who has alarmed the detector should be asked to remove all metal from his or her person and to pass through the walk-through metal detector once again. Should the passenger still alarm the metal detector, an attempt should be made to locate the unaccounted-for metal by means of a hand-held metal detector. Once the metal is located with the hand-held unit, the person should be asked to remove it and walk through the metal detector once more.

It is not enough to act on an alarm indication obtained with a hand-held unit and conclude that all of the metal on a person has now been located. Two examples will serve as illustrations. In one incident at O'Hare Airport, a male passenger alarmed the walk-through metal detector. Upon questioning, he removed a knife from his pocket, indicating that the knife had alarmed the detector. He was asked to walk through once more, and once again he alarmed the device. This time he produced a .22 caliber derringer, saying that the derringer had alarmed the detector. Police then searched the passenger and found another gun and 60 rounds of ammunition. Had the man not been required to walk through the detector again, it is possible that the screening operator might have stopped her search when she was shown the knife.

At a city in California, the screening employee was unable to locate unaccounted-for metal on a female passenger and referred her to the law enforcement officer for further screening. He proceeded to search with a hand wand detector and concluded that the detector was being alarmed by her bracelet. She was allowed to board. Later, it was discovered that she had a pistol in her brassiere. During the screening process she had held her arm so that her hand was near her shoulder, putting the bracelet in such close proximity to the gun in her brassiere that the hand detector could not differentiate between them. Had the officer asked her to hold her arms away from her body during the hand wand search or had he taken the precaution of having her remove the bracelet and walk through the detector, it must be assumed that he would have eventually located the gun. Unfortunately for the airline involved, as soon as the woman boarded the plane, she showed her seat companion how she had gotten the gun on board. Her seat companion was an FAA Security Inspector who was conducting a security survey of the station.[2]

Shortcuts in screening are not only ineffective, but they may be illegal. The United States Court of Appeals, Second District, has ruled that a frisk of a passenger performed immediately after he or she activates a metal detector is not lawful.[3] The court stated the rule as "Exhaust the other efficient and available means, if any, by which to discover the location and identity of the metal activating the magnetometer before utilizing the frisk." The search must be reasonable, but the court felt an immediate search was not reasonable, since other alternatives were available to the carrier. The passenger could have been asked to remove metal from his pockets and walk through the detector once more. He could also have been screened with a hand detector. The court held that the frisk or body search is a last-resort activity and must be utilized as such.

Specific techniques exist for screening a person using the hand-held detector, which is referred to as the hand wand, hand scanner, squealer, and other terms. There is no prohibition against a male screening a female by means of a hand wand, but should the hand unit be allowed to touch the body of the person being searched, you have crossed over, at least technically, into a body search.

Where a screened area is available, its use will contribute to making the passenger feel less self-conscious about the screening with the hand wand; however, such personal screenings have become so commonplace that embarrassment is not often a major consideration, and many carriers no longer arrange for privacy for such checks.

---

2 The word station is sometimes used to describe an air carriers facilities at an airport.

3 *U.S. v. Albarado*, 495 F.2d 799 (1974).

Screening should begin with the person using the detector standing behind the passenger. Ask the passenger to spread his or her arms parallel to the ground. After making a circular motion about the head and neck, continue on down the back, with the unit parallel with the body and about two to three inches from the person. Then pass the unit up the right side under the arm, screen the right arm, and then the right side of the torso to the waistline. Then pass the wand down the outside of the right leg and up the inside to the crotch area, unless the passenger is wearing a skirt, in which case pass the wand over the front of the torso. Repeat the screening for the left side, going down the left leg and up to the underarm, the inside and outside of the left arm.

Particular attention must be given to passing the wand over the crotch area on both males and females, since experience indicates this is a favorite hiding place for weapons when a search is expected. Other area of the body deserving special attention are the waist, armpit, and ankle areas.

Visitors

Of course, nonpassengers or visitors are also subject to search. FAR 107.20 as amended on January 10, 1986, states, "*No person* may enter a sterile area without submitting to the screening of his or her person and property...[italics mine]" It was necessary to make this change in the regulations, since previously the wording referred only to persons attempting to board an aircraft.

BODY SEARCH

If a passenger or visitor cannot be cleared of unaccounted-for metal, he or she should be given a body search. A body search requires a consent, and passengers must be allowed the option of refusing the search.

A private room or well-screened area should be used whenever possible, at least in the United States. In Japan, for example, a significant number of passengers are randomly selected for body searches, and no special privacy is afforded for a quick pat down. Although the private search is never the "up against the wall," spread-eagle search used in an arrest, the fact remains that the individual being searched does have unaccounted-for metal on the body--a consideration that, at the least, suggests the use of caution. A witness should be present during the search, both as a legal precaution and as a safety measure.

It is also necessary for the protection of the passenger. There have been some aggravated situations in which a private pat-down of a female by a male amounted to molestation and resulted in a law suit. My personal feeling is that there should be no screening of any passenger by a member of the opposite sex without a witness present, even if the screening is only with a hand wand. This is particularly essential if the screening is done in a private room.

Performing The Body Search

It is rare that unaccounted-for metal cannot be identified with the use of the hand wand. For the most part, these are cases involving medical or orthopedic appliances. In such cases, it is only necessary to conduct the consent pat-down search in the area of the unaccounted for metal, such as in the case of a leg or back brace. If it is found necessary to conduct a full pat-down, the following guidelines should be applied.

The police officer or employee should stand behind the passenger and begin the search with the head, touching the scalp. Particularly in the case of large amounts of hair, the officer should be alert to the possibility of a weapon being concealed in the hair. Feel firmly between the shoulder blades, moving on down the back to the belt area. From the waist, move your hands up the passenger's sides to the arms. Cover first the right arm (using both hands) and then the right front of the body above the waist. Repeat for the left side. From behind the passenger place one hand on the inside of the left leg, one hand on the outside of the leg, and move down from the crotch to the ankle. Repeat the procedure for the right leg. Because of the embarrassment involved, the place most frequently neglected in a pat-down search is the crotch area. As has been noted, this makes the crotch a common choice for concealing a weapon. Therefore, the crotch area should be included in the search, unless it has been ruled out for the presence of metal with a hand wand.

SPECIAL SITUATIONS

Tests of the effect of walk-through metal detectors upon heart pacemakers were conducted in 1971 at the Aviation Medicine Clinic, Washington, D.C. The tests used the Westinghouse Model WD-4 walk-through detector and two hand-held units: the Federal Transfrisker #6010 and the Rens Squealer. While minor changes were produced in the rate of certain pacemakers, especially the sensitive unipolar atrial, the atrial or the atrioventricular pacing systems,[4] these changes were clinically insignificant for the patient. No clinically significant changes occurred in the rate of the firing pacemaker, nor was there any failure to sense a demand of the pacemaker, or to pace in the appropriate setting.

In early 1974, the Illinois Institute of Technology Research Institute tested ten pacemakers for their susceptibility to the Metor walk-through metal detector and the Solco hand-held unit. The tests of eight widely used U.S. pacemakers and an English and a German unit failed to show any effect on the pacemakers.

---

[4] Nicholas P. D. Smyth, MD; John M. Keshishian, MD; Oliver C. Hood, MD; Archie A. Hoffman, MD; Norman R. Baker; and Edward Podolak: "Effect of an Active Magnetometer on Permanently Implanted Pacemakers," *Journal of American Medical Association (JAMA)*, 221:162-166, July 10, 1972.

Two other studies[5] reported that--with one exception--all pacemakers tested incorporated mechanisms designed to protect the patient against electromagnetic interference. The exception, the pioneer Medtronics Model 5841, ceased production in September 1970, and--based on the average life of pacemakers--should have been out of use by September 1973. Subsequent Medtronics models have the protective circuitry.

A June 1973 bulletin issued by the FAA with the concurrence of the Federal Air Surgeon and the Associate Administrator for Engineering and Development, stated:

> In summary, while we cannot state categorically that there will be no interference to heart pacemakers from active metal detection devices, we believe, based on operational experience and limited testing, that these devices currently in use in the United States are safe for the airline passenger screening program.[6]

A passenger with a pacemaker may still choose to request a body search in preference to walking through a metal detector. In some cases, their doctors may have told them not to allow themselves to be electronically screened. This wish should not be opposed; the option of a body search is available to these passengers, and they should not be pressed to be electronically screened against their wishes. These passengers, of course, must be screened in some manner, or they cannot board the aircraft. Where a carrier does not permit its employees to conduct a body search and the law enforcement officer will not search, the carrier may have to deny the passenger boarding.

In one case, a man arrived at the airport to meet his wife, went through the metal detector, and asked if it would affect his pacemaker. The police officer suggested he contact his doctor for that information. He walked over to a pay phone to call his doctor and died. His pacemaker was subsequently removed and implanted in a dog, which was repeatedly taken through the metal detector with no effect. The final conclusion was that the pacemaker had not been affected by the metal detector and the man had simply died from a bad heart. He had not accompanied his wife on the trip because his heart was not strong enough.

Passengers in wheelchairs present a special screening problem. Sometimes, the passenger can walk through the detector; if so, this is the best method for screening and is probably less tiring than a search. In many cases, the individual is confined to the wheelchair. A hand wand can be used for screening, provided it is possible to differentiate between metal on the person and metal in the wheelchair; it usually is. If a determination cannot be made, a body search is necessary. If possible, this search should be con-

---

[5] Ibid.
H. J. Bisping and W. Irnich:
"The Effect of Interference from Weapons Detection Devices on Cardiac Pacemakers," Research Assignment for the Federal Ministry of Transportation, Federal Republic of Germany, July 5, 1972.

[6] "Possible Interference of Active Metal Detectors with Life Sustaining Devices," from Director of Air Transportation Security to Assistant Secretary for Environment, Safety, and Consumer Affairs, June 1, 1973.

ducted in private, and should include a pat-down of the chair area to make sure weapons are not concealed in clothing or blankets.

Other passengers may be difficult to clear because of metal in or about their bodies. An individual with a steel plate in the head or a steel pin in an arm or a leg will probably alarm the detector. The hand-held metal detector is the most accurate method for verifying the location of the metal, although a pat-down may be necessary on occasion. For example, if a passenger says he has a pin in his hip or thigh and the hand wand indicates the presence of metal at that location, a pat-down search may be needed to confirm that the alarm is not being caused by a gun or knife strapped to the leg in that area.

Orthopedic braces create similar problems. While a hand wand will verify the presence of the brace and even show its location, it cannot tell you if there is other metal at that location, and a quick pat-down search may be necessary.

In any event, these situations should be handled with tact and discretion to avoid embarrassing the passenger. The judgment of the person conducting the inspection is the key to the thoroughness and effectiveness of these "special" inspections.

From time to time, passengers having metal surgically implanted in their bodies will request the airline to provide a special identification card to assist them in clearing the screening. Thus far, no such cards have been provided because the identification card might be misconstrued as authorizing exemption from screening. One of the most novel approaches is that of a woman who presents a miniature transparency of an X-ray showing the metal in her body. Although passengers cannot be exempted from screening, a passenger faced with this problem might benefit from carrying a letter from a doctor confirming that the passenger does have metal implanted and identifying its location. This could be of assistance to the person doing the search and would add to the credibility of the passenger's statements. Of course, letters can be forged and X-rays stolen, and the extent and type of search appropriate remains a matter for the judgment of the screening officer.

In the case of blind passengers, personnel should be prepared to explain what is being done and what the passenger is to do, prior to the time it occurs, so that first-time passengers will be prepared for it. A checkpoint can be a burdensome and even alarming experience to a sighted person, and it must be even more so for the blind and for those with handicaps that limit their understanding of the procedure.

SCREENING AIRPORT AND AIRLINE EMPLOYEES

Formerly, at many large airports, arriving airline employees parked in an employee lot and took a company bus to the terminal on the ramp side, where they entered within a sterile area. Originally, it was necessary to station a guard at that point, set up a magnetometer, and do a hand baggage search. In some airports, the FAA demanded also that a police officer be assigned to that point. One airline was paying $60,000 a year to support the police presence at one such checkpoint. With increased security at terminals, however, family members are almost universally excluded from riding on employee buses that take them into air operations areas or other sterile areas without being screened. Now employees must drop off their families and luggage at the passenger terminal and then go park in the employee lot by themselves, if that is their wish. They can then take the bus back and meet their families at the terminal.

A change was made to the Air Carrier Standard Security Program, to allow flight officers and flight crew members with identification and in uniform to be exempt from screening, regardless of where they entered. This was in effect until September 3, 1987, when new procedures went into effect. The new rule required even airline and airport personnel to go through screening unless approved measures were in place for (1) the limiting of bypass traffic and (2) the positive identification of persons bypassing screening and (3) for screening or searching their baggage or other goods.

Only employees working directly for an airport were considered airport employees. Tenants, concessionaires, or contractors were not airport employees under the new definition. Law enforcement officers in uniform assigned to support airport security were considered airport employees. The new regulations spelled out in detail who must wear identification, what type of I.D. would be worn, and what the provisions were for bypassing screening when that was allowed.

As to airline employees, the screening procedures only applied to persons passing into the sterile area through a checkpoint. When they entered the sterile area some way other than through a checkpoint, such as through a door with a cipher lock on it, they effectively avoided screening. The January 1989 regulations requiring computerized access cards should help. Card readers on all doors leading to the air operations area will effectively restrict access to authorized persons.

SCREENING DIPLOMATS

Both United States and foreign diplomats must be screened. They have the right of any passenger to refuse screening, and they share the same consequences of not being boarded by the carrier.

Foreign dignitaries traveling on U.S. air carriers sometimes wish to be accompanied by their own armed escorts. On such occasions, their government submits the escorts' names to the State Department, their weapons training, and the itinerary. The FAA may then issue authorization for their carriage to the carriers involved, based on their compliance with the requirement for specific training in the use of firearms. These persons can then be legally carried on board the aircraft of U.S. air carriers in an armed capacity, but this is at the option of the carrier.

A diplomatic courier who presents official credentials is entitled to board with his diplomatic pouch unscreened, and the bag is never to be inspected. However, this applies only to the diplomatic pouch; the courier and the other carry-on baggage must be screened.

SCREENING CARRY-ON LUGGAGE

In the *Semiannual Report to Congress on the Effectiveness of the Civil Aviation Security Program* for the period from July 1, 1989 through December 1, 1989,[7] the FAA reported that over 535 million persons were processed through screening checkpoints at U.S. airports. A total of 1464 firearms were detected. X-ray inspection resulted in the detection of 1406 firearms in carry-on items; 28 firearms were detected by use of metal detectors, and 30 were detected as a result of physical searches. In addition, ten explosive/incendiary devices were discovered by X-ray during this period, including 5 grenades, three fireworks, one flare gun, and one tear gas device. There were 746 persons arrested at screening points for the unauthorized carriage of firearms or explosive/incendiary devices. A tabulation of weapons detected during screening from 1981 through 1988 is included in Table 5-1.

The physical search of a passenger's carry-on luggage must be performed with thoroughness and courtesy. Passengers are under strain from the screening process, and business people returning from several days on the road are not delighted to have someone rummaging through their dirty laundry. While economic savings in labor are usually cited as the reason for purchasing X-ray inspection units, the gain in public relations achieved by not having to inconvenience and perhaps embarrass the passengers should also weigh heavily in the company's decision to purchase X-ray units.

The purpose of the carry-on inspection is to prevent or deter the carriage aboard airplanes of any explosive, incendiary, or a deadly or dangerous weapon on or about each individual's person or accessible property. The FAA generally defines carry-on luggage as all articles in the personal possession of a passenger not worn or carried in the passenger's pockets.

This definition yielded one very peculiar result. The passenger who wore a topcoat was directed to walk through the weapon detector, and the result was a perfectly "legal" screening, according to FAA interpretation. However, if the passenger were to carry the coat --same coat, same contents-- the coat became a carry-on item that had to be inspected pocket by pocket. This oddity was resolved in the ACSSP, and coats may now be screened by the walk-through metal detector whether they are carried or worn.

A number of items will probably always require physical inspection. Because the "homesick" Cuban hijackers tended to use a bottle of explosive or flammable liquid as a weapon, any bottle of fluid that is not sealed should be opened and smelled to see if it contains a hazardous fluid. If a camera is not X-rayed, the screener should look through the view finder. If the lens is blocked, the camera should be further inspected. Similar techniques can be used on video cameras if the passenger requests physical inspection.

Special care must be given the inspection of portable radios and cassette players for two reasons. First, there has been at least one case of a disassembled pistol being found by X-ray in a portable radio. Second, most people by now are aware that the explosives that brought down Pan Am 103

---

[7] *Semiannual Report to Congress on the Effectiveness of the Civil Aviation Security Program,* July 1, 1989--December 31, 1989. Department of Transportation, Federal Aviation Administration, Washington, D.C.: February, 1991.

were in a Toshiba radio-cassette player, fully rigged as a bomb, and equipped with a barometric triggering device. A telex sent to airports and air carriers warning about this device cautioned that it would be "very difficult to detect on an X-ray screen." A variety of items designed to perplex screening agents will cross their paths: tool and equipment kits, umbrellas, canes, strollers, flashlights, portable VCRs that will now be rented at some airports, and all type of professional equipment. Between the X-ray and the physical inspection, it should be possible to clear most of these articles.

Passengers will attempt to carry on board an aircraft a variety of items considered to be dangerous. Some of these can be brought on as carry-on baggage, some must be checked, some must be sent as air freight, and some cannot be carried on the aircraft at all. Articles included in the category of dangerous items include aerosol spray paint cans, book safety matches, extra batteries for wheelchairs, cathode ray tubes, dry ice, fire extinguishers, liquid oxygen, mercurial barometers, motor oil, scuba tanks, and small arms ammunition.

Computers--Laptop and Desktop

Computers will need to be turned on and booted (started up) if physically inspected, although the X-ray will not damage the computer memory. (See the expanded discussion of this subject in Chapter 7.) A portable or laptop computer will be relatively easy to boot because it works off of batteries. However, computers other than laptops are carried on board, most notably a variety of Macintosh computers that will fit in the overhead bins or under a seat and are designed for ease of carriage. The passenger with a Macintosh who wants the computer to be physically inspected will need to plug it in to an outlet in order to boot it. Thus, the checkpoint should have an outlet nearby for this purpose.

Since the onset of the war in the Persian Gulf in January 1991 and the accompanying threat of terrorism, electronic devices are being screened much more closely. Laptop computers, tape recorders, and appliances like hair dryers aren't being allowed as carry-on luggage in Germany under current rules. Other countries may have equally stringent rules. In some cases, passengers in the U.S. have been asked to remove the main batteries from their laptop computers and put them in checked luggage. Besides the principal battery pack, virtually all laptops contain a small lithium battery that provides 3 to 3.6 volts of power to the computer's system clock and calendar. This battery is typically soldered into the machine, and removal would damage the computer.

Electronic Devices

Aside from the security implications, some electronic devices may be allowed past screening but should not be used in flight. Among these are portable televisions, personal portable telephones (except for those furnished in flight by the carrier), AM/FM transmitters and receivers, and remote-control toys.

Items that generally may be used on board an aircraft are voice recorders and dictation equipment, heart pacemakers, watches, hand-held calculators, hand-held computer games, hearing aids, shavers, portable tape players (not radios or portable VCRs), portable computers with accessory printers and disk drives. These requirements vary from carrier to carrier, and if there is any question, the passenger should check with the carrier. The restrictions are also generally published in the airline magazine found on board the flight.

THEFTS AT THE CHECKPOINT

Regrettably, there have been instances of theft at the checkpoints. Usually, it takes the form of theft of cash or jewelry from a woman's purse. When reports and allegations of theft are received, the contract security company usually sends in their own female investigators to pass through the screening checkpoint with target or "salted" purses. We know that once thieves steal successfully, they feels they have beaten the system, and the frequency of the thievery increases. Therefore, it is only a matter of time until they are caught. Investigations into these incidents have been highly successful.

According to a July 1990 *Wall Street Journal* article, a passenger carried almost $1 million worth of jewelry with her on a flight from Palm Beach to Manhattan. She watched her handbag containing her jewelry go through the flaps of the X-ray machine and disappear. She and her husband waited for it to reappear, but it never did. The passenger sued the carrier and the security firm. The jury decided she should be reimbursed for her loss, plus interest. The carrier was assessed 65 percent of the loss and the security firm, the remainder.

UNDETECTED WEAPONS

Even though a thorough carry-on search is accomplished, we know from experience that a certain number of weapons will still get through undetected. This most often comes to light when a gun is discovered in a passenger's carry-on luggage, and the detained passenger tells police that he or she had just recently flown on another flight and the gun was not discovered then. In most of these instances, the FAA levies a fine on the carrier who allegedly screened the passenger and the baggage without discovering the gun. These incidents came to be known as ABC cases, because the passenger traveled from point A to point B to point C.

PHYSICAL INSPECTION OF CARRY-ON LUGGAGE

Obviously, carrier personnel should make every effort to conduct a complete and thorough inspection of each carry-on item. This is accomplished by touch as well as sight (which may account for the hesitancy of some inspectors to probe into diaper bags). One should be able to detect a firearm carried in a small suitcase simply by feeling and lifting layers of clothing, if one is alert for the weight and feel of a weapon. This is not to suggest that the sense of touch can substitute for sight in the inspection, but that its importance should not be overlooked.

The bag and all of its contents should be searched; the inspector should run his hand between layers of clothing, and around the sides and lining of the suitcase. He should not be diverted by his preconceptions as was the inspector who pulled a Bible out of a pocket in a minister's attache case, but failed to see a pistol underneath the Bible. Books should be opened and their pages quickly thumbed to ensure they do not conceal a cut-out compartment. Inner bag pockets should be unzipped or unsnapped and searched. Several cases have come to light in which pistols went undiscovered because someone neglected to open the zippered compartment in a woman's purse.

The suitcase or bag should also be inspected for evidence of resewn or reglued lining, false bottoms, locks that are out of alignment, or any signs of tampering. Look for unusually heavy or unusually balanced items or unusual or unexplainable thickness. Plastic explosives can be rolled flat with a rolling pin and laid against the lining of a suitcase. While being screened at the Allenby Bridge going from Jordan into Israel, I noted that bags were emptied and the empty bag was X-rayed.

An unexploded bomb found aboard a Pan Am 747 in Rio de Janeiro on August 25, 1982, was made up of an electronic timer, a barometric sensor, and two AAA batteries. The explosive was a 4 x 10 inch sheet of 1/8 inch thick plastic explosive, weighing approximately 300 grams (2/3 pound).

In February 1986, Israeli authorities discovered a sophisticated suitcase bomb at one of their security screening points. The bomb had plastic explosives molded into the sides, corners, bottom, and top of the suitcase concealed beneath the lining. The bomb had a barometric sensor, a timer, and an electric blasting cap either entirely or partially embedded in the plastic explosives. A connector was provided to attach the batteries for the power source. An arming switch permitted the suitcase bomb to be safely transported.

Neatness is a key to any such search; a bad taste lingers if items are not neatly replaced in their original order. All carry-on items should be inspected in the presence of the passenger, in the presence of only that passenger; a passenger may not want others behind her to see personal garments and articles. A passenger's belongings should not be unpacked and displayed for all to see. A special physical arrangement using ropes or stanchions may be needed to obtain some privacy, but it is a matter that deserves attention. Equally important, the inspection of carry-on baggage should be accomplished without any comment from the inspector regarding the contents, difficult as this may be.

The importance of never inspecting carry-on items except in the presence of the passenger cannot be overemphasized. If this practice is not standard, the carrier has no defense against allegations that cash and other valuables were removed while the passenger's purse or other luggage was being inspected out of view.

Table 5-1 Airline Passenger Screening Results 1981--1988

	1981	*1982*	*1983*	*1984*	*1985*	*1986*	*1987*	*1988*
Persons Screened (Millions)	598.5	630.2	709.1	775.6	992.9	1,055.3	1,095.6	1,054.9
Weapons Detected								
Firearms	2,255	2,676	2,784	2,957	2,987	3,241	3,252	2,773
(1) Handguns	2,124	2,559	2,634	2,766	2,823	2,981	3,012	2,591
(2) Long Guns	44	57	67	100	90	146	99	74
(3) Other	87	70	83	91	74	114	141	108
Explosive/Incendiary Devices	11	1	4	6	12	11	14	11
Persons Arrested								
For Carriage of Firearms/ Explosives	1,187	1,314	1,282	1,285	1,310	1,415	1,581	1,493
For Giving False Information	49	27	34	27	42	89	81	222

Source: Semiannual Report to Congress on the Effectiveness of the Civil Aviation Security Program, July 1, 1988--December 31,1988. Federal Aviation Administration, (May, 1989): Exhibit 7.

Some passengers carry items of extreme value; they include diamond merchants, jewelry salesmen with expensive samples, and persons who, in the course of business, may carry large amounts of currency. It is not only poor security but bad public relations to require such individuals to display items of value in full view of the general public. A private screening area should be afforded these individuals so that they can show the contents of their carry-on luggage without its being seen by persons who need not be aware of the value of the items the passenger carries. At least one major carrier provides an identification card for such persons, to ensure that they will be screened privately and away from public view. This is not an exemption from screening; it merely provides for screening in privacy without long and revealing explanations by the passenger.

X-RAY INSPECTION OF CARRY-ON LUGGAGE

The effectiveness of X-ray screening depends upon the skill of the operator in recognizing objects. Training is necessary, and as the operator's level of experience increases, he or she will more quickly identify objects commonly found in most carry-on luggage. Any mass that cannot be readily identified by the X-ray image on the screen must be physically inspected. One FAA field test found that 10 to 15 percent of the bags inspected were judged as needing closer scrutiny.

Current training requires X-ray operators to make a decision on each image that appears. The operator must decide whether the image represents *no threat,* in which case he or she passes the bag; whether the image represents a *possible threat* and requires that the bag be physically inspected, or whether it is an *obvious threat* and the operator should call for assistance.

If it is an obvious threat, the operator should stop the X-ray belt so that the bag is still inside the machine out of reach of the passenger. Even in the case of a possible threat, where a physical inspection is conducted, it should be done so that the bag is in the view of the passenger but inaccessible to him or her. The passenger's permission should be obtained before conducting a physical search.

Current training directs that until an obvious threat is identified, the passenger has the right to withdraw from the screening process. Therefore, if an opaque substance is observed in an attaché case, the operator should ask permission to physically search the bag. If, for example, he passenger has illicit drugs in the case and had hoped that he would get through the checkpoint without a problem, he or she may refuse to allow anyone to physically inspect the bag and leave the area immediately.

These instructions would appear to be contra to current case law on the subject. Several cases have held that once having entered the screening process, passengers forfeit the right to withdraw, simply because the search discloses items they didn't want discovered.[8] (See Chapter 6, particularly the

---

[8] *U.S. v. Skipwith,* 482 F.2d 1272 (5th Cir. 1973); *U.S. v. Herzbrun,* 723 F.2d 733 (11th Cir. Fl. 1984); *U.S. v. DeAngelo,* 584 F.2d 46 (4th Cir. Vir. 1978), *cert. denied,* 440 U.S. 935, 99

section entitled, "Passenger's Right to Terminate the Search Process," for a discussion of these cases.) Obviously, any item that is not inspected cannot be allowed past the checkpoint.

Once an obvious threat is observed, however, there is no question that passengers have forfeited the right to change their mind. It is required that if the operator believes a weapon or dangerous device has been observed on the monitor, then he or she must call the law enforcement officer. Even though the passenger might demand the bag be returned, that option is no longer open.

Lead-foil film bags are being marketed for the advertised purpose of protecting photographic film from X-ray screening. It is my opinion that such protection is not necessary in the United States. The advertisements also state that the bag will "protect" radios, tape recorders, and calculators, implying that X-ray inspection could damage these items, which is not true. One might surmise that the bag might be used to conceal a pistol, but tests revealed that the gun's outline was still discernible on the X-ray monitor, even though it was enclosed in this lead-foil bag. It is well to remember, though, that not all opportunists pack their guns in such a way as to provide an unimpeded profile view to the X-ray examination, and unidentifiable and unusual masses should be searched accordingly.

The bag may be placed on the conveyer by the passenger or, during peak periods, by someone assigned that task. Passengers have now been sufficiently trained that they place their own bags on the belt, saving the cost of an additional employee. An additional person should be on hand to provide screening for individuals who alarm the metal detector and to manually search suspect bags. Staffing requirements will vary with the arrival rate of passengers, and during slow periods, an individual can be assigned to perform more than one function.

One factor in X-ray inspection should be kept in mind by supervisors of predeparture screening operations: Watching an X-ray monitor for extended periods of time can mesmerize the operator to the point where he or she is unconsciously inattentive to the shapes displayed on the monitor. One remedy is frequent rotation of operators, perhaps as often as every 15 to 30 minutes. This will add variety to the work of those assigned to the screening checkpoint, making their work more enjoyable as well as more effective. Also, the offer of a reward for discovering a weapon will provide added motivation.

For X-ray examination, garment bags should be folded in half or in thirds, so that they do not pulse[9] the machine two or three times instead of just once. (This naturally is not a problem with a continuous-beam unit.) Operators using conveyors should exercise care in processing items through the unit. Sharp edges on the side of the conveyor can tear garment bags or tear the handle off a purse. A shallow box or plastic tray can be provided for vulnerable articles to ensure that they will not be damaged. Care must be taken to use a tray made of material that will not block X-rays. Polyvinyl

---

S.Ct. 1278, 59 L.Ed.2d 493 (1979); *U.S. v. Pulido-Baquerizo*, 800 F.2d 899 (9th Cir. CA 1986); *U.S. v. Edwards*, 498 F.2d 497 (1974).

[9] Activate the X-ray beam.

chloride plastic, for example, is impervious to X-rays. The transmission qualities of material under consideration can be readily tested via the X-ray unit.

A problem noted in regard to many conveyorized X-ray units is that there may be a discrepancy between the width of the belt and the width of the unit's field of view. When the view is narrower than the belt, a suitcase or garment bag may be placed so that a portion of it extends beyond the viewing area and thus will not be X-rayed. This problem has been met by marking on the belt the viewing limits of the screen. Similarly, X-rays are projected in a cone-shaped beam. The upper corners of a large square box of sufficient height may not come within the cone of the X-ray. Such a situation is easily seen on the monitor. If all of the article does not appear on the monitor, the carton must be opened for visual inspection or repositioned and sent through the X-ray until it has been fully visualized.

There have been instances where babies or animals in carriers have been inadvertently screened by X-ray. It's easy to know when this has happened because the operator will usually let out a scream as the form passes by on the screen. Although this will cause no harm to the infant or animal, it is obviously a situation to be avoided. When putting an animal carrier through X-ray screening, make sure there is no animal inside. When putting a baby carrier on the belt, make doubly sure the baby is not hidden in the blankets.

DISCOVERY OF CONTRABAND DURING SCREENING

Handling the inadvertent discovery of drugs or other contrabandContraband is covered more fully in Chapter 6. The screening operator should not detain the passenger, but should notify a supervisor and give a description of the person, the suspected contraband, and the direction of travel. In the case of a concourse with many gates, it would be helpful to ask the passenger in friendly conversation, "Where are you headed?" or some such comment that would help the law enforcement officer to narrow the search to a specific gate.

One of the more common contraband discoveries is that of an attaché case or suitcase full of cash, often a part of the massive drug laundering enterprise. It is estimated that each year, $300 billion in cocaine money moves around the world. In an article dealing with smuggling drugs from Columbia,[10] the author states,

> Within weeks of the cocaine coming north, the heavy suitcases full of cash must find their way back south. You can fit as much as $1 million into an average suitcase and it will weigh perhaps 35 pounds. The money is as illegal as the cocaine itself, and the problems of transporting it without being caught are every bit as daunting.

Several such attaché cases and suitcases have been left at security checkpoints, in rental car trunks, and in checked baggage. One airline employee is

---

[10] Mike Weiss, "Narco Bankers," *San Francisco Examiner*, (22 July 1990), *Image*.

alleged to have opened a suitcase to pilfer from it and found several hundred thousand dollars in cash. He was rather lavish and indiscreet in spending of money, however. So much so, that the owners of the money had no difficulty finding out who took it. His reward was a severe beating that put him in the hospital for an extended period.

In January 1986, San Francisco Airport Police received a $150,000 bonus from the federal government for their part in retrieving a briefcase full of cash linked to drug trafficking. A year earlier, airport officers had stopped a man before he boarded a departing flight because he fit the federal profile of a drug courier. After a trained dog indicated that narcotics were inside the man's briefcase, officers opened it and found almost $300,000. Because the money was believed to have been used in an illegal drug transaction, a federal civil suit was filed for forfeiture of the cash. There were no claimants for the money. Although it would take legislation to accomplish it, money obtained in the form of rewards could be used by the carriers, as well, to pay for screening costs and perhaps to allow them to pay a better wage to screening personnel. It seems like a workable idea, but city fathers should be expected to oppose any such legislation.

SPECIAL SEARCH SITUATIONS

The search of carry-on items seems an easy task until the first passenger arrives carrying a *piñata* or an urn containing the ashes of a departed relative.

During the early weeks of the 100 percent screening program, another problem arose with some frequency. Orthodox Jewish men carry small leather cases about two inches square, with a long leather strap. These cases, which contain small parchment scrolls bearing Scripture verses, are phylacteries (tephillin in Hebrew) and are worn at morning prayers. They are considered to be sacred ritual articles, and taking them apart, as the determined and inexperienced searchers wanted to do, is a desecration that makes them unfit for use.

For these and other carry-on items that create special screening considerations, the FAA has promulgated rules and search procedures that are followed by air carriers. In the case of urns of ashes, for example, if an X-ray unit is available, screening is easily effected. Prior arrangements can also be made with the carrier. These arrangements include documentation to verify the contents and to establish the identity of the person carrying the urn.

While diplomatic pouches are exempt from inspection, as has been noted, other diplomatic carry-on items are not.

Duty free merchandise purchased for delivery aboard an aircraft is exempt from normal predeparture inspection, *provided* that the delivery is made by a bonded, Customs-licensed employee of the duty-free seller, that the merchandise remains in the custody of the duty-free store employee and the U.S. Customs supervisor until delivered to the purchaser, and that the delivery is made beyond the passenger screening point.

The screening of persons carrying U.S. classified material is covered in FAA Advisory Circular 108-3, dated November 6, 1981. Under the provisions of that circular, the individual carrying the material should have a letter of authorization containing the following information:

1. Give the full name of the individual and the employing agency or company.

2. Describe the type of identification the individual will present (e.g., Naval Research Laboratory Identification Card, No. 1234, ABC Corporation Card, No. 1234).

3. Describe the material being carried (e.g., three sealed packages, 9" x 18" x 24", addressee, and addresser).

4. Identify the point of departure, destination, and known transfer points.

5. Carry a date of issue and an expiration date not exceeding seven days from the date of issue.

6. Carry the name, title, signature, and telephone number of the official issuing the letter. Each package or carton to be exempt will be signed on its face by the official who signed the letter.

7. Carry the name of the government agency designated to confirm the letter of authorization and its telephone number. The telephone number of the agency designated shall be a U.S. government official number that is subject to verification for both government and contractor use.

The Armed Forces Courier Services (ARFCOS) has been specifically designated by the Department of Defense to escort courier material classified Top Secret. In most instances, material in the custody of ARFCOS couriers will be of such volume as to require loading in the aircraft cargo compartment. Special arrangements exist providing ARFCOS couriers access to the ramp where they will remain on guard until the cargo compartment is secured. These couriers will generally be the last passengers to board the airplane and the first to deplane.

A related inspection problem is that of items or materials that cannot be inspected, most often because the item is a delicate scientific instrument, often hermetically sealed, that would be contaminated or damaged if opened. I have found that passengers may carry a number of articles that would fall into the category of medical items for humans or animals. I've arranged for the carriage of cow embryos, bull semen, corneas to be transplanted, and so on. In such cases, the same procedures may be followed as are required in the handling of classified documents and packages. Legal evidence also falls into this category. The real key is for the passenger to make advance arrangements. For U.S. carriers, the procedures appear in the Air Carrier Standard Security Program.

CONTRACT SECURITY SELECTION AND TRAINING

In December 1972, when the air carriers were given 30 days to institute a program of 100 percent screening of all passengers and carry-on luggage, few carriers were staffed to accommodate the increased labor burden the screening requirements would entail. In addition, no one was certain how long these requirements would last; it would not be good management to hire a considerable number of new employees only to let them go six months later if the requirements were canceled.

Because of these factors, most carriers elected to go to outside contract security agencies for personnel to perform the required screening. Most of the work was contracted to guard agencies, but this was not always the case. At O'Hare Airport, United Airlines retained an agency that had considerable previous experience in ushering at public events, and which had expanded into supplying uniformed personnel for checking baggage tags at the positive-claim baggage delivery points. This firm employed a high percentage of college-age young people. They could not be legally classed as security guards under Illinois law, which prohibits the use of anyone under 21 as a guard. Most of the personnel supplied would not have qualified under this rule, but they had applied to be, and were used as, baggage inspectors.

Whether a carrier elects to contract out this function will depend on a variety of circumstances. Cost will probably be the primary factor. This includes not only wages and salaries but also fringe benefits such as health and life insurance and the expense of training, uniforms, and liability insurance.

The decision to retain an outside contractor must also be considered in the light of possible repercussions on internal labor-management relations.

The chief advantage to hiring an outside agency is the flexibility this gives the carrier, who is relieved of hiring or firing employees, scheduling vacations, handling pay raises, etc. If the carrier does not like the way an inspector is performing the duties, it can ask the agency to transfer the person from the airport and to assign a new individual to the post.

The carrier should institute some control over the agency people assigned to it at the terminal, as permanency on the part of the people assigned to screening operations is important to competent performance. Considerable training is required in order that the person doing the inspecting or screening will know how to perform the job and will have a good working knowledge of the relevant requirements of the carrier's security program. If the contract security service has a high personnel turnover rate, and most do, severe problems can be created for the carrier.

The carrier may wish to interview any individual scheduled to be assigned by the contract agency before that person reports to work, for the purpose of determining whether to accept or reject the individual before he or she begins the work assignment. Public relations are extremely important to air carriers, and predeparture screening personnel are public-contact people representing the airline. They should have a good physical appearance, dress neatly, and deal courteously with the public. Since agency personnel are not normally highly-paid, the selection process is proportionately more difficult. The old axiom that "you get what you pay for" most certainly has an application here. The carrier who contracts with an agency solely on the basis of the lowest bidder may be doing itself a costly disservice.

Many companies are in the business of supplying contract security services. A number of them are excellent companies, but some may oversell their capabilities and prove to be long on promises and short on performance. The carrier considering an agency contract should visit local sites where the contractor's employees are working. On one rather large contract, I visited the corporate offices of the contractor on the other side of the continent and then spent a week traveling and visiting some of their major accounts. The carrier representative should also arrange for one of the negotiation meetings to be held at the contractor's place of business, so that he or she may see their offices. It is well to ask for a tour of their training facilities but some agencies may balk at that request. A nationwide firm may have a good capability for providing service in one part of the United States, but may be found to be less well-managed in another. Determine what type of supervision will be provided the contract people: will they provide their own on-site supervisor, or is the carrier to do the supervision? Supervision is the key to effectiveness for any type of security force.

Determine what type of personnel screening is used by the security agency under consideration. In some areas of the country, it is notoriously easy to obtain a job as a security guard. Because of the low wages often paid, the qualifications for employment are not high, and legal standards are frequently even lower. In some states, a person who wants to be a security guard carrying a weapon need only be 21 years old and a citizen of the United States. When the people being considered are to be baggage inspectors, not security guards, and there is no reason for them to be armed, the carrier may draw on the agency's guard labor pool for its needs. The carrier should find out if applicants are fingerprinted and what is done with those fingerprints. Are they used to determine if the applicant has a criminal record? How long does it take for this information to get back to the agency? (In some states, the time required for processing these record checks through the necessary channels may consistently be a matter of months.) Does the agency make any attempt to verify the information on the employee's application?

With proper planning and sufficient investigation, a carrier should be able to select a contract service company with whose performance it will be pleased. A contract should be entered into with the agency to be hired. The carrier's legal department is best equipped to draw up such a contract if the contract the agency offers is not satisfactory. Normally, a contract will provide for the agency to have sufficient liability insurance to protect the carrier in case of suit, as well as a hold-harmless clause protecting the carrier against liabilities, demands, claims, suits, losses, damages, causes of action, fines, or judgments including costs, attorneys' and witnesses' fees, and expense incident thereto, for injuries to persons (including death), false arrest, false imprisonment, assault, and/or battery, and for loss of, damage to, or destruction of property. These hold-harmless clauses were partially responsible for two nationwide contract guard agencies pulling out of the airport security marketplace altogether. They feared that passenger lawsuits from planes that had been hijacked might make their continued operation unprofitable.

The security personnel provided should have the factual knowledge needed to do their job, and they should be made aware of the most efficient manner in which to use this knowledge. They will need initial training, and this should include both classroom and on-the-job training. In addition, there must be continuing instruction so that employees will maintain their knowledge and proficiency. The contract should specify who pays for the initial and recurrent training. It could specify, for example, that the contractor will pay for the initial training, but only if the candidate stays with the company for 90 days.

The continuing/recurrent training can take many forms. It may involve refresher classes, a daily briefing session at the time the employees report to work, which covers one small aspect of the predeparture screening program, or the supervisor working alongside the employee for a 30-minute period each week for evaluation and instruction. A method of disseminating new information, changes to the program, and special alerts must be included in each carrier's training program. This can be done through advisory bulletins or, in cases of urgency, through the carrier's Teletype or FAX networks.

One new training method involves video training programs that are transmitted to subscribers with satellite receivers, or more recently, simply preparing video tapes for mailing to their clients, alleviating the need for a satellite receiver. Professional Security Television Network (PSTN) of Carrollton, Texas, is the innovative company that is beginning this new venture. The concept has worked extremely well for law enforcement agencies.

Major security agencies offer training as part of their service, and it is incumbent upon the carrier representative to know the content of that training and how much time is given to it. The FAA is also a resource for training. At present a standardized training program is supplied to the contractor by an FAA approved vendor. The carrier may have a training department that will assist the security department in the preparation of a curriculum. Community colleges are also a valuable resource for such training programs.

ALTERNATIVE SECURITY CONCEPT

In 1987, a concept was considered whereby the Air Transport Association would create an airline subsidiary that would contract on behalf of all the carriers for security services and be responsible for training these personnel. It was felt by some that such an approach would be more economical and would significantly improve security. After considerable thought and discussion, however, the plan was dropped, apparently because it was felt that a renewed emphasis by each carrier on the security training of contract guards would be more effective.

HUMAN FACTORS IN SECURITY SCREENING

Human factors such as low pay, high turnover, hiring problems, and poor training have plagued the larger airports particularly. Much work remains to be done in this area. Because carriers are subject to civil penalties when weapons are not discovered, they often react by terminating the responsible employee. Motivation is a highly significant problem and is all too often accomplished by negative rather than positive reinforcement. In some locations, screening personnel are given monetary rewards for discovering weapons or are designated as Screener of the Month. It is a fertile area for research, and much needs to be done to alleviate the boredom these people feel.

There are other factors to be considered, such as how long an individual can effectively monitor an X-ray screen, training in dealing with the public and upward mobility or promotion possibilities. Many of the more effective screening personnel are older adults, and they tend to be long-term employees. The FAA is spending a considerable amount of money for technical research in the area of explosives detection, and this is absolutely necessary. I think, however, that additional studies need to be accomplished in the area of human factors, so that station managers are not left on their own in that area.

The Report of the President's Commission on Aviation Security and Terrorism The Report of the President's Commission on Aviation Security and Terrorism commented that the quality of training varies from airline to airline and also varies within an airline. The Commission's investigation at Baltimore-Washington International Airport disclosed a screener's failure to detect an obvious explosive device in a briefcase. The screener was incapable of understanding questions posed to him in English concerning the extent of any training he may have received.

The requirement that security officers and screeners be fluent in English is an important one. Too often, applicants must pass a minimal test that usually amounts to whether or not they can communicate with the interviewer. When they are hired and assigned to a security post, they are unable to communicate in person or over the phone and sometimes just hang up rather than struggle through the dialogue. When they are sitting at a reception desk, this is a critical skill.

Similarly, a screener or at least a supervisor at an international airport needs to be bilingual in most cases. If they are in Frankfurt, they need to be able to converse in English as well as German. One screener in Frankfurt, interviewed on TV, stated he memorized the required questions to ask passengers in English, but he didn't know what they meant and couldn't understand the passenger's responses.

One of the trends that has developed in the industry is that the air carriers are utilizing contract security officers for duties other than predeparture screening. The rationale for this is purely economic. Contract security employees are utilized for international greeters, for pushing wheelchairs, for monitoring the security of baggage arriving on carousels and belts, and for assisting skycaps in the loading of baggage onto belts after the skycap has checked the baggage. The ability to shift individuals from one job to another provides variety for the contract employee and makes the job more interesting. Besides, they get tips when they personal services, and they don't when they screen people.

TESTING AND ENFORCEMENT

Frequent tests are conducted of screening personnel in an effort to evaluate their proficiency. The test objects include guns, dynamite, bombs, and other objects large or dense enough to disguise weapons. A different list of simulated weapons is utilized for metal detector, X-ray and physical inspection. The tests are accomplished not only by the FAA, but by the air carrier and the contract screening company. Normally, the response to a failure to detect these devices is to notify the carrier and to take some type of corrective action with the employee who failed to recognize the object.

A civil penalty assessment demonstration program was authorized by Congress in December 1987 for a two-year period. Several months later, the FAA issued a final rule without allowing for a notice and comment period. This civil penalty assessment demonstration program (1) called for an automatic $1,000 fine for failure to detect an FAA test object, (2) raised to a maximum of $10,000 the penalty for a violation of the FARs, and (3) allowed the agency to prosecute and adjudicate administrative penalty actions under $50,000. During the last six months of 1988, the FAA began to file consolidated civil penalty actions against air carriers based on the carriers' failure to detect the test object during the screening of passengers and carry-on baggage. In three separate rounds of filings during this six-month period, the FAA sought $3,977,000 in civil penalties.

By January 1990, the civil penalties totaled $7 million. However, few were being paid as the carriers challenged the FAA's program in court. The ATA filed suit in the U.S. Appeals Court in Washington, D.C. on the grounds that the FAA violated the Administrative Procedures Act (APA) by not issuing a Notice of Proposed Rule Making (NRPM) with the program. On April 13, 1990, the Appeals Court ruled that the FAA could not initiate new prosecutions until it had complied with the procedural requirements of the APA.[11]

Admiral James B. Busey, FAA Administrator, was receptive to a revision of the policy. The same enforcement issues were being raised in the field of engineering and maintenance. The carriers were upset because even when their own audit teams detected violations and reported them, the FAA proceeded with enforcement actions. On March 27, 1990, Admiral Busey announced a new FAA enforcement policy:[12]

> If you discover an inadvertent violation, correct the problem, report it promptly to the FAA, and put in place a permanent fix acceptable to FAA to make sure it will not happen again, the FAA will not penalize you...Period.

---

[11] Aviation Daily (April 16, 1990):103.

[12] Remarks by Admiral James B. Busey, Administrator, FAA, before the Aero Club of Washington, D.C., March 27, 1990.

The initial understanding of the new policy was that every failure to detect a test object would not result in a penalty, but the FAA would retain the 100 percent detection standard.

HEIGHTENED SECURITY

The FAA developed contingency plans in the early stages of the buildup of the threat in the Persian Gulf. These plans were distributed on December 31, 1990 to be implemented in the event of various levels of hostility. The contingency levels are:[13]

Level 1. Normal; the condition prior to Iraq's August 2 invasion of Kuwait.

Level 2. Heightened tension.

Level 2A. Reports of violence or attacks; escalated tensions as a result of a number of possible factors, none of which actually results in violence.

Level 3. Actual violence or attacks; any incident involving actual violence, bloodshed or loss of life on behalf of any participant in the crisis that could be attributed to another participant.

Level 4. Open hostilities; the actual outbreak of sustained hostilities between Iraq and opposing forces in the Persian Gulf Region.

With the threat of terrorism from Iraq, following the start of the Gulf war, the level of security was increased at airports around the world. On January 16, 1991, when the air war with Iraq was beginning, the FAA ordered the airports to go to Level 4 security. In general, this meant the cancellation of all curbside baggage check-in; no baggage check-in at ticket offices or car rental agencies; access to sterile concourses restricted to ticketed passengers; screening and search of all personnel who approach the aircraft for servicing purposes, excluding flight and inflight crews operating the aircraft; safeguarding all checked baggage from the moment the carrier receives it until the passenger receives it at the destination; and intensified the screening of electronic devices, sometimes resulting in the removal of batteries from such devices. In March 1991, the airports returned to Level 2.

Level 4 security had been an expensive experience for the airports.The FAA estimated lost revenue at U.S. airports to be $3 million per week[14] because access to concourses was restricted to ticketed passengers. Los Angeles International Airport estimated the loss to concessionaires and ven-

---

13 Jonna Ferreris, "U.S. Airports Tighten Security during Crisis, "*Access Control*, 34:4, (March, 1991): 29.

14 William E. Schoneberger, "Airports Tally The Cost of Level 4 Security, "*Airline Executive International*, 15:5, (May, 1991): 8.

dors at $130,000 a day. Conversely, at one airport where pickpockets had been a severe problem, reports of this type of crime dropped to nothing.

Chapter 6

SCREENING AND THE LAW

The legal authority for conducting the airport searches is based on FAR Part 108.5 (a), "Security Program: Adoption and Implementation," Part 108.9, "Screening of Passengers and Property," and ICAO Annex 17, Chapter 4.1.1. Annex 17 to the convention on international civil aviation is entitled, "International Standards and Recommended Practices -- Security -- Safeguarding International Civil Aviation Against Acts of Unlawful Interference."

> FAR 108.5 (a). Each certificate holder shall adopt and carry out a security program that meets the requirements of §108.7 for each of the following scheduled or public charter passenger operations:
>
> FAR 108.9 (a). Each certificate holder required to conduct screening under a security program shall use the procedures included, and the facilities and equipment described, in its approved security program to prevent or deter the carriage aboard airplanes of any explosive, incendiary, or a deadly or dangerous weapon on or about each individual's person or accessible property and the carriage of any explosive or incendiary in checked baggage.
>
> ICAO Annex 17, Chapter 4.1.1. Contracting states shall take the necessary measures to prevent weapons or other dangerous devices, the carriage or bearing of which is not authorized for being introduced, by any means whatsoever, on board an aircraft engaged in the carriage of passengers.

SEARCH AND SEIZURE

Of the first ten amendments to the United States Constitution, which are referred to as the *Bill of Rights*, the Fourth Amendment is probably best known to police and security people.

> The right of the people to be secure in their persons, houses, papers and effects, against unreasonable searches and seizures, shall not be violated, and no warrants shall issue, but upon probable cause, supported by oath or affirmation, and particularly describing the place to be searched and the persons or things to be seized.

The legal ins and outs of search and seizure present an endless challenge, for what is legally certain today may be legally uncertain tomorrow, and antihijacking measures have proved to be no exception.

Public reaction to the 100 percent search requirement was positive; the great majority of air travelers gave every indication of gratitude for the searches and the assurance they implied that the traveler could expect a safer trip. Indeed, more complaints were received about inadequate searches than about objections to the screening.

The strongest--or perhaps the best publicized--objection to the screening program was lodged by Indiana Senator Vance Hartke, who refused to be screened on several occasions. Although the Senator was not permitted to board a flight at Phoenix, two other carriers were less prepared to hinder an already irate U.S. Senator and allowed him to board without screening, receiving a fine from the FAA for their pains. Senator Hartke contended that he was being harassed because of his role in an investigation of the FAA. Eventually, he brought suit against the FAA on a slightly different basis, contending that the searches violated his congressional immunity against arrest, on the grounds that the searches constituted a temporary arrest.

THE LEGAL AUTHORITY OF PRIVATE PERSONS TO SEARCH

The legality of airport searches involves a little-known aspect of U.S. case law, the authority of private persons to search without warrant on reasonable cause. This applies whether the search is of a passenger's carry-on baggage or the opening of a package in the air freight terminal.

The law regarding to searches has evolved over the years and began with decisions affecting police searches. An early ruling, *Weeks v. United States*,[1] held that where federal officers have made an "unreasonable" (and consequently illegal) search and seizure, the evidence obtained is not admissible in a federal court proceeding. For many years, this decision applied only to federal cases; indeed, in *Wolf v. Colorado*,[2] the Supreme Court went so far as to state that *Weeks* was not to be applied to illegally seized federal evidence offered in a state court:

---

1 232 U.S. 383 (1914).

2 338 U.S. 25 (1949).

> [I]n prosecution in a state court for a state crime, the Fourteenth Amendment does not forbid the admission of evidence obtained by an unreasonable search and seizure.

A related decision followed shortly. In *Lustig v. United States*,[3] the Supreme Court held that the fruit of illegal state searches was admissible in federal proceedings:

> The crux of that doctrine is that a search is a search by a federal official if he had a hand in it; it is not a search by a federal official if evidence secured by state authorities is turned over to the authorities on a silver platter.

Understandably, this became known as the *silver platter doctrine*.

But time and opinions of the judiciary pass, and a decade later the Supreme Court had moved toward strict construction of the Constitution. The silver platter doctrine was the first to go. In *Elkins v. United States*,[4] the Court held that evidence illegally obtained by state officers and subsequently provided to federal agents would not be admissible in federal prosecutions. In the next year, the Court took another view of the Fourteenth Amendment and decided--in *Mapp v. Ohio*,[5]--that the prohibitions of the Fourth Amendment were were fully applicable to the states under the Amendment's due process clause; evidence obtained by illegal searches by state or federal officers was inadmissible in state courts.

This, essentially is the manner in which the law developed regarding searches by officers of the government--federal, state, county, or municipal. But searches by private persons do not come under these prohibitions, and U.S. case law in this regard has shown little or no change.

The landmark Supreme Court decision regarding searches by private persons is *Burdeau v. McDowell*.[6] In its decision, the Court stated:

> The Fourth Amendment gives protection against unlawful searches and seizures, and as shown in the previous cases, its protection applies to governmental action. Its origin and history clearly show that it was intended as a restraint upon the activities of sovereign authority, and was not intended to be a limitation upon other than governmental agencies; as against such authority it was the purpose of the Fourth Amendment to secure the citizen in the right of unmolested occupation of his dwelling and the possession of his property, subject to the right of seizure by processes duly issued.
>
> In the present case the record clearly shows that no official of the federal government had anything to do with the wrongful seizure of the petitioner's property, or any knowledge thereof until several months after the property had been taken from him and was in the possession of the

---

3 338 U.S. 74 (1949).

4 364 U.S. 206, 89 S.Ct. 1437 (1960).

5 367 U.S. 643 (1961).

6 245 U.S. 465, 41 S.Ct. 574 (1921).

> Cities Service Company. It is manifest that there was no invasion of the security afforded by the Fourth Amendment against unreasonable search and seizure, as whatever wrong was done was the act of individuals in taking the property of another. A portion of the property so taken and held was turned over to the prosecuting officers of the federal government. We assume that petitioner has an unquestionable right of redress against those who illegally and wrongfully took his private property under the circumstances herein disclosed, but with such remedies we are not now concerned.

This position was probably most succinctly stated in *People v. Superior Court of Los Angeles County*,[7] a case involving recorded conversations. The Court stated:

> There are no state standards for "search and seizure" by a private citizen who is not acting as an agent of the state or other governmental unit. Therefore acquisition of property by a private citizen from another person cannot be deemed reasonable or unreasonable.

POLICE PARTICIPATION

On the other hand, courts have developed an eagle eye for the possibility that a private search may have been done for or with law enforcement, and has thus become a governmental search requiring all the safeguards such a search must be given.

Corngold v. United States[8] involved the admissibility of evidence discovered by opening packages for inspection. The packages had been shipped from Los Angeles to New York under what proved to be a false description of their contents. On the arrival of the packages at the destination airport, four Customs officers tested the packages without opening them, using a scintillator. They explained to a transportation agent of the carrier that the scintillator detects the radiation from luminescent watches, and requested that the packages be opened. After checking the company's tariff, the agent opened the packages with the assistance of the Customs officers and found that they contained watches.

The court stated in its decision:

> It would be difficult to justify any conclusion other than that the carrier employee participated in the search solely to serve the purpose of the government. There was nothing else in the record which would indicate that the package was in fact opened for any purpose of the carrier.
>
> The fruits of a search conducted solely in aid of the enforcement of a federal statute, as this one was, are inadmissible when the search fails to meet the Fourth Amendment standards. The search was in substance a federal search cast in the form of a carrier inspection to enable the officers to avoid the requirements of the Fourth Amendment.

---

7 449 v. P.2d 230, 74 Cal Rptr 294 (1969).

8 367 F.2d 1 (1966).

The Court cited a supporting decision, *Taglavore v. United States*,[9] which said:

> [T]he violation of a constitutional right by a subterfuge cannot be justified, and the circumstances of this case leave no other inference than that this is what was done.

A ruling decision in carrier package inspection is *United States v. Pryba*.[10] Here a United Airlines supervisor (and friend of mine) had caused a package to be opened because of "peculiar circumstances" surrounding receipt of the shipment. These were that the package was heavy for its size, and that the individual delivering the package was extremely nervous, evaded questions as to the contents of the package, and admitted that the return address was fictitious. The package was found to contain 30 reels of film that the supervisor concluded were hard-core pornography. He turned the film over to the FBI.

The defendant's argument was that the tariff regulation under which the package was opened was unconstitutional, and for this reason the film was inadmissible in evidence against him. The court decision said that this contention ignored the fact that the initial search was not made at the behest of federal authorities, but was an independent investigation by the carrier. The Court concluded:

> Furthermore, in the present climate of bomb scares, it is difficult to believe that anyone would seriously challenge as unconstitutional the authority delegated to airline personnel to examine the contents of baggage and freight deposited for carriage.

Generally, when air carrier personnel suspect that a suitcase or a piece of freight may contain explosives, they call in the designated military or law enforcement organization that has the requisite expertise to handle such risks. In such a case, the package is opened by a law enforcement agency at the request of the carrier. It is assumed that the courts would hold this to be a valid search, even though it is performed by federal or local authorities. There are a number of grounds for admitting the fruits of such searches: It might be held to be a private search due to its origin; it could be held that the carrier acted in accordance with FAA rules requiring due diligence to prevent the carriage of explosives; or it might be held to be a lawful police search proceeding upon reasonable cause under circumstances that preclude delaying the search long enough to obtain a warrant.

The case of police participation at the request of the carrier was at issue in *Wolf Low v. United States*,[11] and in this case the Court employed the precedents for searches by "private persons." The defendant had checked two overweight suitcases for a flight from Los Angeles to Las Vegas. The airline ticket agent, suspicious of the weight of the bags, held them off the

---

9 291 F.2d 262, 266 (9th Cir. 1961).

10 312 F.Supp. 466 (1970).

11 391 F.2d 61 (1968).

flight. The agent's superior called on a Los Angeles city policeman on duty at the airport to witness the opening of the bags by airline personnel. The contents proved to be 3500 watch movements, and these were turned over to a United States Customs officer who was called to the scene.

The Court stated in its decision:

> The conduct of the airline in turning over to the Government the fruits of its search of the defendant's suitcases, and the conduct of the Government in making use of the information and evidence obtained by the airline's search, fit readily within the doctrine of *Burdeau v. McDowell.*

Another decision involved a search made after law enforcement had called attention to irregularities. In *Gold v. United States*,[12] government agents informed the United Airlines Services Manager at Las Vegas that they had reason to believe a package in the possession of the carrier had a false description of its contents on the air waybill, and that the shipper's address was a nonexistent one. The agents then left the manager's office after declining to reveal what they suspected the true contents of the package to be. The manager opened the package and discovered pornographic film, which he subsequently turned over to the government agents. The Court held:

> We conclude that the initial search of the packages by the airline's employee was not a federal search, but was an independent investigation by the carrier for its own purposes. Unlike *Corngold*, here the agents did not request that the package be opened, and they were not present when it was opened. The agents had the same right as any citizen to point out what they suspected to be a mislabeled shipping document, and they exercised no control over what followed. What did follow was the discretionary action of the airline's manager and was not so connected with government participation of influence as to be characterized as was the search in *Corngold*, as "a federal search cast in the form of a carrier inspection."

By changing the circumstances only slightly, a different rule of law emerges. Consider the case of a misrouted bag opened by airline employees for the purpose of identifying its owner. Within the bag is found what appears to be narcotics, and the police are called. For the interval between calling police and their arrival, the bag is reclosed and returned to the storage shelf. Is the second opening of the bag, done at the request of the police, an illegal search unless done with a warrant? A 1969 decision held that the carrier has a right and possibly an obligation to turn over to law enforcement authorities contraband discovered in a passenger's luggage. In *United States v. Averall*,[13] the Court stated:

> The airline had the right, and perhaps the duty, to discover if its facilities were being used for the commission of a crime. It is an act of responsible citizenship for individuals to give whatever information they may have to aid in law enforcement.

---

12 378 F.2d 488 (1967).

13 296 F.Supp. 1004 (1969).

From that, one might argue that the second opening of the bag was not an actual search, but merely incident to a turning over of the contraband to law enforcement.

Yet, it is well to remember that the courts have consistently held that, where there is time to secure a search warrant, a search should not be made without one. The ruling case is *Trupiano v. United States*,[14] where the Court held that:

> It is a cardinal rule that, in seizing goods and articles, law enforcement agents must secure and use search warrants whenever reasonably practicable. This rule rests upon the desirability of having the magistrate rather than the police officers determine when search and seizures are permissible and what limitations should be placed upon such activities.

This was reaffirmed in *Corngold* (1966), where the Court said:

> We would, of course, agree that if a carrier, while inspecting packages for its own purpose pursuant to a provision in the contract of carriage, discovers contraband and notifies the customs agents of that fact, and the agents then secure a warrant on the basis of this information and conduct a search, the search is entirely proper. This is precisely the procedure which the Fourth Amendment contemplates.

However, also in *Corngold*:

> [M]ere surrender of custody to a carrier did not forfeit appellant's right of privacy. He did not expressly authorize TWA to consent to a search of his package by Customs agents.

From this, it would seem that the best handling of discovered contraband would be to ensure that no subsequent search will be required; when an employee has discovered contraband in a piece of luggage, he or she should leave the contraband exposed awaiting the arrival of law enforcement. Once the luggage has been closed, however, a search warrant should be required, even though the employee's testimony might be later held to remove the need for a warrant. The appellate process is slow and expensive, and it is to everyone's advantage to handle the discovery to law enforcement in an unquestionable manner.

It has been suggested that a narcotics search might be held to have been done under "probable cause" where the suspect package or baggage has been pointed out by dogs trained to detect narcotics. But narcotics do not explode, and baggage can be held long enough for a warrant to be obtained in most cases. In any event, this would appear to be a law enforcement situation, since carriers do not have occasion to screen cargo and luggage for narcotics.

Where public and private police act jointly, the legal technicalities are blurred. In a ruling case dating back to Prohibition, *Byars v. United*

---

14 68 S.Ct. 1229, 334 U.S. 699 (1948).

States,[15] a federal agent who had been invited to accompany a state officer participated in a search that turned up counterfeit strip stamps of the kind used on whiskey bottled in bond. The Court said:

> We cannot avoid the conclusion that the participation of the agent in the search was under color of his federal office and that the search in substance and effect was a joint operation of the local and federal officers. In that view, so far as this inquiry is concerned, the effect is the same as though he had engaged in the undertaking as one exclusively his own.

INEVITABLE DISCOVERY EXCEPTION TO THE EXCLUSIONARY RULE

In *U.S. v. Hernandez-Cano*[16], the defendant was proceeding through screening when the X-ray revealed a large, dark mass in his carry-on bag. The operator asked for and received his permission to inspect the bag. When the operator attempted to pull a large object out of the bag, the defendant objected to any further inspection of the bag. After repeated refusals to permit further inspection, Hernandez-Cano agreed to open the bundle a little. When he did, the operator saw what she thought was white powder.

When he refused to unwrap the bundle further, the supervisor summoned a police officer. The officer told the defendant that if he did not allow his bag to be inspected, he could not pass through the security checkpoint. The defendant left the checkpoint with his carry-on and went to the ticket counter where he had previously checked two pieces of luggage. He said he wanted to transfer an item from his carry-on luggage to his checked luggage, which was done. He then went through screening and told the operator he had thrown away the questionable bundle, which he claimed was dirty clothes.

Two security employees then went to the ticket counter and determined that the defendant had checked the item. Not knowing if the package contained "drugs or a bomb," the ticket agent notified the airline control tower and told them to hold defendant's checked luggage. She then went to the airplane to discuss the situation with the pilot. Because of her concern, the agent proceeded to the baggage area accompanied by the police officer who had asked permission to go with her.

The ticket agent opened one of defendant's checked bags with one of her standard luggage keys because she felt it was imperative to inspect the bag quickly so as not to hold up the airplane. The officer did not instruct her to open the bag and in fact was standing several feet away. As the agent was searching through the bag, the officer, who by this time was looking over her shoulder, reached down into the bottom of the suitcase and pulled out a large bundle containing a white, powdery substance, later identified as cocaine.

---

15 47 S.Ct. 248 (1927).

16 808 F. 2d 779 (11th Cir. Ga. 1987).

The Court said that although the officer's warrantless search was not justified, "we hold that the inevitable discovery exception should have been applied and therefore, the evidence was admissible."

> Under the inevitable discovery exception to the exclusionary rule, evidence is admissible that otherwise would be excludable if it inevitably would have been discovered by lawful means had the illegal conduct not occurred. The Supreme Court first recognized this exception in *Nix v. Williams,* 467 U.S. 431, 104 S.Ct. 2501, 81 L.Ed.2d 377 (1984)....The Court reasoned...that although the government should not be put in a better position because of its illegal conduct, it should not be put in a worse position because of the conduct either.
>
> Therefore, "[i]f the prosecution can establish by a preponderance of the evidence that the information ultimately or inevitably would have been discovered by lawful means...then the deterrence rationale has so little basis that the evidence should be received..."
>
> Had Singleton [police officer] not intervened in the search here, Fleck [ticket agent] inevitably would have discovered the cocaine.

WHO IS A LAW ENFORCEMENT OFFICER?

Because air carrier employees are "private persons" in law and have only the legal authority of private persons to search, together with certain advantageous exemptions, it becomes essential to be able to accurately determine who is a private person in law and who is not.

FAR 107.14 defines a law enforcement officer as one who:

1. Has the authority to arrest without a warrant, while on duty at the airport, for a crime committed in the officer's presence or for a felony, when the officer has reason to believe that the suspect has committed it.
2. Is readily identifiable by uniform and displays or carries a badge or other indicia of authority.
3. Is armed with a firearm and authorized to use it.
4. Has completed a training program meeting the requirements set out in the FAR.

It seems clear that an undeputized store detective, contract guard, or private investigator is not a law enforcement officer because he is not vested with a power of arrest other than that afforded any citizen. Where this authority has been given to a contract guard who meets the above criteria, the guard can replace law enforcement personnel at screening points. In Hawaii, for example, the state legislature enacted a law giving powers of arrest to persons under contract to the State Department of Transportation, and contract guards so empowered perform the law enforcement function at airports in Hawaii.

Properly empowered private guards are fulfilling this function at other airports, and this policy is under consideration elsewhere.

It is worth noting that circumstances can jeopardize the "private person" status. A company guard may lose this status if he is deputized, whether this has any connection with the guard's work or not. It has been held that a police officer working during his off-duty hours as a security guard or a store detective is a police officer at all times, and in *People v. Tarantino*,[17] The Court noted the precedent of *Burdeau (1921)*, but held it inapplicable where the private citizen who obtained the evidence was employed by the district attorney and the police department, worked under the supervision of the police, and was paid out of public funds.

LEGALITY OF AIRPORT SCREENING SEARCHES

This general issue--police participation in private searches--takes on new significance as we consider searches conducted by airline employees in compliance with government regulations. It has been said that these warrantless, nonarrest searches are legal because they are conducted by private persons; but, on the other hand, they are done at the instruction (by requirement) of a federal agency. FAR 108.9 (a) requires each certificate holder to "conduct screening under a security program...to prevent or deter the carriage aboard airplanes of any explosive, incendiary, or a deadly or dangerous weapon on or about each individual's person or accessible property and the carriage of any explosive or incendiary in checked baggage."

In *United States v. Lopez*,[18] which involved a narcotics seizure resulting from a screening search, the Court stated:

> While the abuse in this case was by airline officials, not the government, these employees were acting as government agents insofar as they designated 'selectees' and alerted Marshals. The court's only effective means of controlling official excess is to refuse to admit in evidence the fruits of unconstitutional seizures.

Another authoritative statement on the subject was given in testimony before the Aviation Subcommittee of the Senate Commerce Committee in January 1973 by Assistant United States Attorney General Scott Cramton, who stated:

> Even though private employees of an airline are doing the search, there is no question that where they are ordered to do so by the Federal Government, it really is 'State action,' you can't get around the constitutional question by pretending it is private conduct.

Warrants will not be discussed here because, as the *Lopez* decision states:

> The anti-hijacking system depends upon being able to swiftly sift out potential hijackers for closer scrutiny while permitting all passengers includ-

[17] 45 Cal.2d 590, 290 P.2d 505 (1955).

[18] 328 F.Supp. 1077 (1971).

> ing 'selectees' to board unless weapons are discovered. Exigencies of time precluded the Marshals from obtaining a warrant in this situation.

At issue in *Lopez* was the question of consent. The government contended that the posting of signs advising that passengers and baggage were subject to search was tantamount to "implied consent." The Court rejected this reasoning, stating that consent to a search involves the relinquishment of fundamental constitutional rights and that consent should not be lightly inferred. A subsequent decision, *United States v. Meulener*,[19](1972) supported this view, stating:

> It is clear that in the instant case, the defendant did not give consent to a search of his suitcase. He opened it only after he was ordered to do so by the marshal at a time when he was not free to leave or to avoid the search. Under such circumstances, the search was inherently coercive.

Stressing that consent to a search amounts to a waiver of a constitutional right, the Court quoted *United States v. Blalock*.[20] Here, the decision stated that the requirement of an intelligent consent implies that the subject of the search must have been aware of his or her rights, for an intelligent consent can only embrace the waiver of a known right. The Court, in *Blalock*, imposed upon officers the obligation to advise the subject of his or her constitutional rights as they relate to search and seizure.

In that respect, the *Meulener* case is being followed, as passengers are now advised of their right to refuse to submit to a search. Signs at screening points inform passengers that they may refuse inspection, but that passengers refusing to be screened will not be transported. Contra to that policy is the opinion of the *Lopez* Court rendered in 1971, prior to the introduction of mandatory screening, which said:

> Nor can the government properly argue that it can condition the exercise of the defendant's constitutional right to travel on the voluntary relinquishment of his Fourth Amendment rights.

Nor were airport searches justified as searches incident to arrest by the holding of the Court in *Lopez*. That the search revealed evidence of a violation of a federal statute does not render the search justifiable; a search prosecuted in violation of the Constitution is not made lawful by what it brings to light. There is long-established judicial precedent confirming this, beginning with *Byars* v. *United States*.

There is much discussion in case law of probable cause as justification for searches. Presently, only a limited search for weapons is allowed in the absence of probable cause where the search is not incident to an arrest. As the Court put it in *Terry v. Ohio*:[21]

---

19 351 F.Supp. 1284 (1972).

20 255 F.Supp. 268 (1966).

21 392 U.S. 1 (1968).

> A police officer may in appropriate circumstances and in an appropriate manner approach a person for purposes of investigating possible criminal behavior even though there is no probable cause to make an arrest.

This search must be limited in scope and is permissible only if "a reasonably prudent man in the circumstances would be warranted in the belief that his safety *or that of others* was in danger" (emphasis added).

Courts have applied the tests of these precedents to the profile, the magnetometer, the search of carry-on luggage, and the search of the person. These cases, however, were all decided before the January 1973 imposition of the requirement for 100 percent electronic screening of all passengers and 100 percent search of all carry-on items. The decisions may still stand as precedent, but the overtone throughout most of these decisions was that the slight infringement upon individual rights was to be balanced against the overwhelming need to stop hijacking.

In prior decisions, attempts have been made to justify these searches on the basis of probable cause, with probable cause having been established by correspondence with the profile or by the passenger's having activated the magnetometer. In *Meulener*, the Court held that the failure to make an initial pat-down of the defendant's outer clothing before searching his suitcase violated the passenger's Fourth Amendment right, but this, too, was a "selectee" case.

Decisions in a parallel situation offer some insight into the reasoning of the courts. Following a series of bombings of federal buildings, a precautionary search was established and those entering were required to open packages and briefcases. In a case arising from this requirement, *Barrett v. Kunzig*,[22] the Court stated:

> When the interest in protection of the government property and personnel from destruction is *balanced* against any invasion to the entrant's personal dignity, privacy, and constitutional rights, the government's substantial interest in conducting the cursory inspection *outweighs* the personal inconvenience suffered by the individual...[emphasis added].

The Court was even reluctant to term the process a search and stated:

> The term "search" has been used by the plaintiffs. To the extent that term is applied to more than a casual visual inspection, it has no meaning and is without foundation in this record.

The Court also observed that the inspection could not be held to be accusatory in nature because the inspection was required of everyone entering.

In another courthouse inspection case, *Downing v. Kunzig*,[23] the Court commented similarly, "We cannot agree that such "searches" (if they may be so characterized)...were unreasonable under the Fourth Amendment."

---

22 331 F.Supp. 266 (1971).

23 454 F.2d 1230 (1972).

The argument of *Downing* is most applicable to a discussion of airport searches:

> As in most instances under the law competing values must be weighed and a balance struck. Here it seems to us that the balance is clearly on the side of upholding the very limited sort of search contemplated... .
>
> [I]n times of emergency, government may take reasonable steps to assure that its property and personnel are protected against damage, injury or destruction by resorting to the very minimal type of interference with personal freedom arising from the acts complained of in this case. The "searches" considered here involve no confiscation of legitimate property, no personal inspection of confidential papers, and no undue restraint on freedom of movement.

Consistently, these cases return to the minimal invasion of personal freedom and the reasonableness of the search in the light of the risk. Similarly, in *United States v. Epperson*,[24] regarding the legality of a search by use of a magnetometer, the Court commented:

> The danger is so well known, the government interest so overwhelming, and the invasion of privacy so minimal, that the warrant requirement is excused by exigent national circumstances.

This decision also observed that the constitution does not forbid searches and seizures; it only forbids those that are unreasonable. And again on "reasonableness" of the search, the *Epperson* Court stated:

> The reasonableness of any search must be determined by balancing the governmental interest in searching against the invasion of privacy which the search entails... . [I]t is clear to us that to innocent passengers the use of a magnetometer to detect metal on those boarding an aircraft is not a resented intrusion on privacy, but, instead, a welcome reassurance of safety. Such a search is more than reasonable; it is a compelling necessity to protect essential air commerce and the lives of passengers.

The court went on to state that the use of the magnetometer having been justified at its inception, the subsequent physical frisk was justified by the information the magnetometer developed. Several other decisions on magnetometer searches are in accord with *Epperson.*[25]

From these decisions, it would appear that the courts will continue to support the minimal invasion of privacy inherent in airport searches. It is an added advantage that the acquisition of X-ray devices since these decisions has even further reduced the minimal invasion of privacy that the search of carry-on luggage entailed.

The *Henry* Court[26] differentiated between the X-ray scan and the magnetometer search. They said that the X-ray scan is a more intrusive

---

24 454 F.2d 769 (1972).

25 *U.S. v. Kroll,* 351 F.Supp. 1948 (1972); *U.S. v. Slocum,* 464 F.2d 1180 (1972).

26 U.S. v. Henry, 615 F.2d 1223 (9th Cir. 1980).

search than the magnetometer and that both are a search governed by the Fourth Amendment. The *Edwards*[27] Court distinguished between the two searches, also:

> Thus, one is not forced to choose between flying to one's destination and having one's baggage searched. Rather one may merely consign any baggage he does not want searched to the baggage compartment. The only imposition upon the passenger then is not having the bag during the flight and, perhaps, a little wait at the destination for his luggage. Clearly this is not the same case involved when the only way to avoid search is to forego flying. The conclusion, therefore, must be that a carry-on luggage search, unlike the personal search, may be justified on a consent basis.

Of course, since this decision, checked baggage is inspected by X-ray and other means; flights to foreign destinations receive 100 percent inspection of checked baggage. The courts have not yet considered the mandatory inspection of checked baggage.

PASSENGER'S RIGHT TO TERMINATE THE SEARCH PROCESS

A problematic situation arises when the screener tells a passenger that a carry-on bag being X-rayed must go through a physical search because the screener cannot identify an opaque object in the bag, and the passenger decides not to go through with the search and attempts to leave.

Screening employees are currently instructed that they cannot conduct a search without the passenger's consent, and therefore, they are to let the passenger leave under such circumstances. The ACSSP states that when the object cannot be identified, the carry-on *must* be searched. It leaves no option for the passenger to leave, although it has been interpreted to mean that the carry-on must be searched if it is to pass the screening checkpoint. The law appears to be unsettled on whether the passenger has the right to terminate the search at this point. The following cases support the position that once passengers enter the screening process, they forfeit the right to change their minds and their bags are subject to physical search.

In *United States v. DeAngelo* [28] DeAngelo submitted his briefcase for X-ray examination. The screener discovered an opaque object that he could not identify and told the passenger that the bag would have to be searched. The passenger protested, saying that he would prefer not to take the flight. The bag was searched anyway, and marijuana and hashish were found. The Court stated:

> [W]e believe that the circumstances were sufficiently suspicious to cause a reasonably prudent man to conclude that DeAngelo might endanger the security officers and the other passengers in the airport. The officers

[27] U.S. v. Edwards, 498 F.2d 496 (2d Cir. 1974).

[28] 584 F.2d 46 (4th Cir. 1978), *cert. denied*, 440 U.S. 935, 99 S.Ct. 1278, 59 L.Ed.2d 493 (1979).

> knew that DeAngelo had attempted to board the aircraft carrying a brief case that could not be inspected adequately by X-ray and they were justified in opening it to determine whether he was carrying a gun or explosive device which could pose an immediate danger to persons in the vicinity if the brief case were returned to him... .
>
> We also conclude that the search was reasonable because DeAngelo consented to it... . DeAngelo had a choice of traveling by air or by some other means. The signs in the terminal gave him fair notice that if in the course of the total screening process a physical inspection of his hand luggage should be considered necessary to assure the safety of the traveling public, he could be required to submit it for that purpose. When he then voluntarily entered upon the screening process DeAngelo acquiesced in its full potential scope as represented to him, including physical inspection if, as developed, that should be requested. Allowing him to withdraw his luggage when the X-ray raised the suspicions of the security officers would frustrate the regulation's purpose of deterring hijacking. We therefore conclude that having consented to the search, DeAngelo could not withhold permission after the first step of the process disclosed that he was attempting to carry aboard the aircraft articles that were concealed from X-ray.

In *United States v. Pulido-Baquerizo*[29], where a briefcase was inspected without the passenger's permission, the Court held:

> [T]hose passengers placing luggage on an X-ray machine's conveyor belt for airplane travel at a secured boarding area impliedly consent to a visual inspection and limited hand search of their luggage if the X-ray scan is inconclusive in determining whether the luggage contains weapons or other dangerous objects.

The court recalled that in *Davis*,[30] they had stated that "potential passengers may avoid the search by electing not to fly." They noted that,

> Airplane skyjacking and bombings at airports have proliferated since our decision in *Davis*. Additionally, firearms and explosives can be small and easily concealed. Their detection is difficult if limited to an inconclusive X-ray scan....Given these circumstances, a visual inspection and limited hand search of luggage which is used for the purpose of detecting weapons or explosives, and not in order to uncover other types of contraband, is a privacy intrusion we believe free society is willing to tolerate.

The Court also noted that *Davis* does not specifically hold that consent to an additional search could be withdrawn after an inconclusive X-ray scan if the passenger agreed not to board the plane, nor did it determine at what point in the boarding process a passenger may decide not to fly and thereby withdraw implied consent. The Court added,

---

29 800 F.2d 899 (9th Cir. 1986).

30 *U.S. v. Davis*, 482 F.2d 893 (9th Cir. 1973).

> The requirement in *Davis* of allowing passengers to avoid the search by electing not to fly does not extend to a passenger who has already submitted his luggage for an X-ray scan.

The *Herzbrun*[31] case involved a more determined and forceful attempt by the passenger to terminate the search. Herzbrun placed his shoulder bag on the X-ray belt, and the operator noted a large, dark, unidentifiable mass in the bottom of the bag. Herzbrun told her the bag only contained clothes, and he did not want it opened. She told him it would have to be opened if he wanted to go on the plane. She opened the bag and put her hand inside, touching the mass at the bottom of the bag. At this point Herzbrun said, "Take your hands off. I don't want you to search the bag." Herzbrun then shut the bag on the operator's hand. Two police officers were summoned by means of a silent alarm and told Herzbrun he'd have to have the bag inspected if he wanted to fly. He replied, "I don't want to fly," and retreated toward the nearest exit. He was followed and arrested. His bag was later searched after a warrant had been obtained based on a detector dog alerting[32] to the presence of narcotics in the bag. Over a pound of cocaine was discovered. The court referred to its previous decision in *Skipwith*:[33]

> In *Skipwith* we also dispensed with any notion that a traveler can attempt to enter the secure area and then beat a retreat if the search proves not to his liking. "Such an option would constitute a one-way street for the benefit of a party planning airport mischief, since there is no guarantee that if he were allowed to leave, he might not return and be more successful." We noted that an unimpeded exit would diminish the risk to skyjackers and increase attempts... .
>
> *Skipwith* thus stands for the proposition that travelers who enter airport security areas may be searched on *mere* suspicion... .
>
> We also reiterated that "[t]his Court has made clear that an investigation need not be curtailed simply because a suspect decides not to take a particular flight... . We have expressly decided to reject the right-to-leave argument... ."
>
> Contrary to Herzbrun's contentions at trial and in this appeal, he had no constitutional right to revoke his consent to a search of his bag once it entered the X-ray machine and he walked through the magnetometer.

Other cases suggest that the passenger has every right to terminate the search and leave. The *Albarado*[34] Court, for example, stated,

> Even after activating the magnetometer, the prospective passenger may refuse to submit to a frisk and instead forfeit his ability to travel by air, because this serves the purpose of the whole search procedure, which is

31 *U.S. v. Herzbrun*, 723 F.2d 773 (11th Cir. 1984)

32 A dog "alerts" when he detects narcotics or explosives. Usually he sits and won't continue with the search until his handler acknowledges the alert.

33 *U.S. v. Skipwith*, 482 F.2d 1272 (5th Cir.1973).

34 *U.S. v. Albarado*, 495 F.2d 799 (2d Cir.1974).

> not to catch criminals, but rather to keep armed hijackers from getting on airplanes.

United States v. Meulener,[35] held:

> The defendant's Fourth Amendment rights were violated when he was not told at the time the search was initiated that he had a right to refuse to submit to the search provided he did not board the airplane.... .
>
> If the defendant in this case was not in fact going to board the plane--and he testified that he would not have attempted to board had he been given the opportunity to decline to undergo the search, he would pose no danger to the passengers and crew on the aircraft. Accordingly, the governmental interest which justifies a physical search of the person or hand luggage of a prospective passenger in the process of boarding an airliner is lacking in the case in which the prospective passenger declines to board the plane. To meet Fourth Amendment guarantees, the prospective passenger must be advised that he has to submit to a search if he wants to board the plane, but that he can decline to be searched if he chooses not to board the aircraft.

This case, decided in 1972 prior to mandatory screening, can perhaps be distinguished from the case where a passenger has entered the screening process, having been warned by signs, by common knowledge, and by observing other passengers being screened and can be considered to have consented to the search. It deals with individuals who make the initial decision to enter the screening process, rather than change their minds, once the screener starts to find things they don't want found.

The *Davis* Court[36] justified the searches on the basis of "administrative" searches but still supports the right of the passenger to avoid search by electing not to board the aircraft:

> The essence of these decisions is that searches conducted as part of a general regulatory scheme in furtherance of an administrative purpose, rather than as part of a criminal investigation to secure evidence of a crime, may be permissible under the Fourth Amendment though not supported by a showing of probable cause directed to a particular place or person to be searched.
>
> As we have seen, screening searches of airline passengers are conducted as part of a general regulatory scheme in furtherance of an administrative purpose, namely, to prevent the carrying of weapons or explosives aboard aircraft, and thereby to prevent hijackings. The essential purpose of the scheme is not to detect weapons or explosives or to apprehend those who carry them, but to deter persons carrying such material from seeking to board at all.
>
> Of course, routine airport screening searches will lead to discovery of contraband and apprehension of law violators. This practical consequence does not alter the essentially administrative nature of the screening process, however, or render the searches unconstitutional...

---

35 351 F.Supp. 1284 (C.D.Calif. 1972).

36 *U.S. v. Davis*, 482 F.2d 893 (9th Cir. 1973).

One important caveat should be stressed, however. To meet the test of reasonableness, an administrative screening search must be as limited in its intrusiveness as is consistent with the satisfaction of the administrative need that justifies it. It follows that airport screening searches are valid only if they recognize the right of a person to avoid search by electing not to board the aircraft. (This same *Davis* Court subsequently overruled a portion of this decision in *Pulido-Baquerizo*.[37]) The following dicta supporting the right to leave rather than submit to a search is from the *Davis* Court.

> It is difficult to see how the need to prevent weapons and explosives from being carried aboard the plane could justify the search of a person who had elected not to board. Perhaps it could be argued that a compelled search might lead to the apprehension of a potential hijacker, eliminating or at least reducing the chance that he would try again. Compared to the degree of additional intrusiveness that compulsory searches involved, however, this possibility seems to slight as to be inconsequential. The risk of successful hijacking is not enhanced by allowing a potential passenger to avoid a search on a particular occasion by electing not to fly. Airport screening searches, as carried out in this case and as currently employed nationwide, are not selective. A prospective passenger who elects not to fly on an earlier flight is, like all other passengers, certain to be subjected to a search before he can board a later flight... .
>
> Since a compelled search of persons who elect not to board would not contribute to barring weapons and explosives from the plane, it could serve only the purpose of apprehending violators of either the criminal prohibition against attempting to board an aircraft while carrying a concealed weapon...or some other criminal statute. Such searches would be criminal investigations subject to the warrant and probable cause requirements of the Fourth Amendment.
>
> In sum, airport screening searches of the persons and immediate possessions of potential passengers for weapons and explosives are reasonable under the Fourth Amendment provided each prospective boarder retains the right to leave rather than submit to the search.

Most interestingly, however, because this search occurred before the introduction of mandatory screening, the court "looked ahead" and suggested that, "[I]t may well be that under the present airport screening program [100% mandatory screening had just been introduced before the court rendered its opinion]...the alternatives presented to a potential passenger approaching the screening area are so self-evident that his election to attempt to board necessarily manifests acquiescence in the initiation of the screening process."

In *United States v. Henry*[38] Patrick Henry, using a fictitious name, approached a ticket agent to check his briefcase. He then asked for the briefcase back and took it to the men's room. Because of Henry's nervousness and physical appearance (he was wearing an ill-fitting wig and a extra clothing), the agent contacted his supervisor, who decided to tell Henry that the airline had closed luggage acceptance for the flight and he would have to

---

37 *U.S. v. Pulido-Baquerizo*, 800 F.2d 899 (9th Cir. CA 1986).

38 *U.S. v. Henry*, 615 F.2d 1223 (9th Cir. 1980).

take the briefcase to the gate for checking. The purpose of this was so that the briefcase would have to pass through screening. The screener saw unidentifiable dark objects and asked Henry to open the briefcase. He said he didn't have a key. A police officer told him that if he couldn't open it, he'd have to take it back to the counter to check it. The agent told Henry that he'd get his supervisor and meet him at the checkpoint. The agent kept the briefcase, but when they arrived at the checkpoint, Henry had already gone to the gate.

The agent and supervisor went to the gate and told Henry that unless he was willing to open the briefcase, he couldn't take it with him. Henry said, "I've got to have that bag in Dallas. Okay." He then boarded the plane, and the agent, supervisor, and police officer took the bag to their baggage service area to open it. When they began to open it, they saw a string that they thought might be a fuse. The called the bomb squad, who listened to it with an electronic stethoscope and had the bags checked by two explosive-detection dogs, who alerted on the bag. The bag was opened, and a loaded revolver, bullets, and firecrackers were found.

In its discussion of Henry's consent to the search, the Court noted that when he initially asked for his briefcase back at the ticket counter, it was readily given to him. At the checkpoint, when the screener identified the opaque object, no one attempted to compel him to open the briefcase. He was given his briefcase and told to take it to the counter and check it. The court found that under the totality of the circumstances, Henry freely and voluntarily consented to the search of the briefcase. But note the language of the Court regarding Henry's right to withdraw the briefcase at any point in the search:

> In determining whether Henry freely and voluntarily consented to the search we believe the crucial factor is whether Henry could have freely withdrawn the briefcase and avoided the search.

More interesting and almost in conflict, however, is the footnoted statement, "We express no opinion as to whether submission to the X-ray scan constitutes consent to physical inspection, if requested," referring to *United States v. DeAngelo,* discussed earlier.

In *United States v. Edwards*,[39] the search involved a beach bag carried by Edwards, that caused the magnetometer to alarm when she walked through it. This incident occurred in 1972, and obviously, this procedure of walking through a metal detector with carry-on items is no longer followed. A Deputy U.S. Marshal examined the beach bag and found a package containing 1664 glassine envelopes containing heroin. In this case there were not only signs stating, "Passengers and Baggage Subject to Search," but public address announcements stating that passengers' baggage was subject to search. The court stated:

> [I]t would outrage common sense to suppose that an intelligent woman, neither blind nor deaf nor ignorant of the language, was not aware as the

39 498 F.2d 496 (1974).

> district judge found "that she was as free to step out of the line of passengers (as she had been to enter that line) if she did not want her baggage to be searched," 359 F.Supp. at 767. Miss Edwards simply miscalculated the odds.

If the right to refuse search of carry-on baggage is important, then the question arises as to the search of checked baggage. The *Edwards* Court specifically avoided that issue and stated:

> We are not here required to and do not consider what circumstances may justify the search of checked baggage, a procedure routinely followed by at least some airlines on international flights from the United States.

FAILURE TO CLEAR METAL DETECTOR

It is a well-settled legal principal that failure to clear the metal detector the first time does not justify an immediate pat-down search. In *U.S. v. Albarado*,[40] where the defendant attempted to go through the metal detector with a packet of counterfeit money wrapped in tin foil, the Court went into some detail on this point:

> At the outset it should be noted that any further investigation after activation of the magnetometer is for the metal which did the activation; activating the magnetometer is not a general license to search for anything. Next, it is clear that compared with other methods the frisk is "inefficient" in that a whole body may be frisked only to discover that the keys in the passenger's pocket set the machine off. It would be easier and more effective merely to ask the passenger to remove all his metal items and return through the magnetometer. This procedure is clearly preferred over the immediate frisk because, while still a search, it entails far less invasion of the privacy or dignity of a person than to have a stranger poke and pat his body in various places. Such a procedure is...frequently used. It is proper for several reasons. It limits the search to the metal detected by the magnetometer. If the person passing through the magnetometer the second times does not activate the machine, the metallic items have all been made visible to the officer. If on the other hand, the machine is activated again, the person has failed to divest himself of all metallic items on his own. While this in itself may not provide "probable cause" or even a reasonable suspicion in *Terry* terms, it necessitates that further measures be taken. At this point the only further practical measure is a pat-down, or frisk if necessary. And it is reasonable at this point because no lesser readily available means presently exist for determining what unexplained metal object remains on the passenger's person... .
>
> Of course, the second magnetometer walk-through is not the only alternative available to the authorities. Where airports are equipped with hand-held magnetometers they may be used to locate the metal on the passenger's body and he then may be required to remove it so that they may be assured that it is not a weapon. Other additionally proper possi-

40 *U.S. v. Albarado*, 495 F.2d 799 (2d Cir. 1974).

> bility is to provide instruction to all passengers that metal items should be removed prior to passing through the magnetometer initially and to provide a table or whatever on which these items may be placed so that they may be viewed by an official. These examples are not to be considered exhaustive, but, again, the rule is easy to state: *the frisk* in the typical boarding situation we have been talking about *is to be used only in the last instance.*

CONTRABAND

The introduction of passenger screening to detect weapons also raised constitutional issues regarding the consequences of some searches. Screening resulted in thousands of arrests unrelated to air piracy, most of them for possession of narcotics. Other arrests were made of aliens accused of illegal entry, parole violators, servicemen absent without leave, fugitives, and others accused of local, state, and federal crimes.

Controversy arose when civil libertarians charged that many screening searches were being performed on the pretext of searching for weapons when, in fact, the law enforcement officer was actually seeking drugs or other contraband. The air carriers themselves were, for the most part, only concerned with keeping people moving and selling passenger seats; each discovery of drugs meant delays in processing passengers and time away from the job to testify in court. As a consequence, some local managers instructed their screening agents regarding contraband and emphasized repeatedly that the airline was not in the law enforcement business. If the screening agents found marijuana or other drugs in carry-on luggage, they were to go ahead and board the passenger without confiscation or comment; the carrier's only legal obligation was to locate weapons.

Such methods of achieving efficiency came to a sudden stop in 1973, when a judge of the U.S. District Court in Brooklyn ordered the United States Attorney to seize a Braniff DC-8, which had carried three persons smuggling drugs from South America. The judge was actually within the law; the statute provided for the confiscation of vehicles used in transporting narcotics, but it had not been enforced against commercial carriers. Some months later, the judge suggested seizure of an Avianca 707 on which another drug smuggler had traveled. Although the United States Attorney decided against seizing the Braniff plane, the federal action prompted carriers to re-examine corporate policy regarding contraband items. In February, 1985, another Avianca all-cargo jumbo jet was seized by Customs. Ten months earlier, Eastern had an L-1011 temporarily seized by Customs in Miami. In September, 1985, Air Jamaica paid an $18,000 fine to recover a 727 seized by Customs after 45 pounds of marijuana were found in unticketed baggage.

The U.S. is not alone in these efforts. On May 1, 1987, Air Canada paid a fine of more than $16,500 to retrieve an L-1011 seized by British Customs officers in London, who accused the airline of failing to prevent its use in smuggling drugs from Bombay, India.

A privately owned 707 that was being used to smuggle arms from Houston to South Africa was seized in Houston in May 1981. The two men arrested were charged with violation of the Arms Export Control Act. In 1985, the Fifth Circuit Court of Appeals upheld the forfeiture to the government.

Legislation passed in 1986 imposes stiff penalties on owners and operators of aircraft found to be involved in smuggling or in fraudulent activities. When aircraft are involved in a violation of Customs laws, they can be held, seized, and forfeited, if necessary, to pay penalties. Common carrier aircraft, however, may not be seized in connection with drugs found on passengers, in their baggage, or in cargo found to be properly manifested, unless the carrier had knowledge of the violation or was "grossly negligent" in preventing or discovering it. The aircraft may then be seized and forfeited unless neither the owner or operator, pilot or other responsible employee knew, or through the exercise of "the highest degree of diligence" could have known, that the contraband was aboard. (For further discussion of "the highest degree of diligence," see discussion of drug smuggling in Chapter 9. The task was not made easier by the range of items that are legally contraband. These include narcotic drugs possessed in violation of the law, stolen property, illegal firearms, and counterfeit monies or securities. Federal law prohibits the transportation of these contraband articles, articles that are being or have been illegally imported into the United States, and articles on which appropriate duty or taxes have not been paid.

This opens up the proverbial Pandora's box when it comes to instructing screening agents. Are the pills in the suitcase contraband or are they prescription drugs? Is the white power it the plastic bag an illegal drug? Must we ask the passenger? Is the U.S. currency in the attaché case counterfeit? Must we attempt to find out if it is counterfeit or not? Is the weapon given to us for checking as baggage a stolen weapon?

The answer to all this, of course, is that no carrier is required to embark upon a full-scale law enforcement effort to discover contraband. What is expected is that the carrier take steps to ensure that proscribed articles are not knowingly transported; extraordinary measures to identify contraband items are not required. Here, as in so many other aspects of the antihijacking program, the standard is that of "reasonable judgment." The corporate obligation and responsibility will most probably not be held to extend beyond obvious violations--but it does extend to those.

The problem is exacerbated when various law enforcement agencies see the screening process as a means of investigating and enforcing their particular area of concern. Thus, when DEA agents pass out profiles to ticket agents of people who transport drugs, asking to be notified if someone meeting that profile checks in for a flight, then the screening process, or at least the check-in process, is being used for purposes other than it was originally intended.

It is not an infrequent occurrence that when a piece of carry-on luggage is inspected at a checkpoint, large amounts of cash are discovered. This raises a suspicion that the person may be running drug money. Presumably, it would be beneficial for the IRS, the DEA, and other agencies to know when such discoveries are made. What is the obligation of the carrier in such an instance? Is a large amount of cash carried by a passenger contraband, per se?

The IRS, for example, would like to be notified. The Tax Equity and Fiscal Responsibility Act of 1982 (TEFRA) added Code Section 6867 to expedite jeopardy or termination assessments in cases where there is no known owner of large amounts of cash. Such a circumstance occurs frequently in narcotics enforcement when suspected narcotics dealers, at the time of arrest or in relation to an arrest situation, disavow ownership of a quantity of cash in an attempt to extricate themselves from direct or indirect involvement in the narcotics deal. Acknowledgment of ownership of the cash by a suspected narcotics dealer is tantamount to self-incrimination and is viable evidence for the prosecution at the time of trial. Because the drug trafficking business is a cash and carry trade, paper trails are avoided, and reliance is placed on face-to-face transactions at some point in the deal. Large amounts of cash can be indicative of an actual or intended direct exchange of drugs for money and are frequently found in possession of the drug suspect, with or without drugs with nondrug evidence of trafficking.

The General Rule of Code Section 6867, entitled "Presumption of jeopardy where owner of large amounts of cash is not identified," states that if the individual who is in physical possession of cash in excess of $10,000 does not claim the cash either as his or as belonging to another person whose identity can be readily ascertained, and who in turn acknowledges ownership of such, then it is presumed that the cash represents gross income of a single individual for the taxable year in which the possession occurs. Under such circumstances, the General Rule further presumes that the collection of the tax will be jeopardized or at risk by delay. The General Rule explicitly states that the sum of the cash must be in excess of $10,000 and that ownership must be either denied or unable to be corroborated. It is therefore important that at the time the cash is seized by any agency pursuant to an arrest or other situation, that testimony be documented concerning the ownership of the cash and attempts be made to confirm or deny the defendant's story.

Rules for assessment are also specified in the Code and state that the entire amount of the cash will be treated as taxable income for the tax year in which the possession occurs and will be taxed at a rate of 50 percent. The possessor of the cash will be treated (only with respect to the cash seizure) as the taxpayer for the purposes of the jeopardy and termination assessments. As the taxpayer, the person in possession of the cash is entitled to notice of the assessment and the right to challenge the assessment. The true owner can also come forward and challenge the assessment and will be retroactively substituted for the possessor for all purposes as of the date of the original assessment. Based upon the jeopardy and termination assessments authorized by the new provisions, the I.R.S. Examination Division can assess as tax approximately 50 percent of the cash in question, and the I.R.S. Collection Division can seize any portion of the cash in question to satisfy the tax liability. Coordination of I.R.S. and state and local tax enforcement actions can result in seizure of the entire quantity of cash and can prevent its use by the defendant either at the time of the arrest or at a later date.

The carriers have strongly resisted the efforts of the government to ask them to search for large amounts of cash as well as drugs. The carriers offer a number of valid reasons for this. The Federal Aviation Regulations that mandated the screening program are designed to deter the carriage of weapons and explosives aboard aircraft. There is no indication of any congressional intent that the purpose of the legislation was to look for drugs. Under the law as already set out in this chapter, the courts would undoubtedly consider such expansion of the searches as governmental searches requiring justification under the Fourth Amendment. The *Edwards* Court[41] stated,

> If experience should demonstrate that the Government is abusing its authority and is using the airport search not for the purpose intended but as a general means for enforcing the criminal laws, a means for limitation can be found in the suggestion in Judge Aldrich's dissenting opinion in United States v. Skipwith...[42] [inadmissible under exclusionary rule].
>
> In Terry v. Ohio, 392 U.S. 1, 19, 88 S.Ct. 1868, 1978, 20 L.Ed.2d 889 (1968), the Court said, "The scope of the search must be 'strictly tied to and justified by' the circumstances which rendered its initiation permissible." Thus in the search of carry-on luggage, the right to search, even when based on consent, is not general, but one limited to the search for weapons or explosives. The marshal is not given a license to search for drugs among all airline passengers.

At least one state, Illinois, has enacted a statute[43] that operates as an exclusionary rule:

> Any evidence of criminal activity found during a search pursuant to this Act [crime to board aircraft with firearm or explosive] shall be admissible in legal proceedings for the sole purpose of supporting a violation of this Act and is inadmissible as evidence in any legal proceeding for any other purpose, except in the prosecution of [specified] offenses related to weapons.

Similarly, in *U.S. v. Van Lewis,* the Court stated,

> Although the problem caused by illegal drug traffic in the metropolitan areas of the country are serious indeed, they are quite different from the problems posed by the threat of air piracy when both are viewed within the framework of the Fourth Amendment.
>
> The would-be air pirate, by obstructing the right to travel and endangering life in wholesale fashion, threatens the fabric of the republic

---

41 *U.S. v. Edwards*, 498 F.2d 496 (1974).

42 *U.S. v. Skipwith*, 482 F.2d at 1280-81. "Not only does it seem unfair for the government to profit from a circumstance that was none of the defendant's making, but no hope of striking it rich should serve to lure the agent into a broader search than he would have chosen to make had hijacking materials been his maximum objective. Where special circumstances are allowed to reduce the ordinary conditions precedent to a lawful search there should be special safeguards to see that the opportunity is not abused...I believe the same prophylactic principle that dictates exclusion of property unlawfully seized...should be employed to temper possibly overzealous airport searches. Such a rule would not only limit the decision to make, and the extent of, searches to what the justification permitted in the first place; it would isolate the procedure and guarantee to the public that the government is not using airport search procedure for other purposes, an offensive thought to any innocent person suffering the inconvenience." To date no court has accepted this argument.

43 Illinois Revised Statutes, Chapter 38, §84-4.

> and strikes at the very heart of our legal and moral relationships with one another. To protect public life and liberty, the law has sanctioned the development of special rules governing airport security searches as long as they are "reasonable" within the meaning of the Fourth Amendment. Reasonableness must be tested against the need to search. When the public danger, and hence the need, is as great as it is in the case of airport security, reasonableness within the meaning of the Fourth Amendment requires less factual justification.
>
> As damnable as drug traffic is, its regulation involves the protection of no special public interests like those at play in airport security. Regulation of drug traffic is achieved through enforcement of laws adopted by Congress which are similar to law prohibiting bank robbery, bribery, or conspiracy. Thus, the court perceives no fundamental public interest at stake in routine enforcement of the drug laws which calls for the development of rules unique to airport drug searches. Accordingly, the government's rights in these cases must be tested against basic Fourth Amendment principles rather than by rules derived from an air piracy context.

The carrier could also encounter significant liability problems if it actively worked with the government in looking for drugs. It could be subjected to suits for defamation, false imprisonment, or intentional infliction of emotional distress. Additionally, if the state authorities were involved, 42 U.S.C. §1983 (1976) authorizes a person deprived of any constitutional right as a result of state action to bring a civil suit against the person who deprived him or her of that right. Aside from the legal principles involved, such an expansion of the search might dilute the efforts of the screeners as it relates to their primary purpose.

In Hawaii, the State Department of Agriculture conducts inspections of luggage to determine if "forbidden fruit" is being carried out of the islands. In Honolulu, this is accomplished by connecting TV monitors into the screening station X-ray machines so that inspectors can view the contents of carry-on luggage without having to open it. The state has the authority to inspect for agricultural items, so this latter example can be differentiated from the former. Also, it serves as an additional check for weapons, providing one extra set of eyes looking at the X-ray image. It saves time for the passenger by eliminating a physical inspection by the Agriculture Inspectors.

Significant legal questions would arise out of these requests to use the screening process for other than antiterrorism purposes. if they were allowed to persist (and they have not been). Presumably the courts would not permit the screening to be used for anything other than a search for weapons and explosives,[44] with the caveat that the carrier cannot ignore contraband that it discovers. But not ignoring contraband is much different than actively searching for it. Just how far will the courts go in upholding the dicta in *United States v. Averall*[45]

---

44 Judge Aldrich's dissenting opinion in *U.S. v. Skipwith*, 482 F.2d at 1280-1281.

45 296 F.Supp. 1004 (1969).

> The airline had the right and perhaps the duty to discover if its facilities were being used for the commission of a crime. It is an act of responsible citizenship for individuals to give whatever information they may have to aid in law enforcement.

This has been a problem area for the airlines because the pressure from the law enforcement agencies has been great. When the airlines refuse to cooperate, cases arise of law enforcement officers who have used screening employees as paid informants without the knowledge of the carrier. A clear statement of the law is needed by the U.S. Attorney General with instructions to the various federal agencies involved. The air carriers are put in the middle between the law enforcement agencies and the courts, and that should not be allowed to continue.

Chapter 7

METAL DETECTORS, X-RAY INSPECTION, AND EXPLOSIVES DETECTION

METAL DETECTORS

The electronic metal detectors used for screening passengers are of both the walk-through and the hand-held type. Some carriers purchased their own devices even before the early federal purchase and distribution of walk-through metal detectors. Most of these earlier models were passive detectors known as *magnetometers*. Their greatest weakness was that they did not detect metals incapable of being magnetized, and a significant number of U.S.-manufactured guns are made largely of nonferrous metals for lighter weight.

Passive systems detect metal by changes in the earth's magnetic field. Active detectors operate by creating their own electromagnetic field and alarming when the field is disturbed by metal objects passing through it. Active detectors use various frequencies, usually between 90 Hz and 25 kHz, while hand-held units generally operate from 100 kHz to 1 MHz.

Metals such as aluminum, brass, and copper are highly conductive and hence provide greater signals at higher frequencies. Metals such as iron and steel, on the other hand, produce greater signals at low frequencies. For this reason, a compromise frequency that will detect both types of metal is usually desirable.

High-frequency detectors (which react to highly conductive metals) will alarm at tinfoil from cigarette packages or gum wrappers, and these detectors are more prone to false alarms. Conversely, a low-frequency detector (which reacts to less conductive metals such as iron and steel) can disregard cuff links, pocket change, keys, and small pieces of jewelry. This difference makes the low-frequency unit (below 500 Hz) more practical for walk-through screening of passengers. The hand-held detectors, on the other hand, work at high frequencies, as they are usually employed for a more precise examination of persons who have already alarmed a walk-through device.

The five walk-through metal detectors initially provided by the FAA through the federal program were manufactured by Westinghouse, Metor, Rens, Solco, and Sperry Rand. Now that carriers purchase the detectors there is a whole new list of brand names. Some of the old ones have dropped away and new names are on the scene.

Selecting a Metal Detector

It is useful to know the criteria that should be considered when selecting a weapon detector for purchase.

Sensitivity--The most sensitive metal detector, however invaluable it might be for jewelry manufacturers, is not the best for processing airline passengers. Originally, the FAA's suggested standard called for a maximum false alarm rate of 15 percent; that is, an acceptable detector could alarm on 15 percent of the passengers because they had metal on their person in an amount less than that represented by a gun. Nevertheless, whenever the false alarm rate goes higher than 15 percent, it becomes a problem for the carrier because every passenger who alarms the device must be personally rescreened.

One of the FAA purchased detectors proved to be subject to a particularly high false alarm rate--65 percent in one test conducted by the FAA at a major airport. Airline experience with this detector generally confirmed the 65 percent figure; the device reacted to tinfoil (cigarette packs and gum wrappers) as well as pocket change, keys, and jewelry. The ideal metal detector should detect a gun without fail while passing a person with an ordinary amount of pocket items and jewelry.

Mobility--Some devices can be moved across the concourse, recalibrate themselves, and be ready for use within a matter of seconds after their relocation. Others may require the services of an engineer to recalibrate them if they are moved only a few inches, a process that could take anywhere from several minutes to several hours.

Detection Rate--The FAA has established specific standards that require a high probability of detection of handguns. The acceptable standard is not published, though of course it is well known to those responsible for maintaining it.

External Factors--Metal detectors vary in their sensitivity to different factors varying continuously in their environment. It can make a difference in the operation of a device if the terminal in which it is located is primarily of steel construction or if the detector is located near where aircraft will be docking. One model experienced excessive alarms from sunlight streaming through a window. In another location, mercury vapor ramp lights affected a unit's operation. Other problems have been traced to electronic doors, public address systems, fluorescent lights, motors, switches, relays, and radio transmissions.

Some metal detectors are particularly sensitive to environmental magnetic noise. Even after a checkpoint has been chosen, electromagnetic noise may develop to a point where certain makes and models of walk-through devices cannot operate. In such cases, the only solution is to find an electronically quieter location or to replace the unit with one less susceptible to electromagnetic noise. Experience indicates that the elliptical-coil metal detection unit is the least likely to be bothered by such external electronic noise. Some metal detectors are incompatible with other metal detectors or with certain models of X-ray machines and cannot be effectively operated in their vicinity.

Shielding panels offer some help against interference. These panels are available from most manufacturers of this equipment or they can be constructed by carrier or airport maintenance people. In addition, the unit chosen should incorporate filtering in its circuitry to reduce electronic noise interference.

With the advance in technology in recent years, most of these early problems have been eliminated. Manufacturers can answer questions about their products as they relate to these areas of concern.

Physical Construction--Size must be taken into consideration. Units may be too tall for the best location or there may not be sufficient floor space, particularly when the detector is to operate in combination with an X-ray baggage inspection machine in a narrow concourse. Many of the early units had ramps that required passengers to walk up, through, and down. Because of slip and fall cases and the needs of elderly and handicapped people, these ramps have gradually been eliminated. Size has been reduced to a more reasonable configuration also.

Operation--In evaluating these devices for operational advantages, their ease of calibration, self-calibration, and required frequency of calibration should be considered.

Other Factors--Airline personnel will include in their considerations the aesthetics of a detection unit, because they want a unit that is reasonably attractive. Most of them are.

Cost is a most important factor; at one time three detectors of one manufacturer could be purchased for less than one detector of another manufacturer without appreciable loss in performance.

Before purchasing a specific brand of walk-through metal detector, the potential buyer should contact the security personnel of at least one air carrier that has had substantial experience with that particular unit and possibly with several others to learn their opinion of the unit and its day-to-day performance. It may be possible to contact technical people who are involved in maintaining these units and who are familiar with the potential interference problems at the location.

Walk-through metal detectors are sensitive to the movement of metal near them. This most commonly occurs where a table for the inspection of luggage is located next to the unit, and metal-framed luggage passes back and forth, causing the detector to alarm. Alarms can also be caused by passengers walking beside the unit, by wheelchairs, by baggage tractors or carts moving on the floor below, and even by an aircraft pulling into the gate. One brand of detector, in effect, switches off except when someone is passing through it, which cuts down on false alarms.

As mentioned, screens for shielding can be purchased, or they can be constructed by maintenance people in one of two ways. The simplest is to face a 4 x 8 foot sheet of half-inch or three-quarters-inch plywood with a sheet of 20-gauge aluminum. The same effect can be obtained more attractively by sandwiching the aluminum between two sheets of wood-grained hardboard. The screen should be firmly secured to the detector to avoid its changing the electronic environment when bumped or jostled.

The effects of metal detectors on magnetic tapes were tested at Washington National Airport in 1973 by Eastman Kodak Company. The test results, reported in their Technical Memorandum TM 29-73, indicated that airline passengers may safely permit magnetic recordings to pass metal detection searches without concern for the integrity of the recorded data.

Testing the Metal Detector

All metal detectors are required to meet a specific detection standard set by the FAA. Although the the FAA does not specify a false alarm rate limit, operational efficiency demands that a detector's false alarm rate should not exceed 15 percent; that is, nor more than 15 unarmed passengers out of 100 should alarm the detector, regardless of keys, pocket change, or metal arches in their shoes.

At one time, the FAA had devised what they believed to be a fair test of the detection rate for all brands and models of walk-through detectors. With the Bureau of Alcohol, Tobacco and Firearms, the FAA identified what was believed to be a cross section of the "gun population" of the United States. Test kits containing six specific weapons each were issued to each FAA Regional Office. The test was administered in a manner better seen than described.

To make the test, the FAA Inspector had to walk through the metal detector four times, with each gun in seven different positions and locations on the body--a total of 168 passes through the detector. For computing a test score, each gun had a separate "weighing" factor. Completing the test required about 15 minutes per detector, and was noticeably fatiguing to perform. It did not help matters that a detector set to alarm on the test pistol with the smallest metal content would also alarm on a normal complement of change, keys, and jewelry. The carriers asked the FAA for a simpler test, suggesting that this might use a piece of metal of a specific size, weight, and metallic content, which the carrier could use to calibrate the detector.

Until this test piece was provided, the carriers were in a rather untenable position. A carrier could be liable to a possible violation if a detector was not up to the standard specified, yet there was no way to test the unit short of purchasing a kit of the six designated guns. In practice, detectors were usually calibrated for the day by having the law enforcement officer at the screening point walk through the detector with, and without, his gun; the detector should alarm with the pistol and not alarm without it. At other locations, the calibration "standard" carried through the device as a test was a large wrench or a pair of pliers. One manufacturer provided a piece of metal with which its units could be calibrated, but this was not acceptable to the FAA .

The carrier must test the metal detector weekly if its use is continuous. If it is turned off or left unattended for a significant period of time, it must be retested. The FAA will test the metal detector if it is a new machine, if it is moved to a new location, if it is adjusted by the air carrier, and during the course of a station inspection. If it fails, it must be adjusted or replaced within 48 hours.

The FAA has developed a standardized encapsulated test gun made of 3.7 ounces of stainless steel. Each station receives one of these units, and additional test pieces can be obtained for authorized service personnel.

METAL DETECTORS AND COMPUTERS

The newsletter for one computer user group suggested that walking through a metal detector with a 3.5-inch floppy disk in your pocket would hopelessly garble the data on it. Phil Romine, Product Manager for Del Norte Technology, Inc., distributors for the popular Sentrie metal detector, responded in part as follows:

> 1) The FAA has always regulated the design of these devices to alleviate a threat to implanted electronic devices such as heart pacemakers. European countries also follow the FAA specification or require a similar low field strength in these detectors.
>
> 2) The electromagnetic field allowed from a metal detector (1 Gauss) is about twice the earth's magnetic field you experience every minute of your life.
>
> 3) This letter is written from a 3.5 inch diskette like you describe. It was subjected to 16 hours of illumination while taped inside a working detector with no effect. The computer disc was not impaired in any way. This test is equivalent to 30,000 passes through an airport detector (and several pairs of shoes).
>
> 4) Manufacturers of magnetic tape specify the field required to damage recorded tapes, usually in the order of 600 to 1,000 Gauss. It is also true that unused tapes cannot be damaged by the magnetic field because the tape will be completely scrubbed and rewritten when it is used in a computer.
>
> 5) With millions of passenger boardings every day through thousands of these inspection gates, not one incident of damaged tape has been reported to us.

Because the bomb on Pan Am 103 was hidden in a portable radio, the FAA proposed that all small computers and other electronic devices be banned from aircraft. A hue and cry was raised by journalists, businessmen, and businesswomen who carry their computers with them as tools of the trade. In July 1989, the proposal was dropped.

Early fears that the use of computers on board might interfere with navigation devices have proved groundless, and the FAA permits their use on board. That may not be the policy of foreign carriers, so passengers should check with the flight attendant before turning a computer on. It is against the law to use a cellular phone aboard aircraft because of the interference it creates.

X-RAY INSPECTION UNITS

X-ray units used in U.S. inspection of carry-on luggage are all of the low-dosage or low-radiation type. The low-dosage X-ray source produces a weak image that must be amplified to be seen clearly, often on a closed-circuit television monitor.

Some of the units operate on a "short pulse" principle, with each pulse lasting about 1/120 of a second. The image produced is kept on the monitor by an image storage unit until the next bag appears at the conveyor inspection point or until the operator cancels the image. Other units operate in a continuous X-ray emission mode in which the X-ray source continues to emit radiation as long as the item being inspected remains in the beam path or until the operator turns the unit off.

In one unit, an X-ray beam scans across the object being inspected, generating information from each part of the object sequentially, and then displays the information as an image on the viewing screen. This technique enhances the image by eliminating background radiation scatter. Most of the units in use today utilize digital video storage.

Selecting an X-ray Unit

The purchase of an X-ray inspection device is a considerable investment, and most air carriers must think in terms of many units, making the cost an even more substantial factor. An investment of this size demands prepurchase study and research, as well as knowledge of the factors that ought to be included in purchase considerations.

Picture Quality--Experience has demonstrated that the X-ray unit's effectiveness substantially depends upon the operator's ability to recognize objects on the monitor display. The more items that cannot be identified because of a vague image, the more time that must be spent in opening and inspecting luggage. While the clarity of an image is to some extent a matter of subjective evaluation, objective comparisons are possible. Certain units will have obviously sharper focus, brightness, and contrast, and it is image contrast with shades of gray or color contrast that assists the operator in identifying objects.

Stored Image vs. Continuous Beam---Every X-ray unit in current use at U.S. airports is one of these two types. The stored-image unit is operated by split-second pulsing of the X-ray beam, and it stores the image information produced on a display unit where it can be viewed for as long as several minutes if necessary. Because the pulse is on for only a fraction of a second, these units generally provide an even greater margin of safety than continuous-beam units, although both types are considered to be film safe.[1] Most pictures are now stored by means of digital memory.

In continuous-beam units, the X-ray beam is usually activated throughout the time the bag or package is being inspected (three to six seconds), and the object under examination could receive more radiation than it would from inspection by a stored-image unit.

When coupled with conveyor belt baggage carriage, some continuous beam models show a smear effect from the bag's continuing travel along the conveyor. For such units, it is necessary to stop the conveyor momentarily with the bag in front of the X-ray source, usually no more than two to three seconds, to ensure a clear image. A potential problem with this technique, at least in theory, is that the film-safe radiation limits could be exceeded if the operator were to keep the bag in front of the X-ray source for an unusual length of time. Because of these drawbacks and the increased clarity of the picture, most of the airport X-ray machines in use today are the digital memory, stored-image type.

Horizontal vs. Vertical X-Ray--Some manufacturers have designed their units so that the X-ray beam projects horizontally across the viewing area, while others project the X-ray beam from above, vertical to the target area. A few manufacturers offer either horizontal or vertical presentation.

On a horizontal-projection unit, garment bags and other flat, nonrigid items may present problems but these can be countered. A flat item traveling on the conveyor of a horizontal presentation unit will not X-ray well, but garment bags can be folded in thirds, and flexible briefcases may be leaned against the side of the tunnel in a more upright position. The vertical projection units thus offer an advantage in this particular regard.

Speed of Inspection--The standard minimum rate for X-ray inspection (also spoken of as the *flow-through rate*) has been established by the the FAA as no more than six seconds per item, or 10 items per minute. Most X-ray units average at least 12 items per minute, or one every five seconds. Studies conducted by United Airlines during the early stages of the predeparture screening program indicated that the rate of manual search of carry-on baggage averaged 2.8 inspections per minute per inspector and found that there were 1.2 bags per passenger. Most X-ray units, then, inspect four times as many items as a manual search in the same period of time.

If the X-ray unit chosen has a relatively slow conveyor belt speed, however, processing delays may occur. An independent study reported that the rate of conveyor X-ray examination differed markedly between two manufacturers' units, one requiring 10 seconds longer per item due to its slower belt speed--10 inches per second as opposed to six inches per second.[2] Conveyor belt speeds on some units are adjustable, but others are set to a standard.

---

[1] The effect of X-ray examination on unexposed film is discussed in more detail later in this chapter.

[2] Jose Roman, Jr. and George C. Jackson, Jr., "Analysis of Airport Security Processing" (graduate thesis,Institute of Transportation and Traffic Engineering, University of California, 1974):39

The report also stated that a manual X-ray system required the least time for baggage inspection.[3] This is attributable to the fact that, in manual systems, passengers don't have to wait at the end of the conveyor for their luggage. This advantage is lost in heavy passenger traffic, when the expeditious handling of baggage through small manual X-ray systems often requires additional carrier personnel.

Other factors can affect the speed of inspection. One unit requires a few seconds to formulate a picture, delaying each display by that amount of time. On the other hand, the flow-through can be almost doubled where it is possible to place items side by side (as in vertical viewing) or one after the other (as in horizontal-beam units) so that two items may be screened with one picture display.

Film Safe--All X-ray units used by U.S. carriers are film safe; that is the radiation output level per boarding inspection is so low that undeveloped film will not be fogged or otherwise affected. (See "X-Ray and Photographic Film later in this chapter).

Ability to Detect Explosives--New innovations utilizing backscatter and other technology designed to detect explosives and other organic material have been incorporated into the latest X-ray units on the market. (See Detection of Explosives by X-Ray later in this chapter.)

Safety Considerations--The FAA and the U.S. Bureau of Radiological Health require that every X-ray unit be safe for passengers, for the operator, and for any person within range of the device. The unit must meet not only the Bureau standard, but also the standards of the state in which it is to be registered for use. The system's design, construction, maintenance, and method of operation must all comply with every applicable federal, state, county, and city law, ordinance, regulation, industrial code, and/or recommended practice related to the system, its environment, personnel safety, and working conditions.

Inspection Capability--At present, carry on items are limited to underseat size, which is defined as 13 x 23 x 9 inches. Some carriers allow passengers to board larger carry-on items, which are stored in cabin luggage compartments. If the X-ray is to be used for inspecting checked baggage, it obviously must have a tunnel size that will accommodate all generally accepted boxes and suitcases. The X-ray unit must be capable of processing every type of luggage, including garment bags, duffel bags, wrapped packages and other miscellaneous articles.

A consideration facing the prospective purchaser is whether the unit is to be equipped with a conveyor belt assembly and, if so, whether it should be a long belt or a short one. The means by which the luggage items are to be introduced into the machine and how they are to be retrieved after inspection, must also be resolved.

Mobility--The unit should be mobile, if possible, and mounted on lockable casters or wheels. Mobile units allow for greater flexibility of operation as well as simply facilitating routine floor maintenance. Mobility can be a disadvantage if the unit will require the services of a technician before it

---

3 Ibid:39

can be operated after a move. Before a unit can be moved to another airport, the approval of the FAA Principal Security Inspector must be obtained.

Maintenance--The cost of maintenance and the extent of warranties for parts and labor are important issues in purchase negotiations. For how long are the tubes guaranteed, and what is their replacement cost? Can the carrier train its own people to handle the unit's maintenance, or will it require an outside service organization? What is the cost of a maintenance contract?

The maintenance requirements for major system components should be given in quantitative terms such as "mean time between failures" (MTBF) and "mean time to repair" (MTTR). It can make a considerable difference if a unit is out of operation 48 hours rather than 48 minutes because a repairman was unavailable. Guarantees regarding parts availability must also be discussed, as well as whether or not the vendor can schedule routine maintenance for those hours in which the machine is ordinarily not in use--for instance, between 12:00 A.M. and 5:00 A.M.

It is advantageous if the unit generally uses standard parts and components and needs few parts available only from the manufacturer. As far as possible, subsystems should be modular and connected by plug-ins, rather than hard-wired, for quick replacement of malfunctioning units. What are the unit's power requirements? Can it be operated on the present wiring at the station?

Training--Who will train operators and maintenance personnel, and who will pay for the training? How much training is required to operate the unit? Current units offer more options for the operator and will require a greater degree of training. Operation should be simple enough that an inexperienced contract guard can be taught to operate it within a short period of time.

Relative Cost--Having established the approximate cost of the preferred unit and its most attractive options for a particular application, the carrier must determine whether the expense of X-ray inspection can be justified at the particular station under consideration. Comparing the cost of labor for manual search against the cost of the X-ray unit and the manpower to operate it is probably the best method to use for this evaluation. It is not unusual for an X-ray unit to pay for itself in as little as a year or less from personnel savings alone. The convenience to passengers should also be weighed in the balance. As important is whether manual inspection is contributing to departure delays; the elimination of delays represents a positive operational savings.

Other factors than initial cost and savings should also be considered. One manufacturer's unit may be capable of a faster flow-through than that of another. To achieve the same flow-through rate might require one unit from one manufacturer or two less expensive units from another maker. Maintenance costs, maintenance contracts, and tube replacement can prove a substantial expense and deserve consideration as part of the purchase decision. You may also wish to question the manufacturer about the availability and cost of a lease agreement as an alternative to outright purchase.

Testing The X-Ray Unit

X-ray units must be tested daily, and this testing is the responsibility of the screening supervisor. For X-ray units installed before July 22, 1985, the FAA requires that the unit be capable of distinguishing a 24-gauge insulated copper wire. For units installed after July 21, 1985, the American Society for Testing and Materials (ASTM) step wedge, a piece of metal whose thickness reduces from one thickness to another in several steps, is used.The unit must detect the 24-gauge wire under the fifth step of the step wedge.

The visualization of the strand of wire affixed to an 8 1/2 x 11 inch sheet of paper is a criterion for establishing picture quality and nothing more; if the 24-gauge wire can be seen, the machine meets specification. The rationale for the test may be that this represents the smallest probable gauge of the common lead wire used for blasting caps, but this may not be true; some explosives experts question that explanation. At any rate, the 24-gauge wire visualization test is the required criterion for the earlier machines, although a step wedge can be used on those units also.

Interestingly enough, the brightness and contrast settings that allow 24-gauge wire to be most easily seen are not always the best setting for screening carry-on baggage. Once the operator has demonstrated the system's capability of viewing 24-gauge wire, the unit may then be readjusted to the optimum setting for viewing carry-on baggage.

X-ray penetration varies with voltage. Lower voltage units will show wire and its insulation more readily than higher-voltage units, but the latter penetrate more and thus will discern more objects needing identification. To give one example, a common transistor radio will show up as a black square on the display of a low-voltage X-ray unit. On the other hand, a high-voltage unit less efficient for visualizing wire would display the contents of the same transistor radio in considerable detail. Technical people feel that the best machine for seeing the test wire is not necessarily the best machine for screening packages. They believe that a test more closely related to the overall purpose of X-ray screening might be more appropriate.

This is probably the reason for the introduction of the step gauge or step wedge. An X-ray unit in good operating condition discerns the differences in metal thickness in shades of gray. Another option for testing X-ray units resembles a TV test pattern. Organizations such as the American National Standards Institute (ANSI) and the ASTM suggest other optional standard tests for evaluating the quality of an X-ray picture. One such test is a pentrameter, a bar of lead with holes of decreasing size. The X-ray picture can be rated by the number of holes visible. Bar patterns can be used to represent the system resolution in line pairs per millimeter or millimeters per line pair. Philips Electronic Instruments Company has developed an X-ray test pattern for security systems that utilizes a number of these devices.

When the unit is tested, the supervisor needs to enter the appropriate information on the operational test log of that particular unit. The information generally includes the date of the test, the supervisor's name, and the smallest wire gauge seen.

A Notice of Proposed Rulemaking (NPRM) to phase out old X-ray equipment that does not meet new, higher performance standards is now being developed.

DETECTION OF EXPLOSIVES BY X-RAY

Although a revolver is generally relatively easy to identify on X-ray, a bomb is a bit more difficult. The operator looks for batteries and wires, because on the black-and-white x-ray units, the explosives themselves are extremely difficult to identify. The introduction of E-Scan™, an innovation marketed by EG&G Astrophysics Research Corporation, is of some assistance in that task. The E-Scan can be added on to most of the Astrophysics X-ray units. The color monitor makes explosives and weapons more identifiable by displaying the organic materials in a bright shade of orange-brown and the inorganic materials in a deep shade of blue. Objects too dense to penetrate are displayed in a bright shade of green.

EG&G Astrophysics Research Corporation claims that being able to separate the dense organic materials from the less dense inorganic materials enables the operator to make faster, more accurate judgments on the contents of packages and luggage. Explosives are more dense than other organic materials but not as dense as metals. As a result, they are displayed in a deep orange-brown color, while knives are displayed in the inorganic blue color. Guns are usually too dense to penetrate and, therefore, are shown in a bright shade of green. Regardless of the thickness or density, explosives are displayed in the orange-brown color, as are narcotics such as heroin, cocaine, and marijuana.

In actual usage, the images on the X-ray appear less distinct to the operator in the color mode. An item might show up clearly recognizable in the black and white mode, while in the color mode, it is sometimes more difficult to identify the object. Therefore, some operators tend to operate the unit in the black-and-white mode most of the time, which defeats the purpose of the machine. Generally, operators perform better with the E-Scan if they start out working with that machine rather than be transferred to it after having worked on a machine with the black-and-white screen.

Partially to compensate for this, but also to provide for backscatter imaging, several manufacturers have begun using two monitors: one color and one black-and-white. Objects being subjected to X-rays either reflect or absorb the radiation. Backscatter is produced from reflected X-ray imaging and has proved useful for identifying sheet explosives (plastic explosives rolled into a sheet). Heimann Systems Co. offers systems: HI-MAT and HI-CAT. HI-MAT uses three colors to classify screened objects. HI-CAT uses two monitors; one is black-and-white, and the other is a pseudocolor X-ray for object identification by shape and structure. AS&E (American Science and Engineering) also has two monitors on its Model Z, which utilizes both backscatter X-rays and absorbed X-rays to image contraband invisible to standard X-ray systems.

At the 1990 Annual Seminar of the American Society for Industrial Security, virtually every manufacturer of airport X-ray equipment showed equipment utilizing two monitors. This illustrates the effort made by the X-ray manufacturers and engineers to perfect a technology that will allow X-rays to be successfully utilized to detect explosives. When I look back on where the industry has come since the early green fish eye fluoroscopic monitors, I am amazed at the improvements made by suppliers in the X-ray industry.

In June 1990, EG&G Astrophysics introduced another option for its Linescan X-Ray Baggage Screening System, which they have named *False Image Projection (FIP)*. The primary purpose of FIP is to test operator alertness, but it can also be used to train operators to identify specific threat objects. At a location away from the operator's station, a supervisor can activate FIP and, on a random basis, superimpose a false threat object on the monitor screen showing an X-rayed bag. Thus, if the outline of a Glock 17 Automatic appeared on the monitor screen, the operator would have to press the control panel to acknowledge that they had seen the false threat object. The operator's actions are recorded on a diskette for auditing purposes.

X-RAY SAFETY

The introduction of X-ray baggage inspection into air terminals evoked prompt expressions of alarm from some quarters. Generally, the alarm was traceable to a lack of information about the level and amount of radiation emitted by the units; the protesters' assumed that the inspection units developed the same dosage as medical and dental X-ray machines. Such attention-getting newspaper headlines as "X-Ray Build-Up May Pollute Terminals" and "Scanning Thru X-Rays May Cause Sterility or Cancer as Thousands of Pieces of Luggage Are Processed"[4] were typical.

In August 1973, the Bureau of Radiological Health, a division of the Food and Drug Administration, U.S. Department of Health, Education, and Welfare, published a set of radiation safety recommendations for adoption by state radiation control agencies. The recommended standards were:[5]

1. It shall not be possible to insert any part of the body into the primary beam.
2. Radiation exposure shall not exceed 0.5 mR in any one hour at a distance of 5 centimeters [less than two inches].
3. The control panel shall be equipped with a key lock. It shall not be possible to remove the key while the unit is in the on position.
4. Doors and access panels that prevent entry to an area where the exposure may exceed 0.5 mR in one hour shall be interlocked.
5. Means shall be provided to indicate to the operator when X-rays are being produced.

---

[4] *Airport News*, Vol. 15, No. 1, Av-Sec Communications, Inc., Jamaica, N.Y.: 1

[5] Federal Register, Vol. 38, No. 152, August 8, 1973: 21442.

6. A "deadman" switch shall be provided on the exposure controls. The location of the switch shall be such that the operator has a clear view of the X-ray system.

Three weeks after publication of the BRH recommendations, consumer advocate Ralph Nader entered the radiation controversy. In a letter to then FAA Administrator Alexander Butterfield,[6] Nader urged the Administrator to:

> [D]eclare an immediate moratorium on further operation or installation of carry-on baggage X-ray machines until the public can be assured (a) whether they are needed at all, or (b) whether the highest possible standard of safety will be maintained at the nation's airports if the FAA is lax enough to continue to permit their deployment.

Butterfield responded with assurances that adequate testing had taken place and that, although the federal government did not certify or license X-ray equipment, guidelines had been issued to state radiation control agencies. Nader replied, expressed continued concern, and again asked that current X-ray operations be suspended and no further installations permitted until "proper standards" had been established. The Administrator responded, that "The FAA has no evidence which would support your allegations." Nader and a private group identified as the Aviation Consumer Action Project (ACAP) then filed suit to halt the use of X-ray devices for baggage inspection.[7]

At about the same time, October 10, 1973, the Bureau of Radiological Health published a proposed federal performance standard for cabinet X-ray systems, which expanded the recommendations published for state action, retaining all of the previous requirements. The proposed federal standard added that each cabinet X-ray system should have a permanent floor; that each door was to have a minimum of two safety interlocks; and that each access panel should have at least one safety interlock. Two independent means were to be provided to indicate when--and only when--X-rays were being generated. One of the indicators was to be a visible display stating "X-Ray On." Warning labels would be required, stating "Caution: X-Rays Produced When Energized" and "Caution: Do Not Insert Any Part Of The Body When System Is Energized--X-Ray Hazard."

Originally, the federal standard was to become effective one year from the date of final publication, but a January 11, 1974 proposed revision to the standard suggested an effective date of 15 days from the date of the publication of the final order in the *Federal Register*.

The proposed federal standard posed no grave problems, because several states had immediately adopted the August recommendations, and carriers and manufacturers of X-ray equipment had implemented the federal standard as if it had become effective immediately. Compliance modifications

---

[6] Letter dated August 30, 1973.

[7] Aviation Daily, 209:299, October 29, 1973.

were made first in those states where they were required before the unit could be put in operation.

About six weeks after the publication of the proposed federal standard, the ACAP suit to enjoin X-ray baggage inspection came to a hearing, and Judge B.D. Parker ruled in favor of the consumer group, ordering a halt to the use of X-ray machines throughout the country. The ruling was primarily based on a procedural error and did not address the merits of ACAP's charges that the machines posed a danger of radiation to passengers and operators.

In his decision, Judge Parker held that the March 29, 1973, FAA memorandum authorizing the use of X-ray devices was "an unlawful act" because affected persons had not been allowed to comment on it before it took effect, a requirement of the Administrative Procedure Act. In addition, the judge ruled that the FAA must file an "assessment statement" to the effect that the use of such devices would not harm the environment. Another condition of the decision, and one to which the FAA also agreed, was that, whenever an airline submitted an application to the FAA for installation of an X-ray unit at an airport, the FAA would automatically furnish a copy of the application to ACAP for its review and comment, establishing a precedent that many regard as unfortunate.

The FAA agreed to all conditions, and on April 24, 1974, Judge Parker granted an indefinite stay on the order. Because the court had granted numerous temporary stays to its order in the interim, no actual shutdown of the machines had occurred.

However, another effort to shut down the X-ray units was more successful; at the peak traffic period of five p.m. on Good Friday, 1974, the FAA ordered 55 of the 190 X-ray units in service at 31 U.S. airports to be shut down immediately. The FAA action had been triggered by an advisory copy of a Bureau of Radiological Health defect notice sent to two manufacturers of X-ray machines. In compliance with the provisions of the Radiation Control Act of 1968, the BRH notice gave the manufacturers a specified time in which to reply or take corrective action--but the FAA order was immediate. The carriers objected strongly, pointing out that they had not been given the hearing provided for in the law and charging that the FAA was overreacting because of the ACAP case so recently settled. This the FAA denied.

More to the point were the actual defects and their remedies: one model had been cited because it was a manual unit without a conveyor belt, and thus it was possible to place a part of the body in the primary beam of the X-ray. This was corrected by the installation of an automatic feature that turned off the X-ray if a part of the body was placed in the path of the primary beam. The other unit was cited for excessive radiation, which turned out to be caused by the operators reaching inside the machine and increasing the radiation level to obtain clearer images. That defect was corrected by the installation of lead seals on the unit's service access doors.

Further Rule Revision

To satisfy the court-ordered requirement that affected persons be allowed to comment on the use of X-ray for baggage inspection, an amended notice of proposed rule-making was published on June 21, 1974. The final rule was published in the *Federal Register* on March 5, 1975, and became effective April 4, 1975.

The amended rule, which became a new section of FAR 121, established safety standards for X-ray units and required initial and recurrent training of X-ray operators, the use of film badges, (discussed later in this chapter under Radiation Badges) and radiation surveys every six months. FAR 108.17 now controls the use of X-ray systems, and the inspection requirement has been changed from 6 months to 12 months. The machine must meet the standards spelled out in 21 CFR 1020.40 or guidelines published by the FDA in the *Federal Register* of August 8, 1973 (38 FR 21442).

Also required was the posting of a sign notifying passengers that carry-on items would be X-rayed and recommending that they remove all "X-ray and scientific film" from their carry-on baggage before inspection. That is generally interpreted to include X-ray film and film with an ASA speed of 1000 or more. It should also be noted that film still sealed in factory boxes is not subject to inspection.

Carriers were apprehensive about this requirement, feeling that many passengers would be unnecessarily alarmed about ordinary film in their baggage. One X-ray manufacturer protested that such a sign was "false and misleading" for their equipment which would not damage X-ray and scientific film. The revised rule also stated that passengers could request physical inspection of photographic equipment and film packages, but this provision did not have to be included in the required "X-ray and scientific film" sign.

The sign first issued by the FAA stated only "Physical Inspection May Be Requested," without reference to cameras or film. Predictably, some passengers concluded from this that they could, upon request, have their carry-on items manually searched instead of X-rayed. This interpretation was not generally accepted, not only because of the delay it would produce, but also because X-ray often provides a superior security inspection because of its unique capability of looking inside objects such as portable radios and cans of hair spray. The FAA later changed the sign's wording to limit optional inspection to photographic equipment and film packages.

Radiation Studies

In the midst of showers of charges and countercharges, it is often difficult to arrive at facts. What is the real risk of radiation exposure from a low-radiation X-ray inspection unit?

The FAA's Negative Environmental Declaration, filed with Judge Parker's court on June 19, 1974, stated that the radiation exposure rate limits for cabinet X-ray systems established by the Bureau of Radiological Health are identical with those established for the ordinary television set. The statement also said that the overall risk of radiation exposure to the operator was the same as, or less than, the exposure permissible in any occupation.

The U.S. standard for maximum radiation exposure established by the National Committee on Radiation Protection states that members of the public shall receive no more than 170 millirem per year from all man made radiation sources other than medical exposures. This is in addition to natural exposures; cosmic radiation in the atmosphere contributes approximately 50 millirem annually for each person living at sea level. Persons living at a high altitude, as in Colorado, can receive as much as 200 millirem each year from cosmic radiation. Medical and dental exposure is estimated to average 55 millirem per person per year.

An example of common X-radiation that falls within the maximum radiation exposure standard is color television; these sets generate X-rays that are estimated to contribute 0.2 millirem of radiation annually to the average viewer.

The technical proposal of a major U.S. X-ray manufacturer clarifies this further:[8]

> Even when directly in the beam, a person would receive less exposure from 5,000 inspection pulses than from a single dental X-ray. To an operator in his normal operating position, the leakage radiation exposure is much lower....(I)t is for all practical purposes impossible [for the operator] to encounter sufficient radiation to approach a dose of 1/100th of a Rem per week even if he operated the system 24 hours per day for seven consecutive days.

Radiation Badges

Even before the FAA required the use of film badges for X-ray operators, some carriers were already using them. These film badges are commercially available, and their cost includes analyzing of the badges after use and furnishing the results of the analysis to the customer. A control badge is kept separate, away from the area in which the X-ray machines are operating, to provide an environmental level of radiation exposure for comparison with the badges worn by the operators. Any difference found on analysis is the amount of radiation attributable to leakage from the X-ray machine. In addition, the FAA requires the carrier to maintain a log identifying the operator, any training he or she has received, and a continuing record of duty time for each operator.

The requirement's covering the wearing of badges and the record keeping actually work to the benefit of the carrier in terms of a legal liability. If, for example, a screener delivered a baby who had medical problems of one sort or another, might she presume the disability was caused by the X-ray? It is helpful for the carrier to be able to produce records reflecting low to zero dosage during the time she worked near the X-ray machine.

---

[8] "Technical Proposal for Airline Carry-On Baggage Inspection System", Bendix Aerospace Systems Division, December, 1974, p.3-1.

X-RAY AND PHOTOGRAPHIC FILM

All air carriers using X-ray have heard from irate passengers claiming that X-ray screening damaged their film. Fortunately, the passenger quite often includes the film with the letter. United Airlines has made a practice of tracing the cause of the damage for each roll submitted to them, and has yet to find a single piece of film damaged by an X-ray carry-on baggage inspection unit. Most often, the film sent in is completely black or unexposed. Although the lack of exposure can be caused by a number of errors and camera malfunctions, it cannot be caused by X-radiation, because this fogs or overexposes the film, tending to make the film clear or milky white. This distinction can most readily be seen when the damage claimed concerns the vacationer's color slides.

When a picture is correctly taken, the film that surrounds the sprocket holes is not exposed by the picture taking and remains black. This is readily apparent if the slide is removed from its cardboard frame. Had the slide been damaged because of X-ray exposure, the unexposed film in the sprocket area would be clear or milky-white, as the X-ray unit--unlike the camera--does not limit its exposure to that part of the film that is exposed by the lens of the camera.

There are occasions when checked--not carry-on--baggage is removed from the cargo hold of an aircraft for inspection by a more powerful X-ray unit that is capable of damaging film. However, this is usually only done in response to a sabotage threat, and passengers are always notified so that they can remove film from their luggage before it is X-rayed. Most of the units that X-ray checked luggage are still film safe, however.

A research study published in 1973 by the Film Technical Services Division of Eastman Kodak Company states:[9]

> The results of our tests thus far have led us to the conclusion that a cumulative exposure not exceeding 5 milliroentgens (mR) would seldom cause a noticeable photographic effect, providing the exposure is made in increments of exceeding 1 milliroentgen and the orientation of the film-luggage combination is changed between exposures... . Care should be taken that the luggage containing unprocessed film be X-ray inspected no more than five times *unless the very low level dose units are used.* These units operated in the range of 0.1 milliroentgen (100 microroentgens) are quite safe for film. [Emphasis added.]

Although the radiation dosage will differ according to the X-ray unit employed, no unit used by U.S. carriers generates one mR per inspection. The dosage per inspection ranges from a possible high of 0.6 mR down to as low as 0.15 to 0.2 mR or, in some instances, even less. The majority of the units used today have a maximum of 0.15 mR with an average dose of 0.1 mR. The true risk to passengers' film becomes a matter of calculation: How many inspections can film pass before reaching an exposure of 5 mR (at which the most sensitive, very high speed film may fog), using a worst case average of

---

[9] J.K. Debenham, *A Brief Description of the Effects of X-Ray Inspection on Unprocessed Photographic Film,* Film Technical Services Division, Eastman Kodak Company, 1973.

0.5 mR per inspection. The answer clearly is 10 inspections. At 0.15 mR per inspection, 30 inspections would not be sufficient to fog film. Indeed, one X-ray manufacturer, American Science and Engineering Company, claims its units are film safe for over 300 inspections.

To apply another yardstick, cosmic-ray radiation at 35,000 feet altitude can be as much as 0.1 mR per hour; thus, a four-hour flight could subject airborne film to a radiation exposure equivalent to several low-pulse X-ray examinations. Thus, it would appear that film that can be safely transported by air may just as safely be inspected by any of the X-ray systems used by U.S. carriers.

What can be said so broadly of U.S. carrier inspections cannot be said about all units in use. X-ray units in some overseas locations can cause film damage, and appropriate precautions should be taken where there is any question.

X-RAY AND MAGNETIC TAPE

Another potential claim against air carriers has been in the area of magnetic computer data tapes and cassette tapes. The hypothesis that X-ray inspection could damage computer tapes deserved and got very careful study because the loss of even a small number of of bits or bytes could make gibberish of the recorded data.

On the basis of considerable experimentation, Bendix Aerospace Systems stated:[10]

> Having conducted these tests with tapes containing sophisticated programs and testing under the most severe of anticipated conditions, we are able to conclude, without reservation, [that] magnetic tape may be submitted for X-ray examination by the Bendix Ray Inspection System without any detrimental effect on the magnetic tape or the information on the tape.

No data contradicting this have appeared.

For purposes of illustration, the tests conducted by Bendix in 1973 were as follows:

> *Case #1:* A simple test was performed to determine if data recorded on magnetic tape would be affected by the X-ray pulse generated by the manual [non-conveyorized] Bendix Ray Inspection System.
>
> In conducting the test some standard ALSEP data from the Apollo Space Program was recorded on 3M half-inch mylar-base tape running at 3.75 inches per second with the input data in split phase, NRZ-C at 1060 bits per second. The data was checked visually on an oscilloscope for correctness of waveform and following decommutation by the data processor for completeness of the engineering data content.

---

[10] Letter to the author, December 20, 1973.

The reeled tape, in a standard cardboard box, was then placed in the primary beam of the inspection system and subjected to five (5) pulses, which would be equivalent to five routine examinations by airport security personnel.

The tape was then replayed and no differences were noted on either the oscilloscope displayed waveforms, or in the decommutated output when compared with the pre-X-ray data analysis.

Case #2: A complex test was performed to determine if data recorded on magnetic tape would in any way be altered or degraded by the routine X-ray examination of the tape on the Bendix Ray Automatic Inspection system.

In preparation for the test, three duplicate tapes were generated. The tape (Memorex/MRXIII, one-inch width, 3200 FCI, total surface certified, with approximately 2400 feet per reel) was written on an IBM 9 track dual density tape unit 3420, Model 3 at 75 inches/second--800 BPI. The plot tapes were generated by a FORTRAN IV G Program, using Cal Comp sub-routines, for plots on the Cal Comp plotter Model 763 29 1/2 inch by 120 feet drum plotter. The tapes were prepared on the IBM 370 System at the Bendix Ann Arbor Data Center.

The test, designed to detect and reveal alterations of degradations resulting from error of 1 part in 6125, consisted of an intricate plot registration requirement, involving approximately 58,000 characters (29,000 commands and 29,000 clock items) per plot, with each .00125-inch movement of the plot requiring a command.

Following generation of the tapes, each was read and plotted without incident. Using the Bendix Ray Automated Inspection System, test tape Number 1 was subjected to ten (10) separate X-ray examinations; tape Number 2 was subjected to thirty-five (35) separate X-ray examinations; and tape Number 3 was retained as a control tape and did not undergo any examination, but was otherwise in proximity to tapes 1 and 2 at all times. Replots of all three tapes, post-X-ray, were plotted without error and they were in perfect registration with pre-X-ray plots.

X-RAY AND COMPUTERS

Having worked for Apple Computer, Inc., as the Manager of Corporate Security, I'm a computer junkie who begins to go through withdrawal if I'm kept away from my keyboard and mouse for too long. The prospect of having a hard disk inadvertently erased brings cold chills to seasoned computer users. Of course, everyone should back up their hard disk. My favorite quote on that subject comes from Jean-Louis Gasseé, former President of the Product and Manufacturing Division of Apple: "It's just like your momma always told you. Eat your veggies and back up your hard disk."

Portable or laptop computers are now being reduced to notebook size and some weigh a pound or less. It's common to see passengers working with their computers on airplanes now, and this will increase. Portables are not the only computers affected; I've taken my trusty M&M (Macintosh and modem) on many trips, storing it in an overhead bin or under the seat. Checking a computer as baggage borders on insanity, unless the passenger has the original packing carton. Most carriers would not accept a computer as checked baggage.

There are two possible threats to a computer during the screening process; the metal detector and the X-ray machine. The controversy is much the same as over the screening of film. If a computer is screened at all, it will be done by X-ray or hand search. There is no magnetic field in the X-ray machine other than a very small field in the motor, which drives the conveyor belt.

The FAA states that the X-ray will not damage computer memory but allows a hand inspection if desired, much the same as for cameras and film. That inspection, however, normally requires the passenger to turn the computer on for the person at the checkpoint. Unless you have a battery-powered portable, the computer will have to be plugged it in and booted. As might be anticipated, there probably won't be a table nearby, and you may end up on hands and knees on the floor. It should be mentioned that, because of space limitations some passengers have experienced trouble getting past the flight attendant with something as large as a Macintosh or Mac SE, even though the computer fits easily into the overhead compartment.

Although the FAA states the X-ray will not harm memory, some experts disagree. They believe that though there's only a small chance that the X-ray machine will do any damage, even a little bit of damage can be fatal. Other frequent flyers routinely have their computers and floppy disks X-rayed without problem.

John P. Battema, former Vice President - Marketing for EG&G Astrophysics Research Corporation, believes that X-ray screening poses no threat to personal computers. Over 3000 Astrophysics X-ray machines are in use throughout the world. The company supplies over 90 percent of all the X-ray screening systems used by U.S. airlines. Battema stated:[11]

> [T]he radiation emission used to screen items in a Linescan security screening system will in no way affect personal computers or other magnetic memory devices. The Linescan System One systems used by ________ Airlines utilize 0.1mR (milliroentgens) per inspection.
>
> Tests that we (and others) have conducted indicate that the threshold radiation level to affect magnetic memory devices is at least 10R (Roentgens). This means that a personal computer or other magnetic memory devices could be inspected at least 100,000 times before there was even the slightest possibility of damage.
>
> In another frame of reference, a personal computer flown in a jet at 35,000 feet from Los Angeles to New York will receive at least twenty times more radiation than it received during one inspection with a Linescan System One.

EG&G Astrophysics also claimed that they had subjected floppy disks to 2000 roentgens (2,000,000 mR) in one of their high-dose, industrial X-ray systems without affecting the stored magnetic data in the slightest. They indi-

---

[11] Letter to the author from John Battema dated March 23, 1987.

cated this was equivalent to twenty million inspections with a Linescan system.

PLASTIC GUNS

Significant concern was aroused when the Glock 17, a 9mm automatic pistol made in Austria, was introduced. I have fired both the Glock 17and Glock 19 and find them to be a very fine weapon, used by several law enforcement agencies in the U.S. It was invented by Gaston Glock in the village of Deutsch-Wagram, just outside Vienna. It is not a true plastic gun because the barrel, slide, and one spring are metal. Indeed, it contains more metal than the typical Saturday Night Special. It was feared, however, that the Glock 17 guns would be undetectable by X-ray, particularly if disassembled. Jack Anderson reported in January 1986 that Khadafy was in the process of buying from 100 to 300 Glock 17s which would be difficult for airport security forces to detect.[12]

A study by the Congressional Office of Technology Assessment in 1987 indicated that firearms made entirely of plastic or ceramic materials could be introduced within several years. At least four bills were introduced into Congress that would have banned the introduction of these weapons in the U.S. At the hearings, manufacturers supplemented their testimony with evidence of the Glock 17 clearly shown on X-ray inspection photos. The Glock 17, however, did not appear to be the main concern of the legislators--rather, they were worried about the prospect of future undetectable firearms.

Several aviation industry groups, including the Air Transport Association, Airport Operators Council International, and the American Association of Airport Executives asked the Justice Department in October 1987 to reconsider its withdrawal of support for banning plastic firearms. In December 1987, the Senate voted to table an amendment to other legislation that would have banned firearms not detectable by current airport security technology. Just days later, the FAA told Congress that it was not free to support legislation to ban such firearms because others in the administration formulate the policy, perhaps referring to the Justice Department's stance.

Finally, in October 1988, a bill (H.R. 4445) was passed in the House that would require the DOT to complete a study within six months of available state-of-the-art security screening equipment and to establish a schedule for installation of the equipment at the earliest practicable time, although no funding authorization was provided. The bill would also prohibit the manufacture, importation, sale, or possession of any firearms that are not as detectable as the "security exemplar" by magnetometers and whose shape is not accurately depicted when screened by X-ray machines commonly used at airports. The security exemplar referred to is a test handgun made of 3.7 ounces of stainless steel. Existing firearms were not affected by the bill.

Suppliers of X-ray units have developed a technology that will more readily detect plastic guns. Two such systems are the HiCAT, developed by Heimann Systems Co., and the E-Scan by EG&G Astrophysics Research,

---

12 San Francisco Chronicle, January 14, 1986.

Inc. These techniques involve use of backscatter radiation, previously discussed in this chapter in the section entitled, "Detection of Explosives By X-Ray."

EXPLOSIVES DETECTION SYSTEMS

The predeparture screening program has been highly effective in deterring hijackings. The vulnerability of aircraft to explosives placed in checked baggage is still great, however, and an effective explosives detection system (EDS) is needed that can detect explosives at a fast enough through-put rate to be practical.

A variety of explosive detection systems are on the market. Some are walk-through units and are used at nuclear sites. Others are hand held and can be used to check carry-on items and checked baggage for explosive vapors. In one test[13] of the Ion Track M97 hand held unit, which is standard issue for U.S. Army Explosive Ordnance Disposal (EOD) worldwide, 75.3 percent of the bags were cleared, 2.3 percent received a positive alert and 22.4 percent had to be retested. With carry-on baggage, 81.7 percent were cleared, 0.3 percent tested positive, and 18 percent had to be retested. The checked bags were tested at the ticket counter by a security agent prior to being X-rayed. The security agent was able to spend an average of 20 seconds per bag before another passenger came along. The average time to search a garment bag was close to 35 seconds; a suitcase, about 17 seconds. While vapor detection technology is useful for detecting explosives on people, my personal view is that a hand-held unit such as this might be useful for checking selected or target bags, but is not a practical method of screening a high volume of bags at a major airport. The thermal neutron activation (TNA) method uses nuclear radiation, and as such it cannot be used for screening passengers or carry-on baggage.

The FAA has awarded contracts to a number of firms for the purpose of developing effective EDS units. In mid 1987, the FAA awarded $430,000 to Thermedics, a firm developing a vapor detection device. At about the same time, a prototype of the EDS unit that has received the most attention was being installed at San Francisco International Airport in TWA's bag room for a one-month test. This unit is the thermal neutron activation (TNA) detector manufactured by Science Applications International Corporation (SAIC) with the support of the FAA.[14]

On August 30, 1989, the FAA, under pressure from Congress, issued a final order requiring airlines to install new explosives detection systems at about 15 U.S. and 25 foreign airports. Although the rule did not specify a particular EDS technology, it noted that the TNA had been "shown to have the highest degree of explosives detection currently possible for detecting

---

[13] Report of operational test with the Model 97 explosive detector at Logan International Airport, in conjunction with Northwest Airlines and the FAA, 15-16 February 1989, Ion Track Instruments, Burlington, MA.

[14] The TNA unit was originally marketed under the name SNOOPE (System for Nuclear, On-line Observation of Possible Explosives.

known civilian and military explosives manufactured here and abroad." The FAA had already purchased six of these units at a cost of over $1 million each and had installed one in August at the TWA terminal at JFK airport in New York. In January 1990, Pan Am agreed to install one at Miami. The units purchased by the FAA were equipped with a special X-ray unit called Xenis to provide a dual-sensor system. Other installations were planned for London Gatwick, Frankfurt, Dulles, and a sixth unnamed location.

Considerable controversy exists regarding the TNA. On September 13, 1990, FAA Administrator James Busey classified as secret all new details about the performance of bomb detection devices. He said his order would prohibit further public discussion of the system's false alarm rates, ability to detect small amounts of explosives, and operating methods. "How we are changing the methodology of use of the system, how we are changing its detection capability, how we are speeding up or slowing down its operation, how we are dealing with its known false-alarm capability, those are the kinds of thing I don't want to talk about any more," said Busey.[15] In compliance with those orders, capabilities of the TNA system will not be discussed further. Of course, questions have arisen as to whether the real intent of this order was, in fact, to silence public discussion of a highly controversial issue.

The ATA has commissioned the University of California, Berkeley, Transportation Studies Institute, to conduct a study of the practicability and consequences for U.S. airlines of the introduction and use of the TNA EDS for the inspection of international checked baggage. This study is in furtherance of implementing FAR 108.20, "Use of Explosives Detection Systems" which became effective October 5, 1989.

In February 1991, representatives of the Congressional Office of Technology Assessment (OTA) told the Senate Government Operations Committee that the TNA is flawed and recommended that the FAA not buy more than the six already purchased.[16] The FAA responded that OTA's conclusions were outdated, citing recent encouraging test reports.

Section 108 of the Aviation Security Improvement Act of 1990, signed into law on November 16, 1990, amends Title III of the Federal Aviation Act of 1958[17] by adding the following new section:

> Section 320. Deployment of Explosive Detection Equipment.
>
> (a) GENERAL RULE. No deployment or purchase of any explosive detection equipment pursuant to section 108.7(b)(8) and 108.20 of title 14, Code of Federal Regulations, or any similar rule, shall be required after the date of the enactment of this section, unless the Administrator certifies that, based on the results of tests conducted pursuant to protocols developed in consultation with expert scientists from outside the Federal Aviation Administration, such equipment alone or as a part of an integrated system can detect under realistic air carrier operating conditions the

---

15 San Jose Mercury News (September 16, 1990).

16 *Technology Against Terrorism: The Federal Effort--Summary*. Congress of The United States, Office of Technology Assessment. (Washington, D.C., GPO, S/N 052-003-01226-0): 5.

17 49U.S.C. App. 1341-1358).

amounts, configuration, and types of explosive material which would be likely to be used to cause catastrophic damage to commercial aircraft.

Safety of the TNA

A statement regarding the safety of the TNA system was prepared in cooperation with the California Department of Health Services, Radiological Health Branch, and disseminated by the FAA Technical Center. It reads in part:[18]

> In this system, the luggage passes through a cloud of very low energy ("thermal") neutrons. These neutrons are very penetrating and cause gamma rays to be instantly produced in the luggage. By analyzing these prompt emissions, the presence of explosives is determined. The process of detection of prompt gamma rays does not require the screened materials to be radioactive. Indeed, the induced activity is very minimal, well below natural background, and is unmeasurable within a few minutes.
>
> Even if luggage is returned to the passenger shortly after screening due to flight cancellations, there is no possible risk to the passenger, since all bags will be monitored after screening and any bag with a level above naturally occurring background will be held for the number of seconds necessary for it to reach background again. The dose level applied to the baggage is orders of magnitude lower than that which has already been approved for medicine and foodstuffs in other commercial avenues. Aluminum is the most likely common material to become slightly radioactive, but even several pounds of it will only produce a small fraction (about 1%) of the constantly present naturally occurring background within a few minutes after screening. It is absolutely safe to wear jewelry, belt buckles, and other items carried in luggage immediately after obtaining your baggage. There is no cumulative build-up of radiation no matter how frequently your luggage may be screened and there is no risk to the fetus of pregnant women when they carry screened luggage, nor to animals or their fetuses if they fly in close proximity to baggage in aircraft cargo holds. Animals will not be screened by this system. The system is completely film safe and test measurements on the film radiation exposure effects have been made by the American Society for Testing and Materials.
>
> In short, during most flights, the radiation dose routinely received by passengers at flight altitudes due to cosmic rays is many times greater than the dose received, if any, by handling baggage screened by the TNA system.

Research continues into various devices for the detection of explosives. EG&G Astrophysics is working on a machine called T-Scan, which they claim is lighter, faster, more efficient, and less costly than the TNA. Several X-ray manufacturers market a backscatter imaging X-ray that they say can detect types and amounts of explosives known to be a problem. Manufacturers of vapor detection devices such as Thermedics and Ion Track Instruments believe that their technology, in combination with X-ray and

[18] "Safety of Baggage Explosive Detection System" undated.

TNA, would be a good approach. As the report of the President's Commission stated, "Only a massive effort now will bring our technology ahead of the destructive devices of terrorist adversaries."[19]

Taggants

One area where progress continues is that of "tagging" explosives so that they can be identified. On January 12, 1990, a special subcommittee of the ICAO Legal Committee completed work on a draft treaty to require the addition of taggants to explosives manufactured by contracting states. The taggants would render explosives detectable by gas-analysis methods that are currently available. On April 12, 1990, the United States participated as one of the 68 countries represented on the full ICAO Legal Committee in the preparation of a new international convention on taggants. The Legal Committee presented the draft convention to a Diplomatic Conference on Air Law under ICAO's auspices for consideration in early 1991. The United States signed the Convention on the Marking of Plastic Explosives for the Purpose of Detection on March 1, 1991.

---

19 *Report of the President's Commission on Aviation Security and Terrorism,* May 15, 1990,(Washington, D.C.: U.S. GPO , 1990): 63.

Chapter 8

WEAPONS, PRISONERS, AND RESTRICTED ARTICLES

WEAPONS

The emergency order requiring 100 percent screening of all passengers and carry-on luggage stated in part: "The certificate holder shall not permit any passenger to board its aircraft unless the carry-on baggage items are inspected to detect weapons, explosives, or other dangerous objects"

The term "dangerous weapon" occurs in the Air Transportation Security Act of 1974:

> Whoever, while aboard, or while attempting to board, any aircraft in, or intended for operation in, air transportation, has on or about his person or his property a concealed deadly or dangerous weapon, which is or would be, accessible to such a person in flight, or any person who has on or about his person, or who has placed, attempted to place, or attempted to have placed aboard such aircraft any bomb, or similar explosive or incendiary device, shall be fined not more than $1,000, or imprisoned not more than one year, or both.
>
> This subsection shall not apply to law enforcement officers of any municipal or State government, or the Federal Government, who are authorized or required within their official capacities to carry arms."

Others have had difficulty in defining a dangerous weapon. Over the years, the courts have held that bare hands, a fist, a lighted cigarette, a bottle, and numerous other items were specific cases, to be considered deadly or dangerous weapons. Indeed, few would argue this. Still, with or without definition, some guidance must be given the contract guard or airport employee who must decide what items are to be taken from the passenger and what items the passenger is to be permitted to take on the aircraft.

Concealed weapons can take many forms, and those responsible for predeparture screening should be alert to even the most unlikely possibilities. Notices disseminated to carriers have reported knives disguised as belt buckles and daggers or knives disguised as the handle of a cane, a fountain pen, or other innocuous object. A list of all the different hidden and improvised weapons that have been found would be virtually endless, and it is not my purpose to provide a guide for those aspiring to conceal a weapon. It is sufficient to say that those conducting the inspections should not hesitate to take a closer look at any item that arouses their suspicions. For example, twisting the handle of a cane or umbrella to see if it comes apart, is sometimes an instructive procedure. In this problem area, X-ray has been invaluable, and X-ray operators should be alert to this and similar possibilities.

All objects generally recognized as weapons should be checked as baggage and not allowed in the cabin of the aircraft. This naturally includes imitation or toy weapons, starter pistols, and compressed air guns. Firearms are acceptable as checked baggage only under the following specific arrangements:

1. The passenger has notified the carrier before checking the baggage that the weapon is in the baggage and is unloaded.

2. The weapon is in a case suitable to protect it under normal baggage handling conditions.

3. When the firearm is other than a shotgun, rifle, or other firearm normally fired from the shoulder position, the baggage in which it is carried is locked, and only the passenger checking the baggage retains the key or the combination.

Whether or not knivesare to be permitted in the cabin is a decision made on the spot. Many men carry pocket knives, and it soon became evident that a standard criteria was needed. The Air Carrier Standard Security Program (ACSSP) established that knives "with blades four inches or longer" were not to be carried in the cabin, but also required that knives must be legal in the local jurisdiction. One airline was cited by the FAA for permitting a passenger to board in California carrying a spring-loaded switchblade knife with a three-inch blade; switchblade knives are illegal in California without reference to blade length. All larger cutting weapons, such as sabers, swords, machetes and hunting knives, should be checked.

Any blackjack, billy club, nun-chaku sticks, or bludgeon should be checked as baggage, provided their possession is legal in that jurisdiction. In many areas, such weapons are illegal and must be confiscated by the law enforcement officers. A "bludgeon" can be a souvenir reproduction of an ancient mace or of a "morningstar"--a spiked ball on the end of a chain.

Any device or object, whether or not it is commonly thought of as a dangerous weapon, should be picked up if its possession supports a reasonable presumption that it could be used as a weapon; this includes ice picks, straight razors, and elongated scissors. Passengers have been required to check pocket nail clippers, too; because each decision is a matter of judgment, both extremes and every point in between can be seen at security checkpoints. Bearing in mind that the flight crew will be giving each passenger a steak knife with the meal, a fairly usable general rule can be derived that one should not confiscate any items less dangerous than the eating utensils to be furnished by the carrier.

FAR 107.21 was amended on January 10, 1986, to substitute the words "deadly or dangerous weapon" for the word "firearm." The regulation had stated, "No person may have a *firearm*, an explosive, or an incendiary device" (italics mine) on or about the individual's person or accessible property when presenting himself or herself for screening or when entering or in a sterile area." This change broadened the regulation to include such items as chemical Mace™ or knives. These have always been unacceptable at the screening checkpoint, but this change permits the FAA to take administrative action against the person carrying these items.

PERSONS AUTHORIZED TO CARRY WEAPONS DURING FLIGHT

FAR 108.11 (a)(1) states that persons authorized to carry a weapon during flight include:

> (i) An official or employee of the United States, or a State or political subdivision of a State, or of a municipality who is authorized by his or her agency to have the weapon; or
>
> (ii) Authorized to have the weapon by the certificate holder and the Administrator and has successfully completed a course of training in the use of firearms acceptable to the Administrator.

It further specifies in §108.11 (a)(2) that:

> The person having the weapon needs to have the weapon accessible in connection with the performance of his or her duty from the time he or she would otherwise check it...until the time it would be returned after deplaning.

U.S. carriers now require nonfederal officers to present an official letter from their agency, attesting to their need to be armed on the flight, specifying the trip itinerary and the particular segments of the flight for which the officer needs to be armed. An officer traveling somewhere to bring back a prisoner probably doesn't need the weapon on the outbound flight, nor does an officer going on vacation need to be armed.

Federal law enforcement agencies submit to the FAA a copy of their policy under which employees under their jurisdiction are required to be armed during the flight. Thereafter, federal officers of that agency do not need to present a letter to the carrier. They will be asked to sign the form, however, because that is the carrier's method of keeping track of who has a weapon on the flight.

Persons who are not law enforcement officers and who wish to carry a weapon aboard must (1) have completed a firearms training course within the preceding 12 months, (2) be approved by the carrier, and (3) be approved by the FAA Administrator. Approvals are granted by the FAA for no longer than six months at a time. The carriers are extremely reluctant to give their approval, and most carriers require that the approval be given by the president of the airline. Without some convincing and compelling need, the odds of this portion of the law being met are slim to none.

Persons carrying a weapon cannot be served alcoholic beverages. Armed passengers are to be known to one another and to crew members. Each armed law enforcement officer on the plane must be aware of the presence and seat location of every other officer on the flight. This is to prevent one officer from becoming alarmed if he or she notices another individual on the flight who is armed. When armed law enforcement officers are boarded, the crew members, and especially the Captain, are to be advised of their seat locations and the fact that they are armed. The Captain will probably want to meet the officer so that the Captain will recognize the officer if an incident does occur.

Passengers should be advised to keep the weapon concealed while on the plane, which usually means they cannot take off their suit coats or jackets if they are wearing the weapon. (Some will elect to carry their weapon in an attaché case.) Many officers prefer not to board as an armed officer, if for no other reason than to be able to have a drink on the flight. The officer-passenger should be advised that his assistance--even in the event of a hijacking--is not expected unless asked for. United Airlines conveys much of this information to the armed passenger by means of a handout notice that says, in part:

> Discharge of a firearm aboard an aircraft could cause a situation far more dangerous than the original disturbance--and this includes hijacking. Wounding or killing innocent bystanders is always a possibility, but aboard an aircraft the innocent bystanders might be the pilots and without them the flight will end in disaster. Behind the walls, under the floor, and above the ceiling there are many fuel lines, control cables, electric wires, and hydraulic systems, all essential to safe flight and all subject to damage or destruction by a stray bullet or ricochet.

I have observed law enforcement officers going through screening on a number of occasions. I agree that it should not have to be said, but if a law enforcement officer is exempt from screening, it means just that. He or she is *exempt*. That means they don't have to be screened. The screening is for the purpose of detecting weapons, and the officer has already admitted to having a weapon. I have observed screeners on one or two occasions review all of an officer's paperwork and then require him to pass through the walk-through metal detector and even go over him with a hand wand.

One conflict that was resolved during the early years of the antihijacking program was that some captains refused to allow law enforcement officers to have their weapons on board and would personally relieve them of their weapons. That only happened successfully with one FBI agent before the Bureau let the airlines know it would no longer be tolerated and that agents were under orders not to surrender their weapons. One showdown came when an agent refused to surrender his weapon and also refused to leave the aircraft at the captain's request. The captain left instead, and it was necessary for the carrier to get another pilot for the flight. After a few such skirmishes, that issue now seems to be well settled.

Another case involved an FBI Special Agent who was also a U.S. Air Force reservist pilot who flew refueling tankers for the Air Force. When the captain of the airline asked the agent for his weapon, the agent said, "O.K., you can have the gun and play FBI Agent, if you let me be the pilot and fly the plane," and he could have done it.

ARMED POLICE OFFICERS IN THE STERILE CONCOURSE

Yet another problem has been presented by armed police officers visiting a sterile concourse area. The officers are there to meet a flight or to see someone off; they are not in uniform, but they are armed. Indeed, many police departments require their officers to be armed at all times whether on duty or off. The problem of the visiting officer, which was formerly the responsibility of the airport, acting in agreement with the law enforcement authority on duty, is now handled uniformly under the ACSSP.

The procedure worked out by the FAA and the carriers states, in essence, that the armed officer who is not a passenger should be referred to the Checkpoint Security Supervisor. The visitor presents credentials and verification of identity to the CSS, whose responsibility it is to authorize the armed officer to pass the screening point without going through the inspection procedure.

Armed Flight Crews

Although U.S. flight crews could be armed under FAR 108.11, provided the Administrator and the air carrier were to agree to authorize weapons, no U.S. carriers are known to permit their crews to be armed. Some crews have been armed in the past. On May 8, 1964, a suicidal passenger shot and killed the flight officers of a Pacific Air Lines F-27 shortly after takeoff from Stockton, California. All 44 persons aboard were killed when the plane crashed. Shortly thereafter, pilots for Pacific Air Lines and Bonanza Air Lines carried sidearms, as much for protection against would-be suicides as against hijackers. At one time, it was estimated that as many as 10 to 15 percent of pilots were carrying weapons despite carrier bans, but this proportion has almost certainly decreased with the decline in hijacking attacks.

CARRIAGE OF PRISONERS

The carriage of prisoners aboard commercial airlines is a relatively recent practice. Although the carriers might prefer not to carry prisoners, there is a considerable amount of revenue is involved, so most carriers accept the business.

Under the rules of some carriers, a prisoner who so much as expresses an objection to going on the flight within the hearing of carrier personnel will not be boarded. No carrier willingly risks a struggle with an unruly prisoner at an altitude of 35,000 feet. Prisoners who are violent or who physically resist escorts or guards, will rarely be boarded. Some--but not all--carriers refuse to accept prisoners in handcuffs or leg irons because of the visual impact of a restrained prisoner on the other passengers. In such cases, it sometimes occurs that two or three armed law enforcement officers bring a prisoner to the boarding gate, where they remove the handcuffs just prior to boarding. As soon as the plane has landed and the prisoner and escort deplane, the cuffs are put on again. Most law enforcement officers would prefer to transport a prisoner in handcuffs or some kind of restraint if the prisoner has a record of violence.

Many carriers now prefer to have the prisoners restrained in handcuffs attached to a waist belt, but they leave the final decision to the escorting officer. Safety questions have been raised regarding the use of leg braces or leg irons, but their use is not specifically excluded.

A serious question arises, though, as to the advisability of an unrestrained prisoner being seated next to an armed officer; should the prisoner succeed in obtaining the officer's weapon, the lives of all aboard would be placed in jeopardy. In one case, escorting law enforcement officers went to sleep, and the flight attendant saw the prisoner eying the gun of one of the escorts. She woke the escorts, who did not appreciate her concern on the grounds that there was nowhere for the prisoner to go. The airline has since made a practice of impressing upon prisoner escorts their responsibility to stay awake for the entire flight.

The air carrier is required to board the escort with the person in custody before all other enplaning passengers board and to deplane them last. The carriers seat these persons in the rearmost available passenger seats that are not located in any lounge area or next to or directly across from an aircraft exit. I investigated one case where while the airplane was taxiing, the prisoner ran forward a few rows, opened the emergency exit and jumped to the ground.

FAA Advisory Circular 108-2, dated October 16, 1981, concerns security rules for the carriage of weapons and escorted persons. No regulatory requirements exist in FAR 108 governing the escort of persons by *unarmed* individuals. With regard to armed law enforcement escorts who have in their custody a person being transported aboard an airplane for which screening is required, the FAA requires the officer to:

(1) Notify the certificate holder at least one hour before departure, or in an emergency, as soon as practicable, of the identity of the escorted person and the flight on which that person will be carried.

(2) Determine and notify the certificate holder as to whether the escorted person is considered in a "maximum risk" category by the government entity having custody of the person. Two LEOs [law enforcement officers] are required for the escort of one maximum risk category person and only one "maximum risk" person may be carried aboard an airplane. One LEO may escort no more than two persons in a risk category less than that of maximum risk. Where a question arises concerning risk category, the governmental entity responsible for the person in custody should be relied on for such a determination. [The FAA believes that the determination of whether a person is considered a "maximum risk" is properly one to be made by the professional law enforcement organization having custody, and no attempt will be made herein to provide a general definition.]

(3) Assure the certificate holder that the individual in custody has been searched and does not have on or about the individual's person or property any article which could be used as a deadly or dangerous weapon.

(4) Be equipped with adequate restraining devices to be used in the event that a determination is made that restraint of the person in custody is or becomes necessary. It should be noted that some certificate holders do not permit restraining devices to be used while in flight due to perceived safety/liability concerns that might arise in the event of an incident or an accident. The escorting law enforcement jurisdiction and the certificate holder should have a clear understanding of this issue prior to initiation of the intended air travel.

(5) Assure that the escorted person is accompanied at all times and kept under surveillance.

(6) Be familiar with the requirements of the certificate holder.

Rule 5, that the prisoner must be accompanied at all times, extends to occasions when either the prisoner or the escort must use the rest room (or the "blue room," as it is called in the industry).

A violation of this rule which had severe consequences. During a July 1976 flight, a mental patient under escort by two male nurses was permitted to go to the rest room unaccompanied. He savagely attacked a six-year-old child he found there who had not known how to lock the door. Though the patient was then subdued and sedated, upon landing he was able to seize the gun of a police officer and shoot him. The carrier had not been advised that any passenger on the flight was under escort, let alone that a dangerous mental patient was aboard.

If a prisoner is determined to be dangerous or a "maximum risk" by the law enforcement agency, two escorts are required. Because most law enforcement agencies are under budget constraints, they are reluctant to classify any prisoner as dangerous, thereby avoiding the cost of two escorts. This could create added risks for the passengers.

Some carriers have reportedly had problems with military prisoners who refuse to open personal property containers for predeparture inspection. Because the requirements for predeparture screening are certainly no less stringent for prisoners than they are for the rest of the traveling public, the containers must be inspected or checked as baggage.

Unescorted juvenile wards of the court flown from one location to another can present problems similar to those of transporting prisoners. In these cases, though the carrier is advised, and arrangements are made for a juvenile officer to meet the passenger at the destination, and the juvenile is unescorted during the flight. At least one hijacking attempt was made in 1975 by an unescorted juvenile runaway. The authorities who put her on the plane had not advised the carrier that the juvenile had a history of mental problems, or that she was being returned to a mental institution. Surely these juveniles should be escorted, at least when they have a history of aggressive anti-social behavior. The airlines cannot supervise what are, in effect, juveniles in custody.

ALIENS

Another category of unescorted persons who could create problems is that of aliens being deported. Aliens who are being deported for criminal conduct are under escort; however, in the case of a normal deportation, the alien is placed aboard the aircraft, and the Immigration and Naturalization Service (INS) escort gives the alien's papers to the flight attendant to be given to the immigration official meeting the alien at the destination. The very few problems that have resulted from the transportation of deportees have primarily been attempted escapes.

One of the tasks assigned to the air carriers by the INS is the detention of aliens who arrived in the U.S. without proper documentation and are awaiting deportation. This may be a matter of a few hours, or in the case of an alien who appeals the deportation, it could stretch through many months while the alien awaits hearings. The INS feels it is proper to charge the airline with this responsibility, on the theory that it was the airline that allegedly failed to check for proper documentation. Sometimes, that documentation involves sophisticated counterfeit passports, which are not readily detectable. The air carriers feel, once again, that they are being placed in a law enforcement function and that they are not equipped to deal with it.

In the most common case, the guarding of an alien will involve putting him or her up in an airport hotel overnight with a contract security guard to make certain the alien does not escape. There have been some unfortunate incidents arising out of these situations. Security guards have been attacked and severely wounded, and security guards have attacked and raped the individual they were supposedly guarding.

When an alien decides to appeal the deportation, the air carrier is still responsible for 24-hour protection while he or she awaits the hearing process. Airlines have repeatedly asked Congress to assign to the government the lodging and food costs of aliens, estimated to be around $5 million a year. In July 1985, the House Appropriations Committee approved legislation directing the INS to report to Congress on its current policy of requiring airlines to pay detention costs for aliens denied entry into the U.S. pending an immigration hearing. The committee said it was concerned about the possible ramifications of detention of aliens by airline personnel or their agents who are not law enforcement officials. They also asked the INS to assess the impact of a requirement that no alien be kept in airline custody for more than 24 hours or that the INS pay detention costs when airlines have made good faith efforts to screen aliens for admissibility prior to departure.

In January 1989, the INS published a proposed rule that made the INS responsible for the physical custody of potentially excludable aliens. If an alien subsequently is ordered to be deported, carriers will be responsible for providing transportation to the last foreign port of embarkation.

RESTRICTED MATERIALS

Inspectors must also be alert for certain items that are classified as hazardous materials (HAZMAT) and that cannot be carried in the cabin or checked as baggage. The most common articles carried by passengers that come under this classification are gases, chemical agents, and lighter fluid. Canisters of propane gas (propylene) or Mace (chloroacetophenone omega) are also normally prohibited unless packaged and labeled in accordance with the Restricted Articles Tariff. This proviso is more demanding than it sounds at first reading; for example, it might cost $100 to ship a $5 can of Mace. To meet the Restricted Articles Tariff, the can would have to be packed in a special container with specialized packing around it and be conspicuously marked with hazardous materials labels appropriate to the contents.

Since hazardous materials cannot be checked as baggage, passengers are presented with a dilemma; they cannot take the item with them, but they usually don't want to lose it. Often, they will ask the carrier to store it for them until their return, to mail it to them, or simply to dispose of it for them. Each of these options may present legal ramifications. In some jurisdictions, it is a misdemeanor or a felony to carry Mace. Besides the many restrictions on the mailing of hazardous materials, there are frequent risks in storing them. Claims are also a problem. Not only have carriers received claims for items held at the request of the passenger that could not be found on the passenger's return, but passengers occasionally submit claims for reimbursement for the value of items that have been confiscated. If the airline disposes of an unshippable item from a passenger, the passenger should be made aware that he or she will not be reimbursed by the carrier. It is not the carrier's responsibility; the passenger has attempted to board with an uncarryable item, and its preservation, if it is not confiscated by law enforcement as illegal, is the responsibility of the passenger. That the passenger acted in ignorance of the regulations does not alter the case; here, as everywhere, ignorance of the law creates no exemption from it.

The carrier must know the local law with respect to possession of such items and determine whether it has a legal obligation to refer such passengers to the law enforcement officer on duty. The ACSSP requires that the law enforcement officer be notified of all weapons, explosives, incendiary devices, or dangerous objects, "as appropriate." In most cases, the best course is to tell the passengers to dispose of the item themselves or to agree to dispose of the item for them, with the understanding that they will not be reimbursed.

The FAA does not consider butane cigarette lighters or cigarette lighters containing liquefied petroleum gas to be hazardous, and these need not be confiscated from passengers unless the carrier feels that an unusual situation exists. But this exemption should apply only to lighters; butane or liquefied petroleum gas in any other containers (usually a lighter refill container) is prohibited from carriage aboard a passenger-carrying aircraft. Nor can refill cans of ordinary lighter fluid be carried in the passenger cabin, though they may be carried as baggage in quantities of a quart or less in the unlikely event that the containers are packaged and labeled in accordance with the Restricted Articles Tariff.

Passengers carrying clear plastic reservoir lighters should be warned not to use them at altitude, as the pressure differential may cause them to overflow, and numerous cases of burns due to lighter overflow have been reported. Actually, one clear reservoir lighter has been developed that will not overflow at altitude, but it is not likely that passengers will know whether theirs is the one safe model.

All types of explosives, ammunition (except small arms ammunition for personal use), incendiaries, and fireworks are also prohibited from carriage aboard passenger-carrying aircraft. They can be carried in the belly pits, but only if shipped as air freight meeting the specific requirements set out in the Restricted Articles Tariff.

When an item is taken from a passenger for carriage as checked baggage, the item is in the care, custody, and control of the carrier. The significance of this was memorably illustrated when a screened passenger seized the opportunity offered by an accessible pistol at a boarding desk, where it was being held following its confiscation from another passenger. The weapon was subsequently used in a hijacking attempt on that carrier.

Previously, a rifle or shotgun could be brought into the cabin as carry-on luggage. This practice arose from the fact that hunters prize their weapons highly--often for good monetary cause--and do not want to expose them to the handling of the baggage crews and possible damage, loss, or theft. The concern is understandable, but the practice was discriminatory; passengers who had been thoroughly screened, and perhaps had been relieved of a slingshot purchased for a grandson, were not pleased to watch another passenger board with a rifle or a shotgun.

In 1973, many carriers changed their tariffs to prohibit the carriage of sporting firearms in the cabin of an aircraft. Several carriers developed improved containers for packing rifles and shotguns for carriage as baggage. In some cases, passengers were offered the choice between a "giveaway" carton or a more rugged container that they could purchase for a few dollars.

With the restriction against carrying firearms in the cabin, requests from passengers that their weapons be carried in the cockpit of the aircraft increased to a point that pilots rebelled, and in many cases, they were successful in motivating the carrier to stop accepting weapons for carriage in the cockpit. The problem ended with the issuance of FAR 121.585(b)(d), effective July 20, 1975 (now FAR 108.11[a] & [d]), which prohibits the carriage of a checked weapon in the flight crew compartment because in a hostage situation the weapon could become available to a hijacker.

Certainly the responsible security person should know many regulations, but knowing the regulations is not always enough, since regulations sometimes appear to be in conflict. FAR 103 regulates the carriage of dangerous articles aboard civil aircraft of United States registry. Part 103.1(c)(1) specifically exempts "small arms ammunition for personal use" from these regulations, and Part 103.7 provides for shipping small arms ammunition other than for personal use. FAA guidelines for identifying dangerous weapons and dangerous objects state that all types of ammunition are to be prohibited from carriage in the cabin in the interest of air security. The regulations of one carrier provide that a maximum of 11 pounds of small arms ammunition for personal use may be checked as baggage if it is packed in the manufacturer's original box or securely packed in fiber, wood, or metal boxes to prevent movement of the cartridges. Specific regulations applicable to individual carriers should be available from the carrier at the time the reservations are made. This type of information is available to reservations sales agents through their computer.

Chapter 9

BAGGAGE, CARGO, AND EXEMPT FLIGHTS

In this chapter, the discussion of baggage and cargo handling is generally confined to their relation to protecting the aircraft against incendiary or explosive attacks. It would be folly to contend that security controls on baggage and cargo are or will be foolproof, yet the risk can be and is being substantially reduced.

FAR 108.13 (b) requires that baggage carried in an airplane be checked by a responsible agent and that identification be obtained from all persons, other than known shippers, who seek to ship goods or cargo aboard the airplane. Section (c) requires the carrier to ensure that cargo and checked baggage carried aboard the airplane are handled in a manner that prohibits unauthorized access. The ramifications of failing to accomplishing the requirements of section (c) are great.

For example the President's Commission on Aviation Security and Terrorism noted in its report that a baggage container destined for Pan Am 103 sat open and unguarded for 30 minutes at Heathrow. The Executive summary stated, "Requiring that all baggage containers be fully secured would have prevented any tampering that may have occurred with baggage left in a partially filled, unguarded baggage container that was later loaded on a flight at Heathrow."[1]

Once carriers have accepted the baggage, they must keep it under continuous control. In normal circumstances, baggage is sent down a conveyor belt to the bag make-up area where it is routed to the place where baggage for

---

[1] *Report of the President's Commission on Aviation Security and Terrorism*, May 15, 1990 (Washington, D.C., GPO): i

a specific flight is being assembled. There might be relatively long periods of time when the baggage is not under observation, and indeed, this is where some of the pilferage problems occur. Baggage carts are rarely sealed.

Regulations do call for the carrier to have procedures in effect that will restrict the access of unauthorized people to baggage once it has been checked. This is generally accomplished by locking bag make-up areas. These areas are always located in a part of the terminal that is not accessible to the public.

One of the first control measures introduced require the passenger to have a ticket in order to be able to have check a bag. Looking back, it seems almost unbelievable that an individual could drive up to the curb, give a couple of boxes to the skycap, ask him to ship them to New York, tip him, and then drive off, all without showing a ticket or paying for the shipment. Obviously, the system was abused, sometimes by passengers and sometimes by employees. The purpose of requiring a ticket, of course, is to prevent people from checking a bag on a flight they will not be on. Conversely, the mere possession of a ticket is no guarantee that the passenger will in fact board the flight. Less important than the fact that this ploy is an inexpensive way of shipping goods, "passenger-unaccompanied baggage," as it is termed, presents an opportunity to the terrorist bomber.

A related but different problem is presented by explosives in luggage boarded by a passenger who intends to die in the resulting crash.Prior to the epidemic of hijackings, there were several instances of suicidal acts aboard aircraft. The 1964 Pacific Air Lines murders was one such incident. In another, more fortunate case, a suicide set off a bomb in the plane's rear rest room during a flight from Las Vegas to Los Angeles, but the pilot was able to land safely.

Many, if not most, of the suicides appeared to have been influenced by the hope of leaving their dependents substantial amounts of flight insurance. The Air Line Pilots Association (ALPA) vigorously protested the insurance vending machines available in terminals and campaigned for greater care in the writing of flight insurance. The suicide attacks eventually stopped, however, probably because of the wide publicity given the fact that, in each case, the cause of the crash and the identity of the perpetrator were promptly established, and no monetary reward was ever achieved.

Additionally, acts of homicide have taken place in an attempt to collect on insurance policies. On November 1, 1955, a United Airlines DC-6B on a flight from Denver to Portland disintegrated in flight 11 minutes after takeoff. The explosion, which detonated in the plane's Number Four baggage compartment, was traced to the luggage of the mother of Jack Gilbert Graham, who had not only packed his mother's suitcase, but had also taken out a large flight insurance policy on her life. Graham was executed for the crime.

Curbside check-in has traditionally been a competitive marketing feature among carriers. Carriers emphasize the speed and convenience of one-stop check-in, which may include computerized ticketing, computerized seat assignments, and--most certainly--curbside baggage check-in.

A problem that arose from curbside check-in is the security of baggage tags. Although the problem obviously is not limited to the curbside check-in operation, the risk exposure appears to be greatest at that point. The baggage tags are usually in an unprotected, easily accessible rack, past which goes a fair amount of foot traffic. Besides the potential loss that fraudulent claims by means of stolen tags can represent, the baggage tag could be used to ship an unaccompanied suitcase with relative ease. Because of this, skycap carts at most airports have now been modified to add a locking cover over that portion of the cart containing the baggage tags, providing the skycap with the ability to secure the tags without requiring constant attention to them.

For international flights, it is no longer possible to check bags with the skycap. He or she will assist the passenger with the bags but will put them on a cart and wheel them in to the ticket counter for check-in, rather than allow the passenger to check them at the curb. All international bags are also X-rayed or physically inspected.

PASSENGER/BAGGAGE RECONCILIATION

Passenger/baggage reconciliation is a system for making certain that a bag is not transported without the corresponding passenger on board. As of December 19, 1987, on all international flights departing, entering or operating outside of the U.S., the air carrier must have security control of checked baggage. The carrier may use any one or a combination of the following: passenger/bag match, x-ray, physical inspection, or explosive detection. For flights from specific designated airports, these requirements are more strenuous. By the end of 1990, 100 percent passenger/bag match was required for all international flights leaving from U.S. airports.

EXPLOSIVES AND SCREENING

In recent years, the terrorist threat has shifted from hijackings to bombings. Targets have been both aircraft and terminals. Some of the bombings, of course, have not been terrorist related. For example, a bomb exploded in a self-service baggage locker in August 1974 at Los Angeles International Airport, killing three people injuring 35 others. The perpetrator, who acted alone, was later found to be legally insane. His grievance apparently had no connection with airports or airlines.

After this explosion, self-service lockers were relocated within sterile terminal areas in many cities to reduce the risk factor. At least, the items placed in the lockers would have been screened. The FAA strongly recommends that such lockers should either be removed or relocated past screening points within sterile concourse areas. These airport locker bombings have also resulted in the redesign of baggage lockers by manufacturers. The FAA's recommendations had not been implemented everywhere, however, when, on December 29, 1975, a bomb planted in a similar self-service baggage locker exploded in a baggage claim area at LaGuardia Airport in New York, killing 11 persons and injuring 57 others.

Both of these incidents occurred before the FAA announced the institution of a new requirement for the selective screening of baggage checked by passengers. Under this program, the carrier is responsible for identifying those individuals whose baggage should be screened. Passengers who are so identified, are asked to permit a search of their baggage; searched baggage will be boarded. Should a passenger refuse, the carrier could choose to hold the passenger's baggage off the flight until he or she is boarded. However, not only did this option pose severe logistic and security problems, but carriers were generally reluctant to board the baggage of passengers who refused permission for a baggage search. It is interesting to note, however that a few carriers expressed their willingness to board such passengers without their baggage. Other airlines were less willing to retain suspect bags.

Signs were posted that read, "Passengers with checked baggage are advised of recently enacted FAA regulations which may require: Inspection of checked baggage, passengers to present identification. Passengers may refuse inspection. Passengers refusing inspection may not be transported."

In addition to the selective screening of baggage, the FAA announced it was stepping up research and development efforts to determine the feasibility of (1) using the X-ray fluorescence concept as an explosives detection technique, (2) using nuclear magnetic resonance to detect different types of explosives, (3) applying explosive vapor detection techniques to the inspection of hold baggage cargo in the aircraft, and (4) adapting a thermal neutron detection concept for the examination of checked baggage in cargo. The latter was a forerunner of the thermal neutron analysis machine in use today. The goal of a current FAA program is the strategic deployment of explosives-detecting dogs at a sufficient number of U.S. airports to ensure that no aircraft in flight will be more than one hour from an airport so equipped.

A number of cases have been reported of explosives being carried by "dupes" who were unaware that they were carrying explosives. During December 1983 and January 1984, a British woman unknowingly carried a bomb concealed in her suitcase from Athens to Tel Aviv to London and back to Athens. The suitcase had 1/8 inch sheets of plastic explosive concealed inside the lining. The bomb's triggering mechanism contained an electronic timer and a barometric sensor. The bomb failed to detonate as designed and was recovered by the Greek Police.

On April 17, 1986, an Irish woman attempted to board an El Al flight at Heathrow Airport in London. She was discovered to be unwittingly carrying a functioning bomb in her handbag. The bomb's detonating mechanism, including the initiator, a small amount of plastic explosive, and timer, was contained in a fully functioning calculator. The calculator was lying on the bottom of the bag, and concealed inside the false bottom of the bag were approximately three pounds of plastic explosives.

And on June 26, 1986, a Spanish man was duped into thinking he was transporting illegal drugs in his suitcase. The bomb began to smoke while the suitcase was open and was being inspected by a member of the El Al security team at Barajas International Airport in Madrid. The man was arrested, and a Palestinian associated with the Abu Musa group was later apprehended. If the bomb had escaped detection and the timing device had functioned properly, it would have exploded two hours after takeoff.

To combat the use of dupes to carry explosives, the screening program has been expanded in many airports to include a number of diagnostic questions for the passenger. These include, "Did you pack your own bag? Has it been out of your sight since you packed it? Did anyone give you a package or a gift to carry that you did not wrap?" Several years ago, while traveling abroad, I received the following notice from my hotel:

> We should like to draw your attention to the fact that it is not advisable to receive from strangers (persons other than your family and personal friends) parcels in any form, whether as a gift or for delivery to someone abroad.
>
> These gifts and parcels which might be offered to you during your sightseeing tour or at the hotel at which you are staying could be of a dangerous nature.
>
> You are also advised to check your bags before departure from your hotel and from the country in order to be sure that no such item is included in your belongings, endangering the plane.

At the risk of redundancy, I'd like to quote a sign I saw recently at Sky Harbor International Airport in Phoenix, Arizona, which I thought was excellent.

DON'T BE FOOLED

> TAKE EXTRA PRECAUTIONS to help us insure the safety of air travel.
> KEEP CLOSE WATCH on all your carry-on packages and baggage.
> NEVER LEAVE anything unattended, even for a moment; it might be tampered with.
> NEVER CARRY an article, package or baggage for another person, especially for someone you do not know well.
> WATCH for unattended articles and packages anywhere in the airport or on the airplane, and REPORT anything unusual to security personnel or to the airline crew.

SCREENING CARGO

Care must also be taken in accepting shipments, particularly those from unknown shippers. Virtually every carrier of any size offers a service for the rapid delivery of small packages under such names as Small Package Dispatch (United), Dash (Delta Airlines Special Handling), Priority Parcel Service (American), Proud Bird Express (Continental), and Next Flight Out (TWA). Those packages are handled differently from air cargo because they are forwarded as baggage at the passenger terminal or in some cases, from the air freight office. The proliferation of facsimile machines on the scene has cut into this business to some degree.

One of the changes brought about by the Air Carrier Standard Security Program has been the development of guidelines to deter or delay shipments from unknown shippers. Generally speaking, known shippers are those who regularly offer business to a carrier, and their shipments may still be handled by the traditional "first flight out" procedure if there are no unusual circumstances connected with the shipment. Provisions exist for accepting or refusing shipments from taxi drivers and unknown shippers and forwarders. These involve such things as requiring certain kinds of identification and possibly holding such shipments for 24 hours prior to dispatch. Not unnaturally, the option of holding a suspect package for 24 hours does not appeal to most carriers. Additionally, it doesn't catch all bombs set with a timer, and it has no effect on barometric bombs that detonate at specific altitudes. At least one carrier places its cargo in an atmospheric pressure chamber and simulates the altitude of the flight to check for barometric bombs.

Besides all the other provisions, shipments from unknown shippers or forwarders may be accepted after adequate inspection, either by X-ray (if received where X-ray inspection equipment is available) or by opening the shipment and physically examining the contents.

There are, of course, two important exceptions to inspecting packages. The carrier has no right to search U.S. mail; postal authorities should be notified of any question about a package in the mail. The other exception is any case where an employee suspects that the package may contain an explosive device. Opening such a package is not recommended, and an established procedure should be followed to bring explosives experts to the location to deal with the possible problem.

SCREENING MAIL

One of the vulnerabilities to sabotage of aircraft is the carriage of mail. Pan Am 103 carried 43 bags of mail and over 20 tons of cargo. Standards for the security of mail are different from those for air cargo, which in turn are different from those for checked baggage. Yet, it all goes in the belly of the aircraft.

In fiscal 1989, the U.S. Postal Service (USPS) paid carriers over $1 billion to carry mail. That represents big business for the carriers. Yet, there are no routine procedures for screening of the mail. Carriers receive the mail from the Postal Service already sealed in bags and marked for the destination, and this is the way they are loaded on the planes.

The USPS has taken the position, for legal reasons, that X-ray or other screening of mail "sealed against inspection" cannot be undertaken by the airlines without first obtaining a search warrant, except in extraordinary circumstances. Section 3263 of Title 18, U.S. Code, requires the Postal Service to "maintain one or more classes of mail for the transmission of letters sealed against inspection." It also states that "no letter of such a class of domestic origin [those sealed against inspection] shall be opened except under authority of a search warrant authorized by law." Postal Service Regulations, Part 115.4 and 115.5 state that no person may "open, read, search or divulge the contents of mail sealed against inspection" without a warrant unless extraordinary circumstances create a reasonable suspicion to an inspecting authority that a letter or parcel could be dangerous. And finally, 18 U.S.C 1702 imposes a criminal penalty on anyone who delays the mailing of a letter or parcel "with design to obstruct the correspondence, or to pry into the business or secrets of another, or opens, secretes, embezzles, or destroys the same." Therefore, the FAA has relinquished to the USPS the responsibility for air mail security. However, the law provides that all USPS regulations pertaining to air mail must be consistent with the Federal Aviation Act of 1958, which could give the FAA more leverage in this area.

On November 15, 1979, a bomb on board an American Airlines flight exploded about 30 minutes after the flight left Chicago. The bomb was in a wooden box in a small bag inside a metal postal container. When it detonated, it blew a hole in the side of the box and started a fire in the metal container. The aircraft landed safely at Dulles International Airport, Washington, D.C. This incident prompted a Memorandum of Understanding dated December 11, 1979, between the FAA and the USPS. It required all direct and indirect carriers, including the USPS, to implement an air parcel security program with procedures to prevent, detect, and deter the introduction of any unauthorized explosive or incendiary device into air mail parcels. The USPS adhered to the position that the airlines would never be allowed to screen the mail and that the USPS could screen sealed mail only under very limited circumstances. The Memorandum of Understanding was not workable and was soon disregarded by both parties. One of the recommendations of The President's Committee on Aviation Security and Terrorism is that the USPS should effect a regulatory change redefining the category of mail "sealed against inspection" to include written materials and those parcels below a certain weight. They further recommend that the air carriers must be initially responsible for any screening of mail.

Following the initiation of Operation Desert Storm in January 1991, the Postal Service announced a plan to ship most packages on all-cargo carriers instead of passenger airliners as a safety precaution. Passenger carriers took in an estimated $300 million to $500 million in revenues from the Postal Service's package business, and the loss of this business cost them millions of dollars a week. This was not a welcome development to the airline industry, which had already suffered losses because of rising oil prices and fears of terrorism.

INDIRECT AIR CARRIERS

An indirect air carrier is a company that is in business for the purpose of accepting and shipping items on commercial airlines. They are known as "air freight forwarders." A package can be forwarded through several indirect air carriers before it reaches the airline. There is generally no practical way for the carrier to know who the original shipper was. Conversely, from the standpoint of the terrorist, it is extremely difficult to pinpoint a specific flight or even a specific air carrier when going through a freight forwarder, therefore the target will be totally random. If a terrorist does specify a particular flight, it immediately raises suspicion. Terrorists are inclined to leave their cargo or package with the specific carrier they are targeting, if that is their plan. The same logic applies to mail shipments.

FAR Part 109 was promulgated in 1979 as an emergency rule. It requires indirect air carriers to develop and file for FAA approval a security program designed to "prevent or deter the unauthorized introduction of any explosive or incendiary device into any package cargo intended for carriage by air" The purpose of the rule was to place the requirement for screening at the "point of acceptance" rather than with the airline.

One of the problems with the Part 109 program has been that the FAA does not know the identity of most of the indirect carriers. In testimony before the President's Commission on Aviation Security and Terrorism, an FAA representative stated that there were from 4000 to 6000 indirect carriers in operation and subject to Part 109 security requirements. The FAA had identified only 408 of these indirect carriers.[2] While some benefits accrue from this program, it would appear to be relatively ineffective, and it does not appear to be the final answer.

DRUG SMUGGLING

On September 26, 1986, twenty-two Eastern Air Lines cargo and baggage handlers were arrested in Miami and indicted on drug smuggling charges. Eastern fired 50 other employees as an aftermath of the arrests. Investigators charged that the ring had smuggled tons of cocaine past Customs over the previous three years. Federal authorities said they believed that Eastern airport crews were paid as much as $250,000 to clear each cocaine shipment through the airport. The temptation of large sums of fast, easy money makes the drug problem difficult to fight. An accomplice in Eastern's communication control room directed planes to be unloaded by handpicked crews. When the drug-laden planes arrived, the baggage handlers removed all bags except those containing drugs, which were left in the cargo bays. After the plane was moved from the international zone to load passengers for domestic flights, the baggage containing drugs was removed from the plane.

Also in September 1986, the security chief of Avianca, Colombia's largest privately owned airline, was murdered by drug traffickers two days after he had seized $5 million worth of cocaine that had been smuggled aboard a Miami-bound cargo jet. Avianca had tried for two years to install a sophisticated X-ray machine at its facilities in Bogota to detect smuggled packages of cocaine. As of 1986, the system was still not operational because of threats and intimidation of employees and the sabotage of the equipment.

In March 1987, federal agents and New York Port Authority Police arrested 40 people, including 16 current or former Pan Am employees, two Delta Airlines workers, and eight Eastern Air Lines workers. They were accused of smuggling almost $1.5 billion worth of cocaine into JFK Airport aboard Pan Am flights from Brazil. The head of Pan Am Security in Brazil is reported to have assisted another Pan Am employee who was the courier, in getting the cocaine-laden suitcases through Brazilian Customs. The ring transferred much of the cocaine to domestic flights and distributed it in New York,

---

[2]Ibid., p. 52.

San Francisco, Los Angeles, Las Vegas, Denver, Boston, Chicago, St. Louis, Atlantic City, Miami, and Montreal.

Eleven percent of all the cocaine smuggled into the United States arrives by commercial air. The Omnibus Drug Act of 1986 has forced many carriers to work with the government in attempting to slow the flow of drug traffic. The act got the attention of the carriers by getting to their pocketbooks. It increased fines from $50 to $1,000 per ounce ($35,000 per kilo) for cocaine and from $25 to $500 per ounce for marijuana. It also required that carriers exercise the "highest degree of diligence" to prevent their aircraft from being used for drug smuggling.

As an example of the impact of this regulation, Tampa Airlines, a small Colombian-owned freighter with only four aircraft, has invested $1 million in upgrading its security force and purchasing computerized scales, X-ray equipment, and CCTV equipment. During the same time, fines totaling $4.5 million have been imposed on the airline, with $3.7 million of that sum reduced to $370,000 by negotiation. The remainder is being appealed. The airline, which is headquartered in Medillin, Colombia, probably has a more sophisticated detection system than most of the other carriers in the world. The argument of the airline is that in Colombia the rule is, "Cooperate or die." Tampa argues that every man has a price, whether it be money or his life. The traffickers will threaten to blow away your kids, and do it, if you don't cooperate. How can a company be blamed for what one man does under such pressure? The carriers believe they are being made to act as a police organization, but without authority. They are caught between the drug producers and the end users. The phrase, "highest degree of diligence" has been of particular concern to the airline industry. Industry members feel that the government is passing on its responsibility to the private sector. Customs holds the carrier responsible for drug smuggling whether it was involved or not.

It has been suggested that the strict liability standard of Section 1584 of the Omnibus Drug Act of 1986 is invalid because it conflicts with provisions of the Warsaw Convention of 1929.[3] The convention makes it clear that it is the responsibility of the consignor of the goods to the airlines, and not the airlines, to account for the nature of the goods that will be shipped. It is argued that placing the burden for erroneous representations of the air waybills on the carriers rather than the consignors might also be contrary to (1) the standard IATA air waybill (which directly refers to and relies upon the Warsaw Convention), (2) IATA Resolution 600(b)(2) of IATA's Manual of Traffic Conference Resolutions, (3) Annex 18 to the Convention of International Civil Aviation of 1947 (commonly called the "Chicago Convention"), (4) the International Civil Aviation Organization's Technical instructions regarding hazardous materials, and (5) IATA's Dangerous Goods Regulations. All of these appear to recognize that in cases where controlled substances such as cocaine are smuggled in the cargo, it is it is the consignor of the goods who is responsible for the declarations in the air waybills.

---

[3] Howard G. Slavit, , "Airlines Battle Cocaine Traffickers...Are the Laws Stacked against Them?" *Airline Executive*, (March 1985): 32.

CHARTER FLIGHT SECURITY

Due to various factors, security considerations for charter flights are different from those of scheduled commercial flights. Charter groups are usually all members of a single organization or are made up of individuals generally known to one another.

There are two types of charter flights: public and private. Screening is required for public charters but not for private charters. "Public charter" is defined in FAR 108.3 as "any charter that is not a private charter." The question then hinges on the definition of a "private charter." FAR 108.3(c) states:

> "Private charter" means any charter for which the charterer engages the total capacity of an airplane for the carriage of:
>
> (1) Passengers in civil or military air movements conducted under contract with the government of the United States or the Government of a foreign country; or
>
> (2) Passengers invited by the charterer, the cost of which is borne entirely by the charterer and not directly or indirectly by the individual passengers.

Thus, if an individual buys a ticket for a chartered flight to Hawaii from Happy Hawaii Vacations, the flight is a public charter, even though all the passengers are customers of Happy Hawaii Vacations. However, if XYZ Widget Corporation charters a flight to Hawaii for a sales incentive reward trip for its employees and their families, it is a private charter as long as the entire plane is reserved and paid for by the company. If the employees of the XYZ corporation charter the same plane but pay their own way, it is a public charter. When the San Francisco Forty Niners fly back to San Francisco after winning the Super Bowl, it is a private charter.

Modifications made in charter flight regulations, such as the group inclusive tour (GIT) charters and the one-stop inclusive tour charter (OTC), have removed the group affinity rule, but retained the requirement for prepayment of fare as much as 90 days in advance. The lead time required to organize and arrange a charter is also regarded as a factor discouraging to the aspiring hijacker.

It is because of affinity and the requirements for advance arrangements that charter flights are generally exempt from predeparture screening, unless they are boarded from a sterile concourse. Another instance in which charter passengers are screened is when there are passengers on the plane other than the charter passengers; this is generally true of GIT charters. If the group charters the entire aircraft, passengers need not be screened provided they enplane from an unsecured (nonsterile) area of the terminal.

At the destination station, passengers who were screened at origin can deplane into a sterile concourse. If the passengers were not screened at the origin station, they may not deplane into a sterile concourse unless they are escorted through the concourse by employees of the carrier or its authorized agents. FAA Advisory Circular 108-1, dated October 16, 1981, suggests that the ratio should be one air carrier escort for each 10 passengers.

If passengers are to be escorted through a sterile concourse upon landing, they should be advised of this over the aircraft public address system while taxiing in. Stops at the bar or a rest room cannot be permitted unless the carrier can provide a one-to-one escort. If passengers must patronize an establishment within the sterile concourse, they should be escorted beyond the checkpoint, and re-enter the sterile concourse after being screened.

MAINTAINING THE SECURITY LEVEL

Consistent follow-up provides vital management control for any delegation of responsibility and thus is a necessary element in any company security program. Besides the constant supervision given by the operational people, the program needs occasional inspection by the security department to ensure compliance with established procedures, the carrier's security program, and government directives.

The following is a set of guidelines which can be used as a basis for the survey or inspection of a carrier's aircraft security program.

I General

- Size, location of airport, hours of operation, concourses and gates used or shared, number of departures and arrivals.
- Type of metal detectors in use and quantity?
- Type of X-ray in use.
- Badge/I.D./access card system in use.

II Screening Procedures

- Are there any problem areas? (Talk to people doing the screening and to law enforcement officers on duty; they know the problems and, sometimes, the solutions.)
- Are carry-on items searched in the presence of the passenger or visitor?
- Have there been reports of thefts during the screening process?
- What procedures are used for people who alarm the walk-through metal detector?
- Have problems arisen with the sensitivity or false-alarm rate of the metal detector?
- Is the hand scanner used properly?
- How and where are pat-downs conducted?
- Are both passengers and visitors allowed on the sterile concourse?
- With what frequency are unidentified objects on the X-ray identified through physical search?
- Are there any technical or passenger flow problems with the X-ray?
- Would rearranging the metal detector or X-ray be beneficial?

III Personnel Conducting Screening

- Is screening done by a contract agency? Which one?
- Has there been a frequent turnover of contract screening companies? If so, why? (poor service or going with the lowest bid)
- What is the turnover rate of contract employees?
- What is your observation of their performance in screening operations, including thoroughness and courtesy?
- Describe the quality of supervision.
- What training or retraining is given? Type and duration? Who conducts it?
- Are training records kept (FAA requirement)?
- Do employees wear dosimeters?
- Results of any tests or attempts to get by screening with simulated weapon.

IV Law Enforcement Support

- Agency involved? What personnel are supplied? Is a policewoman available?
- What cooperation is received from the agency? Under what conditions? What is their response time?
- Are there any liaison problems with other law enforcement agencies at the airport (such as U.S. Customs, DEA, FBI, Department of Agriculture, etc.)?

V Access Control

- Describe the access control system in use.
- Review records for procedural controls regarding badge issuance, cancellation, lost or stolen badges, etc.
- Check the overall concourse and ramp access security.
- Check the ease of access to the ramp and to the sterile concourse from the ramp.
- Were you challenged at the checkpoint? Challenged without badge in secure areas?

VI Aircraft Security

- Investigate the security of attended and unattended aircraft.
- Is there overnight storage of aircraft? Number and type. What protection is provided?
- Prior incidents of vandalism?
- Is adequate lighting available and are planes parked in lighted areas?
- Are ventral stairs retracted and jetway doors locked?
- Are there special threat situations requiring greater protection?

VII Baggage, Cargo, and Small Package Security

- Is the emphasis on protection against planted explosives?
- Are bags in make-up area secure against penetration attempts?
- Is a procedure for passenger/bag match in place? Is it workable?
- Are questionable packages X-rayed or inspected?
- Do ticket agents know the profile for selective baggage screening?
- What security procedures are in place for accepting small package dispatch items?
- Are identification procedures being enforced with regard to known versus unknown shippers at the cargo terminal?

VIII Bomb Threat, Sabotage, and Hijack Procedures

- Is a command post set up?
- What is the availability of the bomb squad? What agency responds? Army EOD? Police Department? What is their response time?
- Is mobile X-ray available?
- Are dogs available?

IX Station Management Liaison

- Comment briefly on contacts with management.
- Record the results of any presentation to management.

X Liaison Contacts

- Identify the people talked with by name and title.
- Briefly state any significant comments.
- This is an excellent opportunity for liaison contacts with local FAA, FBI, police sheriff, and airport management.

XI Evaluation

- In a short paragraph, state the conclusions reached and the evaluation of security as it pertains to antihijacking program.

XII Recommendations

- Briefly summarize recommendations made in the body of the report.

XIII Attachments

- Drawings, maps, and sketches of the terminal and airport with runways laid out and sketches of the screening setup can be helpful as resource material, particularly in case of a bomb threat or hijacking. These materials should not be big charts but something that can be filed with the report.

- Photographs of facilities such as the passenger terminal, cargo terminal, vehicle gates, ramp, fences, fuel farm, and screening setup might be useful in making a point or refreshing your memory, particularly where the facility is some distance away from your home base. These photos can be useful in writing the report and, in some cases, can be used in the report.

Chapter 10

BOMB THREAT RESPONSE

According to the State Department, bombings accounted for 63 percent of all international terrorist incidents during 1990.[1] In the United States in 1988, property damage resulting from bombings was $2.3 million, down from $4.2 million the previous year. During 1988, one explosive bombing, at a utility apparatus, accounted for $1 million in damage, making that category the highest ($1,007,250) in total damages. Damages to homes and other residential properties were valued at $193,162; business and other commercial operations suffered $617,430 in damages; vehicles, $222,135; public safety and governmental targets (including schools and U.S. Postal Service facilities such as collection boxes), $66,158. These five targets account for approximately 94% of all the damages reported.[2] (See Table 10-1.)

HANDLING THE BOMB THREAT

Threatening destruction or damage to an aircraft and conveying false information about an actual or threatened sabotage attempt against an aircraft, are federal offenses.[3] FAR 108.19 deals with bomb or air piracy threats. It states

---

[1] Patterns of Global Terrorism: 1990, April 1991, U.S. Department of State (Washington, D.C., GPO): 37.

[2] Patterns of Global Terrorism: 1989, April 1990, U.S. Department of State (Washington, D.C., GPO): 5

[3] 18 U.S.C. 32, Destruction of aircraft or aircraft facilities; 18 U.S.C. 1951, Interference with commerce by threats of violence.

that upon receipt of a bomb threat against a specific airplane, the carrier is required to conduct a security inspection on the ground before the next flight or, if the airplane is in flight, immediately after its next landing. If the airplane is in flight, the Captain must be notified of the threat of all pertinent information regarding the threat so that necessary emergency action can be taken.

When a bomb threat or sabotage threat is received, the decision as to the action to be taken requires evaluating the threat in light of all available information to determine whether it is to be handled as a hoax or taken as a serious threat. Criteria for evaluating such threats are set out in the ACSSP, are not discussed here for obvious reasons.

Each carrier should designate a specific individual or the person occupying a specific post as the airline representative whose responsibility it will be to make such a determination. The decision should not be left to the individual receiving the call, unless he or she has the necessary background information and experience to evaluate the threat. When the text of all such calls is filtered into the one designated position, that person will gain the considerable experience essential to an accurate evaluation of the threat's probability. Generally, the decision will be made in the operations section of the airline, with the advice and counsel of the security division.

The person who makes the decision will need to know the exact text of the call, the nature of the threat, the amount of detail furnished by the caller, flight number (if given), the time or altitude of detonation claimed if a bomb is involved, extortion demands, background noises, vocal characteristics, sex and approximate age of the caller, the evaluation of the threat made by the person receiving the call, and any other information that might be of value in assessing the seriousness of the threat.

Attempts to trace calls are rarely fruitful. Bomb threat calls are typically short, lasting only seconds, while a trace can take as long as 45 minutes to an hour. Further, noises on the phone arising from the trace attempt would frighten away the rare caller who has remained on the line. The telephone company can be of assistance to you in determining your policy on this point; however, there is probably no utility in including a trace attempt as part of your bomb threat response procedures.

Where the threat is made in person, the evaluation becomes a much easier task. In such cases, the individual is available for further questioning to obtain additional details helpful in resolving whether what was heard as a threat was actually an innocent or ill-advised remark or a delusion or whether a danger actually exists.

Because of the frequency of threat calls, each carrier should have a standard procedure for their handling. Essential to this is an up-to-date list of persons to be notified of a bomb threat; the method of notification should also be specified--telephone, Teletype™, or radio. One of the first persons notified must be the Captain of the flight. In the case of serious threats, the FAA and the FBI should also be notified immediately. In some cases, it will be appropriate to notify the airport's crash, fire, and rescue facility. The Captain should be given the facts of the threat and advised of the evaluation that has been made. If the threat meets the specific criteria established by the FAA, a search will be required, and in such cases, no further evaluation or consideration is taken.

Table 10-1 Bombing Incidents, 1978--1988

Year	*Total*	*Actual Explosive*	*Actual Incendiary*	*Explosive Attempts*	*Incendiary Attempts*	*Property Damage (in thousands)*	*Persons Injured*	*Deaths*
1978	1,301	768	349	105	79	9,161	135	18
1979	1,220	728	305	104	83	9,273	173	22
1980	1,249	742	336	99	72	12,562	160	34
1981	1,142	637	315	92	98	67,082	133	30
1982	795	485	194	77	39	7,203	99	16
1983	687	442	127	77	41	6,343	100	12
1984	803	518	127	118	40	5,619	112	6
1985	847	575	102	113	57	6,352	144	28
1986	858	580	129	101	48	3,405	185	14
1987	848	600	104	102	42	4,201	107	21
1988	977*	593	156	161	40	2,257	145	20
Total	10,727	6,668	2,244	1,149	639	133,458	1493	221

*Includes 27 incidents involving combination devices.

Source: Federal Bureau of Investigation,
Bomb Summary, 1988.

Whether the flight is on the ground or in the air, the final decision and ultimate responsibility for deplaning passengers and crew rests with the Captain, subject to such regulations regarding bomb threats as may be in effect at the particular airport. Every effort should be made to assist the Captain of an airborne flight in determining where to land for the search.

BOMB THREAT SEARCH PROCEDURES

Deplaning Passengers

In determining the timing and manner of deplaning passengers, the seriousness of the threat and the safety of passengers and crew members must be considered. On occasion, the use of emergency evacuation slides has been necessary, but these should not be used unless there are no equally safe alternatives. It is not unusual for nonfatal injuries to result from the use of these slides, and while there should be no hesitancy in using them in an actual emergency, they should not be deployed unless their use is appropriate to the risk evaluation.

Normally, there will be time to deplane passengers through a jetway or by means of mobile stairs. The FAA recommends that passengers should be moved at least 100 yards away from the threatened aircraft and from the flight's baggage and cargo.

The decision whether to tell the passengers the true reason they are returning to the airport, or landing somewhere other than their destination, should be a matter of carrier policy, but flight attendants must be informed so that they will be prepared to expedite passenger deplaning.

Where the airport authority permits a threatened aircraft to deplane passengers at the terminal, many carriers prefer to do so for reasons of speed and passenger convenience. Those terminals that do not permit this usually designate a specific airport area to which the aircraft is to taxi for deplaning. One major airport, which requires a plane on the ground to taxi away to an unobserved area of the airport for deplaning, states that this procedure reduces the number of hoax calls, apparently because the caller cannot see the airplane at the terminal and assumes the plane has taken off despite the threat.

Even if the airport does allow passengers to deplane at the terminal, the threatened aircraft should be moved to a more remote area of the field for the bomb search. Passengers should be requested to take their carry-on articles with them. As the passengers deplane, flight attendants should inspect under seats, in overhead bins, and in the blue rooms to make certain no packages have been left on board; any packages found must be regarded as suspect.

Search of Baggage and Freight

In most cases, it is necessary to bring the passengers to the point to which checked baggage has been taken so that passengers can be matched with their luggage before the luggage is searched. Some carriers only match passengers and baggage; in the absence of unusual circumstances, they do not search the bags. The purpose of the matching is to determine if there are any bags not claimed by a passenger and to have the passenger present when the luggage is searched. Any unclaimed bags will, of course, be treated as if they did contain explosives. Mobile X-ray units for bomb-threat baggage inspection are available at several locations throughout the United States. Where such X-ray units are available, the need to match passengers with baggage is not as critical, because all baggage will be X-rayed. An advantage of the X-ray inspection is that it allows passengers to remain in the terminal in comfort, protected from the weather, while their bags are being X-rayed. A mobile X-ray unit can screen 200 passengers' luggage in about an hour; a 747 can normally be searched in a little more than two hours. However, if the bags are to be X-rayed, and if the X-ray being used is not film safe, passengers should be notified so they can remove their film. Most units in use in the U.S. are film safe, however.

A scenario to be avoided in the event of a passenger/bag match following a bomb threat is to herd all the passengers into a room with their baggage so they can select which bags are theirs. If there really is a bomb, the entire roomful of passengers will be at risk. It is better to send them to identify their luggage one at a time.

Freight must also be removed and inspected. Mail should be taken to the airport post office or air mail facility. The appropriate postal officials there must be notified that the mail comes from a flight identified with a bomb threat and that the mail shipments should receive particular scrutiny. The FAA suggests that cargo and mail should be kept at least 100 yards distant from the passengers and the aircraft.

The carrier has the right under its tariffs to inspect baggage and should refuse to carry any baggage for which the passenger will not allow inspection. Any piece of baggage suspect because it is unclaimed or because a passenger has raised objections to or has refused to allow its inspection should be turned over to the military or law enforcement agency responsible for conducting examinations of suspected bombs. Seasoned business travelers may know that the flight is going to be delayed a number of hours and will demand their bags back immediately so that they can take another flight, or the passenger may not want to fly on an aircraft that has received a bomb threat even though the plane has been searched, and will want his bags returned in order to take another flight. Ordinarily, these bags are suspect and should not be released to the passenger without having been searched. If however, only a passenger/bag match is being conducted, they can be released in the same manners as bags to other passengers.

Baggage may be X-rayed by a mobile unit if one is available, but suspect luggage should not be brought into the terminal for inspection by a carry-on baggage X-ray inspection unit because the risk to passengers in the terminal is too great.

Inspection of passengers' luggage may be conducted by air carrier personnel, but if a bag becomes suspect for any reason, it should be left alone and turned over to law enforcement or military demolition experts. Every carrier should be aware of which law enforcement or military organizations are available to respond to calls for explosive ordnance disposal (EOD) assistance at each airport the airline uses, and it should establish liaison activities with that agency prior to an emergency.

Search of Aircraft

The search of an aircraft is best performed by aircraft maintenance people. Their knowledge and experience equip them for the task, and they should be given primary responsibility, though passenger service personnel can assist in the search of the cabin's seats, tray tables, overhead racks, seat pockets, and so forth.

Every carrier should establish detailed procedures for the search of each type of aircraft in its fleet. The more manpower available, the quicker the plane can be searched, but the search must be organized and specific areas of responsibility assigned. To search a B-747 takes about seven man-hours, a B-737 requires about three, a DC-10 five, and a DC-8 about four man-hours.

Before releasing the aircraft to maintenance for a search, waste containers should be searched and lavatory sewage tanks drained and flushed. Although the list of places to inspect on an aircraft is necessarily long and detailed, the same general rules apply to the search of an aircraft as apply to the search of a building; the searchers are principally looking for something that does not belong where it was found and that does not fit its surroundings. Nevertheless, two actual cases have indicated that the object need not be unusual or out of place. Of the two TWA aircraft threatened in March 1972, dogs alerted to explosives in a briefcase in the cockpit of one. It had been ignored by searchers because it had a crew member tag on it. Investigation of the wreckage of the second plane, which blew up on the ground in Las Vegas, indicated that the explosives had been placed in a first aid kit that had been resealed.

When a suspect item is found, it should not be moved until it has been inspected by local authorities. Once the search is completed, the searched baggage and cargo may be reloaded. Security should be maintained on the aircraft to ensure its sterile condition until takeoff.

USE OF SEARCH DOGS

The FAA sponsored a program funded by a discretionary grant from the Law Enforcement Assistance Administration (LEAA), Department of Justice, that provided a 21-week course for the training of explosives-detecting dogs and their handlers at Lackland Air Force Base, San Antonio, Texas. The first class of six law enforcement handlers and their German shepherds was graduated on March 16, 1973. The officers were from police departments in Los Angeles, Cleveland, and Houston. As of December 31, 1988, FAA-sponsored detection dogs and their handlers are now available for bomb-threat search of aircraft at 31 locations.[4]

The dogs are very good at their given task of detecting explosives. For the first 12 weeks of their course, the dogs go through obedience training and the demanding training required for law enforcement patrol dogs. During the last nine weeks of training, which are devoted to explosives detection, the dogs are trained in techniques similar to those employed in detecting marijuana and other drugs.

The dogs are so valuable an asset to the search of an aircraft and baggage that they should be used whenever they are available. A number of factors, however, can significantly limit the effectiveness of the dog and the handler in some cases. The presence of other dogs, particularly female dogs, and any other animals who may have been in the baggage compartment of the aircraft, will affect their ability. In the search of baggage by dogs at one station, the dog made a positive identification of one bag three times. The owner of the bag, who opened it for a physical search, explained that she occasionally carried her cat in the bag. Humidity and wind may affect the dog's capabilities. If dogs are used, the baggage search should be conducted in an area protected from the wind. Other factors that affect dogs cannot be discussed here for security reasons.

The length of time the dog can work effectively is probably the greatest limiting factor. This is usually about 40 to 50 bags, or 20 to 25 minutes. After that, the dog reaches a saturation point and is "all sniffed out" and must be allowed to rest. In such a situation, it is sometimes possible to alternate dogs, with breaks of about ten minutes between work cycles.

At one demonstration I attended, the dogs were quite effective in light of the items they were required to detect. Very small quantities of explosives were placed in sealed bottles or jars inside suitcases, but the dogs still managed to sniff out almost all these items in repeated tests. It was evident that a dog would probably sense a bomb containing several sticks of dynamite or plastic explosives immediately upon entering the room.

The dogs are particularly effective in inspecting the interior of aircraft, and they can do it much faster than their human counterparts. Here again, their capacity is limited, and a single dog may not make it through a wide-body aircraft before reaching the end of its efficiency. Nonetheless, the dogs are effective; in 340 tests, the dogs found 340 explosives. They did not miss one.

I observed one test where an individual walked onto the aircraft and placed some C-4 plastic explosive in the blue room at the rear of the airplane and then left the aircraft. The dog was then let on board to search. When the dog alerted about half way down the aisle, those present thought he had failed the test. It was not until later that we found out that the individual who had

---

[4] Including Atlanta, Baltimore, Birmingham, Bismarck, Boston, Buffalo, Colorado Springs, Dallas, Detroit, Houston, Jacksonville, Kansas City, Los Angeles, Memphis, Miami, Milwaukee, New Orleans, Phoenix, Pittsburgh, Portland (OR), Salt Lake City, San Diego, San Francisco, San Juan, Seattle, St. Louis, Tucson, and Tulsa.

handled and planted the explosive had placed his hand on the back of the seat at the location where the dog alerted.

If dogs are needed for a search, the local handlers should be among the first notified when the threat is received because these dogs are used for other duties in some cities and may not be kenneled at the airport. Sufficient notice must be given the handlers to allow the team to reach the airport in good time for the search.

Dogs may not be the last word in sniffing out explosives or drugs. It was reported on a news broadcast in June 1991, that a pot bellied pig named Rookie was being trained in the State of Washington to detect drugs. Pigs have been used for hundred of years in Europe to sniff out truffles and have a good sense of smell.

Section III

AIRPORT SECURITY

Chapter 11

PERIMETER SECURITY AND LIGHTING

The first step in security planning is to identify the areas that must be protected so that each can be given the attention it requires. For each area so identified, protection starts with establishing an appropriate degree and kind of access control.

POSTING

At a minimum, sensitive areas--and particularly air operations areas--should be posted with signs announcing that the area is restricted to authorized persons. Although such signs will not stop the determined invader, they define the restricted area and turn back those who might stray into the area idly or accidentally. To be effective, the signs must be large enough and repeated frequently enough to ensure that they will be seen by anyone entering or within the restricted area. The FAA recommends as a minimum that these signs be posted at intervals of no more that one every 100 feet.

FENCING

Where chain-link fencing is used as a perimeter barrier, the fencing should be at least eight feet high and topped with a three-strand barbed wire "outrigger" or top guard extending outward at a 45° angle. (Oddly enough, a considerable number of airport fences have this top guard pointing inward, as if the intent was to prevent escapes, rather than intrusions. Indeed, some local ordinances even forbid top guards from extending outward.)

Although the FAA will accept fencing of 11-gauge wire, the chain link should be nine-gauge or stronger with a two-inch mesh or smaller. The barbed wire strands of the top guard should be a minimum of six inches apart. Where the risk of access is particularly critical, such as around a fuel farm, double outriggers--one angled outward and the other angled inward--should be used. Razor wire should be considered in some geographical areas.

The bottom of the fencing should be within two inches of the ground. However, where the soil is loose or sandy, the fencing should extend two inches into the ground around critical areas. Alternative solutions to the tunneling problem are a six-inch cement apron under the fence line and--if a further precaution is appropriate--embedding the fence fabric into the cement. Frame piping used at the top of the chain-link can also be installed along the bottom of the fence. These precautions are not ordinarily required for the entire perimeter of the airport, but may be justified or even essential around critical installations, particularly if these are located in isolated areas.

Usually, the top guard is made up of three strands of barbed wire at six-inch intervals, for a total outrigger length of 18 inches. However, because the top guard is at a 45° angle to the fence, it only increases fence height by one foot. Understandably, a top guard should never be installed on fences less than six feet tall because of the risk to persons walking near the fence.

The standards pertaining to fence height can vary as to the airport and location of the airport. The FAA can provide guidance here. In some locations, greater protection may be required, and in others, less protection may be appropriate. Because hijackings can occur at any airport, whatever its size, air carriers prefer to see an eight-foot fence with a top guard around every airport perimeter. The FAA guidelines have in the past implied the smaller the airport, the shorter the fence required. A four-or five-foot fence can be scaled with relative ease and cannot possibly offer meaningful protection against intrusion at any airport.

Whatever the fence height, the points through which the public can enter must be protected by more than signs. Two incidents at one airport where FAA standards had been met illustrate the access problem. In one instance, an intoxicated woman drove off the access road onto a ramp where three airplanes were parked. When challenged by a ground employee, the driver explained that she thought she was at a motel. Two weeks later, ticketed passengers late for a flight persuaded their cab driver to drive onto the ramp and let them out at the steps of the aircraft. Where access to a ramp area can be gained easily, the airport is far from secure. These incidents occurred early in the antihijacking program, and security has improved considerably since then.

A barrier such as fencing can be overcome by anyone determined to do so. The purpose of fencing is to deter penetration attempts, to force intruders into the open so that they can be seen climbing, cutting, or tunneling, or--at the least--to cause intruders to leave physical evidence that an attempt has occured.

The location of the fence is also important. Adjacent to an airport in the Midwest that was served by a regional carrier was a "chip and putt" public golf course. Although the airport was fenced, the golf course was *inside* the fence; anyone who paid the course fee had access to airport runways and taxiways. In a critical foreign airport, a private golf course is located between two parallel runways.

For fencing to be as effective as possible, a clear space of at least 20 feet should be kept on each side. On the airfield side, this can often be provided by a frequent mowing of the grass near the fence. Nearer the terminal, it may mean restricting the parking of cars and prohibiting the stacking of items at or near the fence. The protection offered by a fence is nullified when materials, cartons, or crates are stacked on either side of the fence so as to enable intruders to climb over the fence.

The number of gates in the perimeter fencing should be limited to the minimum number necessary for efficient operations. Gates can be guarded by a security officer, or they can be opened or closed electronically, by either key card actuation at the gate or remote control via a switch at the central guard post from which the gate is observed by CCTV. The most elementary method of securing a gate is to padlock it and issue keys to those authorized for access at that gate. Alternatively, where several people or departments need access, each can apply his or her own padlock and make it a part of the "chain." Then, any one of these people can unlock their own padlock and gain access. In practice, keys and padlocks are an insecure solution; gates are often left unlocked.

The unmanned electronic gate relocks automatically, but the security offered depends upon the users making sure that unauthorized individuals or vehicles do not follow them through the gate before it has time to close and lock. An additional limitation of the unmanned, card-operated electronic gate is that it closes within a fixed interval of time. If the gate is timed for use by trains of mail or cargo carts, this interval may be too long for security after the passage of a single vehicle. Gate hinges should be so constructed (or adapted by spot welding) that the hinge pins cannot be removed to permit surreptitious entry through the hinged side of a locked gate.

The airport perimeter may incorporate buildings, masonry walls, or natural boundaries such as lakes or rivers. Depending upon the degree of security appropriate, the perimeter should be fenced at natural boundaries if access to the field might be gained by boat or other means. Culvert or similar openings should generally be secured by a barrier that stops entry without impeding the purpose of the opening, usually drainage. Any opening greater than 96 square inches is considered large enough for a person to pass through. Openings larger than this should be secured by such means as hardened steel bars of three-eights of an inch to a half inch in diameter, spaced four inches apart; 9-gauge chain link fencing; or expanded metal. Fastened securely to the opening, these should successfully keep intruders out, but the intrusion barrier chosen must not collect debris and create drainage problems.

Concertina razor wire or razor ribbon is utilized in locations where maximum security is required. It can be laid on the ground in several rows in front of the fence, and it can serve as a top guard for the fence. Possible liability issues exist with this type of fence. In August 1988, a homeless man bled to death after he slashed his arms trying to scale a fence rimmed with razor wire, at a Transit Authority storage yard in Brooklyn. He was trying to get out of the yard, which was encircled by a double set of chain-link fences, each topped with razor wire.

EXTERIOR ALARM SENSORS

Fence and perimeter alarm systems ranging from adequate to excellent are available, but the cost must be balanced against the threat. A fence alone is at best a minor deterrent to trespassers. If it can't be watched or monitored in some way, its value is minimal. A fence with lighting is far more effective than one without, and the addition of fence or perimeter alarms provides just that much more protection. Using a combination of two or more sensors significantly increases the probability of detection (Pd). Then, there is CCTV or concertina razor ribbon. At some point, the line must be drawn as to how much protection is necessary, based on some type of a rational decision-making process.

This is an area where one really needs professional advice. A number of factors such as weather, terrain, distance and area to be protected, physical interference, electromagnetic interferences (microwave transmitters, radars, power transmission lines) must all be considered. For those who want to consider alarms, however, the following are some of the types that can be utilized effectively:

> *Electric Field.* Known as E-Field and other brand names. System consists of a field generator, that excites a long field wire and a sense wire, which run parallel to the electric field wire. An intruder who penetrates within a few feet of the electric-field fence, momentarily alters the level and characteristic of the signal.
>
> *Buried Line.* Buried-line intrusion detectors (BLID) detect small variations in stress exerted on the surrounding soil by anyone walking over or near the sensor. The sensor can be a cable, a geophone, or a pair of pressurized tubes. These detectors are covert, follow the terrain, and have a low false alarm rate. These sensors, also referred to as "seismic sensors," have been used effectively by the U.S. Border Patrol to detect illegal immigrants crossing the border at night.
>
> *Ported Coax.* Also known as the "leaky cable intrusion detection system." This detector senses changes in the electromagnetic field surrounding two or more leaky coaxial cables. One cable is the transmitting device, and the other cable is the receiving device. They are buried about a foot deep and six feet apart. Changes in the field caused by intrusion attempts are processed as alarms.

Microwave. Microwave detectors are useful where line of sight and level ground--clear of trees, tall grass, and bushes--are available. They are generally unaffected by environmental factors. The device consists of one or more transmitters and receivers, a power supply, alarm relays, and tamper switches. In the bistatic type, a transmitter is aimed at a distant receiver. The electronic field is roughly cigar shaped and therefore most effective midway between the transmitter and the receiver. Blind spots may exist near the transmitter and receiver; their presence should be compensated for by overlapping transmitters and receivers. In the monostatic type, the transmitter and receiver are in one location, and the Doppler shift is used as the detection mechanism. The monostatic detector is generally not used for long-range perimeter applications.

Fence Disturbance. Vibration detectors are mounted on fence posts at twenty-to thirty-foot intervals. They will detect attempts to cut the fence or to climb it. The three basic types are mechanical, electromechanical, and strain-sensitive cables.

Taut Wire. The sensor mechanism consists of a series of closely spaced horizontal wires which are pulled taut. Barbed wire is used more often than not. Any attempt to cut the wires, climb on them, or spread them apart to crawl through the fence will trigger an alarm.

CCTV. With adequate lighting or even low-light level cameras, a fence can be monitored from a central security post by means of closed circuit television (CCTV). Cameras or monitors can be equipped with motion detectors so that if there is activity where none is anticipated, a visible and/or audible alarm can alert the guard, switch to a specific monitor, record the activity, and allow appropriate action to be taken, such as dispatching a guard to investigate. A clear zone should be maintained on both sides of the fence if CCTV will be used. I have seen it used very effectively to monitor car clouting[1] and car theft in an airline parking lot. Several arrests were attributable to the monitoring with CCTV.

CCD. The charged coupled device (CCD) is an improvement over the CCTV camera. It is sometimes known as the "chip" camera, and has a small, photo-sensitive, solid-state unit that has been developed to replace the imager or tube in a closed circuit camera.[2] Some of the advantages of the chip camera appear to be color capability, life expectancy, low heat output,

---

1 Breaking into cars, generally to steal radios or items left in the car.

2 Charles Pierce, "CCTV Technology" *Access Control*: V34:5, (April 1991):1.

lack of "blooming" problems, and higher density of horizontal lines of resolution.

Active Infrared. This is the photoelectric system with which most are familiar. Alarms are triggered by the breaking of an infrared beam of light sent from a transmitter to a receiver. This system has an effective range of about 500 feet, limited to line-of-sight transmission.

Passive Infrared. Passive infrared (PIR) sensors detect heat energy radiated by warm objects. They are sometimes used to detect the presence of someone approaching a door from the inside of a building and to shunt the alarm.

LIGHTING

Ideally, fencing and lighting should be closely tied together. Despite the fact that an unlighted perimeter fence can offer little protection against nighttime intrusion, many airports do not light their perimeter fences, or they light them only near populated or high-activity areas. On the back side of these fields, the perimeter fence will probably be dark. Even when the perimeter is completely lighted, frequent patrols or remote observation is needed if intruders are to be detected. Of course, lights are now available that are tripped or turned on by any type of alarm triggering device. They are so common now that many homes have outdoor lighting that turns on automatically if an individual approaches. Any such lights should be the type which come on immediately and require no warm-up period before the light reaches full strength.

It has occasionally been suggested that security observation could be done with field glasses by control tower personnel. This is not practical; tower operators are usually too busy with the safety function of air traffic direction to assume added responsibilities. Although most tower personnel promptly report to airport police any incidents they do observe, tower operators cannot be expected to perform a continuing security function.

At the same time, the tower's advantage of overview points to a possibility that should not be overlooked. A member of the airport police force stationed three or more floors above the ground commands a field of view no mobile or foot patrol can equal. At most airports, however, if such a vantage point were available, it would more frequently be used to observe the airport parking lots than the ramp where aircraft are parked. This is not to suggest that airport security does not recognize the criticality of the protection of aircraft in the air operations areas; it only recognizes that the airport security force necessarily divides a limited number of officers into a known frequency of problems. Because the airport receives more complaints and files more police reports on parking lot thefts and attacks than on any other security "failure," parking lots tend to receive more of the limited surveillance time available to the force than does ramp security.

Fence lighting must include the clear zone on either side. The purpose of the clear zone is to make it possible to observe individuals near or at the fence and to ensure that no parked vehicle or stored items assist an intruder in scaling the fence. If the clear zone is not well lighted, its value during nighttime hours is reduced if not virtually eliminated.

Perimeter lighting is particularly important around terminal areas. Intrusion over fencing near ramp areas is probably most attractive to the type of intruder--hijacker or terrorist--of greatest concern to security.

Light Placement and Protection

It must be remembered that lights can blind as well as illuminate. FAR Part 139 requires the applicant to "show that all surface apron, vehicle parking, roadway, and building illumination lighting...is so designed, adjusted, or shielded as not to blind or hinder air traffic control or aircraft operations." Careful planning is also necessary to achieve the most advantageous use of light. Lights should be directed toward a potential intruder and away from a guard post, for example.

In most circumstances, it is better to mount floodlights near the roof of a building so as to shine down on the outside area than to locate the lights on the ground, shining up at the building. Landscape architects, particularly, appear to prefer the latter practice, not realizing the inherent security vulnerability--lights on the ground can be more easily reached and extinguished. Both methods of mounting illuminate the building exterior to the advantage of passing patrols, but ground-mounted lights blind any observers within the lighted building, just as lights pointed toward a guard post will do. I much prefer to have the lights positioned to blind the intruder rather than the security guard. The security officer's patrol path should be behind the perimeter lights. Lights also must not be directed upward at such an angle or in such a direction as to impair the view of air traffic controllers or of pilots approaching from any direction whether in the air or taxiing.

Perimeter lighting used in conjunction with a chain-link fence should project light toward the fence from within the perimeter. The illuminated zone being observed should extend outside the fence for a minimum distance of 20 yards, although a distance of 50 yards is better. The light standards should be from six to ten yards inside the fence line. Any obstructions will create shadows that might aid in concealing an intruder. Raising the height of the light fixture will sometimes reduce these shadows.

Parking lot lighting is a crime-prevention measure of considerable importance in reducing the incidence of muggings, other physical attacks, car break-ins, and car theft.

Everywhere, effective patrolling by police or security vehicles requires adequate lighting. Dark areas form natural hiding places for thieves seeking to avoid the patrols; full illumination improves the patrolling officer's productivity and the effective coverage of airport property along the patrol route. Where dark areas remain, the officer may have to leave the patrol vehicle and investigate the shadows with a flashlight, an alternative that is time-consuming and dangerous. Consistent illumination also eliminates the dark areas that are so inviting for stashing stolen cargo items for later pickup, an indirect but significant benefit.

Energy shortages have caused some airports to reduce lighting, but marked reduction in illumination is unwise in areas that require nighttime visibility for reasons of security. If there must be a reduction in lighting, it should be made within the terminal where police protection is adequate and there is considerable foot traffic.

Planning should ensure that the area of illumination provided by each lighting fixture overlaps that of adjacent fixtures so that a burned-out or broken light will not black out an area. Also, security patrols should immediately and routinely report any lights observed to be out, and these should be put back into service with the least possible delay.

Lights must also be protected from vandalism. If a light can be eliminated by a well-aimed rock, its value as a security measure is questionable. Fixtures can be protected from vandalism in a number of ways. Certainly they should be located as high as is consistent with the overall illumination plan, and all should have impact-resistant conduits. Failing that, wiring should be run high enough to reduce its exposure to tampering.

FINANCIAL LIMITS

A major and continuing obstacle to the acquisition and implementation of security measures called for by the Federal Aviation Regulations is the airport's lack of discretionary dollars with which to meet these requirements. To a large extent, the problem is the same for all U.S. airports, but the smaller airports are the most constrained in this respect.

Most U.S. airports are owned by local or regional governments, and operated by authorities or commissions responsible to elected officials of the particular governmental unit. Whether the airport regularly operates at a stable deficit, regularly breaks even, or regularly returns a surplus to the governmental unit, any major increase in costs is an immediate financial problem. Three unattractive solutions are available to the local body: increase airport revenue (raise fees), reduce airport services, or draw directory or indirectly on general revenue of the governmental unit. Hence, all airports, but particularly the smaller ones, operate within financial limitations ranging from moderate to severe.

Another factor is the actual cost of fulfilling the new requirements. To give a single example of implementing one requirement, enclosing Spokane International Airport (Geiger Field) required seven and one-half miles of fencing.

Chapter 12

IDENTIFICATION FOR SECURITY

Until the requirements of Part 107 went into effect, personnel identification programs received little attention or effort from airports, except for the badging programs of a few larger terminals. Traditionally, the primary means by which a person was authorized for access to air operations areas has been the wearing of whatever constituted the ramp uniform. Anyone clad in the airport's or the carrier's standard coverall was regarded as approved for unlimited access to ramp areas.

Clearly, dress can no longer be safely accepted as identification. The uniforms of flight officers and the overalls of flight mechanics are too easy to acquire. Flight crew uniforms have been stolen, and outdated flight attendant uniforms are sold. Authentic embroidered emblems or patches of the various airlines are offered for sale in many airport gift shops. One can also buy a baseball cap with the shield of the California Highway Patrol in any one of a number of service stations in California, and a similar situation exists in most states.

A particularly tragic incident highlighted the need to control employee identification. David Burke was fired from USAir on November 19, 1987, after being caught on camera stealing $69 in liquor receipts. On December 7, 1987, he bypassed screening, apparently by using airline I.D. that he had not turned in when terminated. This allowed him to carry on board the pistol with which he shot his former supervisor and the cockpit crew, causing the aircraft to crash. Forty-three people on board were killed. The family of the pilot sued and was eventually awarded nearly $3.9 million in a wrongful death lawsuit. Liability was not an issue at the trial because both sides had agreed that only the matter of damages would be tried.

This incident prompted a Notice of Proposed Rule Making on March 11, 1988, for the establishment of a computerized access control system at 269 airports. Although that system was not called out, the specifications could only be met with a computerized system. A relatively short 60-day comment deadline was given and the DOT rejected requests for extensions.

The regulations required the access control system to meet the following specifications:

1. It must enable only those persons authorized to have access to secured areas to obtain that access.

2. It must immediately deny access at the access point to individuals whose access authority has changed.

3. It must have the capability of zone coding, so that it can admit or deny access by area.

4. It must have the capability of time coding, so that it can admit or deny access by time and date.

FAR 107.14 took effect on February 9, 1989, but gave airports six months to specify the system that would be used and an additional 18 to 30 months, depending on their size, to install and implement the system. Any new airport beginning operation after December 31, 1990, also had to have such a system.

These regulations applied only to those airports conducting passenger operations in aircraft that seat more than 60 people. The first category consisted of airports that screened more than 25 million people annually; second, more than 2 million; third, between 500,000 and 2 million, and fourth, less than 500,000 people.

Of the 122 written comments received by the FAA, most objected to the proposal. Most airline and airport organizations--such as Airport Operators Council International (AOCI), the American Association of Airport Executives (AAAE), the Regional Airline Association (RAA) and the Air Transport Association (ATA)--opposed the legislation, claiming it was overkill and overly expensive. Meeting its requirements would cost far more than projected by the FAA. The FAA projected a cost based on 128 access points per Category 1 airport. Larger airports such as John F. Kennedy (JFK), Newark (EWR), and La Guardia (LGA) estimated they had 1500, 1000 and 750 access points respectively.

Some industry people suggested that more time be given to the study or that a test be run at a limited number of airports. The FAA stated that, "while worthy of merit under less compelling circumstances, the implementation of any of these recommendations would result in the postponement of a security measure intended to promote the safety of air transportation and therefore must be balanced carefully against that goal."

The FAA estimated the cost for the entire program at $169.9 million and anticipated that the airport operators would not be unduly burdened because of the availability of Airport Improvement Funds (AIP). The AOCI pointed out, however, that those funds would come out of already limited discretionary funds at the expense of other important projects.

An effective personnel identification program requires issuing personal identification badges to all personnel who have access to air operations areas, and an enforced requirement that the badge be worn at all times when at the airport. In addition, the badges must be identified and controlled by a distinguishing number and retrieved whenever an employee is terminated or transferred.

With computerized access control, it is possible to eliminate the individual's access the moment he or she is terminated or resigns, whether or not their badge and access card are turned in. Nonetheless, for a variety of reasons it is still important to get the badges back. Some access cards are reusable and are expensive to replace. The Schlage proximity cards cost around $10.00 each, and many others are equally expensive. In at least one sterile concourse, employees are required to insert their card into a badge reader before being permitted access. Thus, the employee is not only screened, but a check is made to make certain they are a current employee authorized to have access.

Airline employees may be a reluctant to turn in their I.D. cards because they can be used to obtain discounts at hotels and other travel businesses. It is necessary to have a formal program in place for employees who are being terminated or being furloughed, or are resigning, retiring, or leaving the company for any reason. Included in the regulations should be a checklist that requires the manager to pick up the I.D. badge, the access card, keys, corporate credit cards, telephone charge cards, medical insurance cards, parking lot passes, and any other items issued to the employee. Aside from all of the other good reasons for recovering a badge, the failure to do so in one case cost the lives of 43 people and cost the airline almost $3.9 million in damages to just one plaintiff.

Anyone who receives advertising circulars from vendors of security supplies is aware of the wide variety of identification badges and access control cards available. It is first necessary to differentiate between access control cards and photo I.D. cards, even though it is possible to combine them in almost all systems. Some photo I.D. badges are permanently affixed to the access control card, and some intentionally are not. The theory in keeping them separate is that if the access card is lost, the finder won't know what access the card might provide because it won't have any company identification on it. The fallacy in this reasoning, however, is that employees will most probably keep the badges together on the same clip, so that if one is lost, both are lost.

ACCESS CONTROL CARDS

Access control has become a science of its own, and much has been written on the subject. If you are selecting a system, you'll need to study the problem in far more detail than is covered in this book. However, for purposes of familiarization, some of the more commonly used types of access cards are described:

Hollerith Card. Probably the earliest of the access control cards, the Hollerith card utilizes the rectangular punched holes of the IBM card. It is easily duplicated and not to be used for high-security applications.

Embossed Card. This card utilizes optical character recognition (OCR) to read carbon impressions from plastic credit cards. It is rarely used as an identification card.

Bar Codes. Although bar codes are coming into more widespread use, they are not a high-security solution because it is simple to generate bar codes with readily available software and they can be photocopied. Some companies put a bar code on badges for use in cafeterias, fitness centers, and company libraries and for property pass applications, where high security is not demanded. Versions of this card for security applications encode the information in nonvisible code or cover the code with a colored film through which the code can be read by the infrared beam but can't be seen.

Magnetic Slugs. The slugs are arrayed on a sheet of material that is laminated in plastic so they are not visible. The card is read by passing it through a row of magnetic sensing heads. This type of card was very popular during the late sixties and early seventies.

Wiegand Technology. In this system, data are encoded in a wire that is twisted under tension, is heat-tempered, and has a magnetic snap action. Wiegand wires (normally 26 of them) are embedded into positions in the access control device to create the codes. It is one of the most widely used cards, along with magnetic stripe and proximity cards. The cards are difficult to duplicate and the readers are immune to the elements, especially moisture, heat, and cold. The system is named after its inventor, John R. Wiegand.

Magnetic Stripe. A magnetic stripe is applied on the surface of these cards and the code is recorded magnetically using conventional magnetic tape-recording technology. The cards can be easily erased or altered, and the technology is well known. They are relatively inexpensive. Data can be encoded and the cards can be used in conjunction with another device such as a keypad. (One of the problems with these cards is that the magnetic stripe is vulnerable to damage by being placed near magnetic media or other magnetic strip cards. I have had to return a couple of mag stripe credit cards to the issuer because they had been placed next to other mag stripe cards in my wallet and the code had become scrambled. My last ATM card came from the bank with a protective envelope in which to carry it and a notice that it could be damaged by contact with magnetic stripes of other cards or by "close proximity to magnetic fields associated with security cards, x-rays, garage door openers and some automobile glove compartments.")

Magnetic Sandwich or Barium Ferrite Card. These cards use barium ferrite for the sheet of magnetic material. A row and column array of spots on the magnetic sheet can be magnetized or not, creating the coding that is read by a row of magnetic sensing heads. This card uses the insertion type reader only. Cards are factory encoded.

Proximity Card. The cards can generally be read from two to four inches from the card reader, but some manufacturers claim from 28 inches to several feet. Cards can be kept in a billfold, purse, or coat pocket and still open the door or gate. They are difficult to duplicate, but relatively expensive. The system employs radio receiver/transmitter implants. A wall-mounted reader transmits low-frequency RF energy. A card entering the field receives the RF signal through its own antenna and interprets the transmission. The system has a higher through-put rate than other cards. The coding and access can be changed without having the card in hand--a distinct advantage.

Laser Card or Optical Memory Card (OMC). These cards consist of an optical recording stripe encapsulated between multiple, protective, transparent card layers. This product uses the technology of writing and reading with light, similar to that in much larger optical systems such as read-only-memory (ROM) or write-once-read-many-times (WORM) products. It has ample memory to store digital photos, signatures, voice prints, fingerprints, hand geometry, extensive biographic data or combinations of these and other personal identifiers.

Smart Card Technology. The laser card is one type of smart card. Smart cards have one or more integrated circuit chips capable of storing and interpreting information. They have been used as debit cards, whereby the cost of a purchase is deducted from the balance in the card and a new balance is entered. At present, the cost ranges from expensive to prohibitively expensive, but as with all new technology, prices are expected to decline.

Because a computerized access control system records the time of entry or departure, unions in some locations have objected to they systems as intrusive. In some cases, management has negotiated this issue with the unions and agreed not to use the system as a time card. At the same time, however, employers must reserve the right to access these records for internal company investigations, because they can provide significant and helpful information.

One of the more innovative uses of access control cards is at Neuvic, a prison in France. Every prisoner carries a personalized "smart card," which must be used to enter or leave any part of the prison. The cards tell the central computer where all the prisoners are at every moment.

All of these various access cards share one failing, of which you may already be aware. An access card or a card reader does not identify a specific individual. It merely tells you that a specific piece of plastic was placed in or near a reader and admittance was granted to the bearer of that piece of plastic. It is hoped that the individual assigned to that card is the one who used it.

PERSONAL IDENTIFICATION AND VERIFICATION--BIOMETRICS

The next higher level of security from an access card alone is a personal identification and verification (PIV) device . In most cases at an airport, that extra level of security may not be needed, but applications for it probably do exist, such as in a cashier's office, computer center, or similar high-security area. A keypad is the most common type of security device used in addition to an access card. I don't regard a keypad as a PIV, because the same person who loans out an access card might also give out a code to a keypad. Someone may also write down the code on a calendar pad or even on the wall near the keypad. The only true PIVs are those that are biometric in design.

The Hirsch keypad deserves mention because it helps to defeat the individual who observes the code while standing near the keypad. The Hirsch unit scrambles the location of the numbers after each use. For example, if the numeral 1 was in the upper left-hand corner, it may change to center right on the next pass. The keypad has a light baffle, which prevents the numbers from being seen from anywhere other than in direct line with the display. It also defeats the old practice of learning the combination by identifying the numbers on the keypad with the paint worn off of them. It has a duress alarm, where an individual can insert an extra digit in the code and send an alarm, and it has physical tamper and code tamper alarms. The Hirsch keypad is being used in several airports.

A biometric test is one that is based on the presence of personal characteristics, either biological or biologically based learned characteristics. These characteristics include fingerprints, voice prints, hand geometry, retinal scans, palm prints, signature dynamics, and keyboard rhythms. Biometric tests are often used with an access card or keypad with a personal identification number (PIN). They can be used to grant access to buildings, rooms, or to only he authorized user. The science of biometrics has progressed to the point where it is a very practical and reliable method of verification and identification.

On June 19, 1990, it was announced that Identix, Inc., of Sunnyvale, California, had been awarded a five-year contract by the California Department of Justice to provide 600 of its fingerprint devices to state, county, and city law enforcement authorities. The contract is estimated to be worth $40 million. The Identix unit captures fingerprints with electronic scanning rather than ink and paper. The electronic image can be reproduced with a laser printer and is more accurate than an ink impression, which is apt to be smudged. An additional advantage is that the fingerprints can be transmitted electronically to Sacramento to verify a suspect's identity, a process that is expected to take only minutes instead of the several days or even weeks. In this instance, law enforcement has grabbed onto a technology originally developed for the security profession.

References to Type I errors and Type II errors are common in discussions of biometric technologies. A Type I error is the false rejection of an authorized individual. A Type II error is the false acceptance of an imposter. Factors to be considered in the selection of a biometric system include the verification error rates (Types I and II), resistance to counterfeiting, throughput rate, cost, storage requirements and enrollment time.

PHOTO IDENTIFICATION BADGES

Before deciding on a badge system, the security administrator will profit from talking with a number of vendors and with actual users of each badge system under consideration. In addition, the security administrator should be acquainted with all the options each system offers and the costs involved.

Certain criteria and information should be incorporated in the program regardless of the type of badge system chosen, and these essentials must be identified before selection. Once the basic badge system is chosen, decisions must be made as to what information is to appear on the badge, and where it is to appear.

As discussed earlier, the airport is responsible for issuing the I.D. card so the FAA recommends that the badge be a minimum of 2 3/4 x 1 3/4 inches in size and that it be tamper resistant and wearable by clip, pouch, chain, or some other means. As a minimum, the front of the badge should have the name of the airport; color coding for access zones; U.S. Customs area authorization, if needed; a color photo of the employee; the employee's name; and the name or logo of the employer or air carrier.

The applicable domicile or station designation is noted on the badge for most employees, but flight officers and flight attendants are usually an exception. These personnel are in numerous airports during any given week or month. The inclusion of their domicile location on the face of their badge could restrict their access at many locations. Usually, a two letter code, such as "FO" for flight operations, is sufficient. If not printed on the front of the badge, the back should include an expiration date and a badge serial number. The airline code usually involves a five-letter designator; the first three letters indicate the station, and the last two, the department or organization. Thus, the designator SFOCS, might designate an employee who works in Customer Service at San Francisco.

Other items that can be included on the badge or in a computer database include the individual's social security number, date of birth, physical description and signature and the issue date. After it has been determined which of this information the badge will carry, the next step is to establish the layout of the badge, particularly which information will go on the front and which will go on the back.

Because the photograph affords the primary means of on sight verification of identification, the FAA's suggestion that the photo be in color and as large as 50% of the badge face or larger is well worth following. Some information should not be placed on the face of the badge. Employees have strong feelings as to what should be visible. In one instance, date of birth and weight were included on the front of flight attendants' badges and strong objections were voiced. Another problem is that some employees who don't like their official identification picture are apt to wear their badges reversed to conceal the photograph. It's worth the extra time required to get a picture the employees are happy with before they leave, or the badge may not be worn often.

I confess to having a preference for the individual's name in large block letters so that it is easily readable from a distance. Aside from the social benefits of knowing who you are talking to, it also has certain security benefits. It's quite frustrating and almost impossible to get close enough to read a person's name on a badge if it is printed in small, faded type. I've always had a desire to put the phrase "If you can read this, you're standing too close" on one of these badges to see how many months it would take for someone to notice it. The return address should be on the reverse of the badge in case it is lost.

Under FAR Part 107, each airport operator is responsible for (and, hence, alone determines) the badge used at the airport, subject only to regional FAA approval as part of the airport's master security plan. As each airport's badge requirements may differ, it is virtually impossible for the carrier to develop a uniform badge system that will satisfy the badge requirements of every airport at which carrier personnel may visit or be stationed. Even so, a number of carriers have successfully established system-wide uniform badge programs. Where airport badge requirements are incompatible, carrier employees may wear two badges--the carrier's and the airport's.

VIDEO BADGES

The most recent innovation in the field of identification badges is that of video badging, another blessing of the computer age. This system offers a number of advantages. First, the photo is stored in the computer's memory, perhaps utilizing CD ROM or WORM drive as a part of the storage and retrieval capability. With the photo is stored whatever background information is desired, including name, address, employee number, blood type, vehicle type and license number, work phone, and building location. Employees who lose their badges no longer need to go to the security office and have a new photo taken. The old photo can be called up and another copy printed. This is a particularly good feature for those who renew contractor badges on a regular schedule, such as every six months.

Additionally, the employee's photo can be called up on a color monitor for authentication of identity. This can be done by typing in an employee number, scanning the bar code on the badge, or as an adjunct feature every time an employee presents an access card. Consider, for example, a turnstile operated by an access card system but guarded by a security officer. If, when the card is presented to the reader, the employee's color photo pops up on a 19-or 21-inch color monitor as the employee is going through the gate, it is a simple task for the security officer to note whether or not the photo matches the person passing through the turnstile.

It is also helpful for identifying employees who claim to have forgotten their badges. When a security officer is monitoring a number of gates from a central control station, the system can assist the officer in authenticating the identity of the employee.

In its more exotic forms, the video badge makes it technically possible to bring up the photo of an employee with whom you are talking on the phone. This project was shown to be technically feasible at Apple Computer, where everyone has their own Macintosh, but rejected for reasons relating generally to invasion of privacy. The video photo does raise some human resource and legal issues. The plans of the security department should be cleared with the responsible departments within the company prior to embarking on an expensive project. It may be desirable to restrict the locations and posts that have access to the photo files on the computer. Other issues revolve around what type of information is stored in the data base and what type of notification is given to the employees at the time their photo is taken regarding how their photo will be used and who will have access to it.

It is important to establish security safeguards regarding access to printing these photos to prevent the duplication or counterfeiting of badges. I have seen one situation where badging supplies were not locked up and employees made duplicate badges on their own.

A new federal statute concerning commercial driver's licenses requires that, by 1992, every state have a data base accessible to all other states and to federal authorities. States that do not comply will lose a percentage of their federal highway funds. The purpose of this legislation is to monitor long-haul truckers. This statute is expected to boost the sale of video photo systems. Any firm contemplating a new badging system should most certainly investigate the video systems and the advantages they offer.

BADGING OF VISITORS AND CONTRACT EMPLOYEES

Special badges should be provided for contractor personnel (construction workers, repairmen, vendor servicemen, etc.) who have temporary, irregular, or infrequent need for access to air operations areas. These badges should be of a distinctive color, serialized, and issued by number to the workers every time they come, rather than issued as a block to an employer or to a contractor supervisor for distribution to the crew. By assigning the badges to individuals, the airport operator retains reasonable accountability (and traceability) for lost and unreturned badges.

A register should be established for visitor control, and each person issued a visitor badge should be signed in and out by time and date. One of the problems with a visitor register, however, is that vendors and other visitors can review the register while signing in and see which of their competitors have been there that day or that week. Also, the register is a slow process because people must wait to sign in until the person in front of them is finished.

One remedy for this is to use 3 x 5 inch cards to register visitors. That way, they can't see who else has registered that day. Additionally, if five people show up at once, they can all fill out their cards at the same time, as is done at a hotel registration desk.

The better solution, however, is to use a computer to sign people in and out--indeed to allow them to sign themselves in and out, provided a security officer is present at the desk. I've used such a system and it works, plus it give you a number of other significant advantages. A Macintosh will be needed with instructions on the screen to use with the mouse to "click" on the appropriate button. Individuals must identify themselves as visitors, contractors, or employees who forgot their badges. They might also record whether they are bringing in or taking out property. Depending on which button is clicked, the appropriate screen will appear for the individual's name, company, and whom he or she is there to see. The computer will automatically insert the proper date and time. It might also be possible to incorporate touch screens into such a process.

It can also automatically print a visitor, contractor, or employee badge that creates a far better impression than a cloth badge with the name written in ball-point pen. A printed badge can be typed in any font and any size desired, so that the name is readable from a distance. All of this information is recorded in a database providing a record of who visited and when. Equally important, it provides a database of property taken in and out of the building. Managers can be furnished a periodic printout of what property was removed from their department and who removed it. In the system developed at Apple, we utilized a HyperCard front end with a 4th Dimension database. I've also utilized a simpler system with FileMaker 4.0 software. Most programmers worth their salt can come up with a program to accomplish these tasks. There is probably also some software already on the market that can be utilized effectively.

LOST BADGES

A standard procedure should be established for reporting the loss of any badge. Upon a report of loss, a notice should be distributed to guard posts, instructing security officers to watch for the missing badge. A new badge with a different serial number should be issued in replacement. Procedures should also be established for follow-up on unreturned contractor badges and reusable badges. Computerized access cards can be eliminated from the data base with a simple entry, and that fact should be made known to employees to encourage them to report lost access cards immediately.

ENFORCEMENT OF IDENTIFICATION PROGRAM

Badge identification programs quickly deteriorate into ineffectiveness unless constant attention is given to enforcement. Some employees will resist wearing badges and will leave their badges at home or carry them without wearing them. If they find they are still admitted to their work area and are not challenged by other employees or rebuked by their supervisor their success will inevitably spread the practice to other employees.

Certainly, all airport security officers should challenge persons not wearing badges, but they cannot be expected to catch every violation; the effective enforcement of a badge program depends on supervision and other employees. An employee on the ramp without a badge should be challenged by other employees, no matter how well known the unbadged employee may be. Some airports have encouraged employee awareness by offering cash rewards to those who detect unauthorized persons on the ramp. On the other hand, the employee who fails to challenge apparently cannot be disciplined unless there is a provision covering this in the union contract or company regulations.

Ramp workers should be instructed about the action they are authorized to take in regard to any unauthorized individual found on the ramp. In some cases, unauthorized persons have been discovered, challenged, and allowed to leave--only to board an aircraft with the intent to take command of it. A tough "up-against-the-wall" approach isn't necessary. It is easy to ask questionable persons if you can be of assistance, but simply giving the individual information or pointing him in the proper direction is not sufficient. The challenged person should be escorted from the sensitive area to a law enforcement or security officer, depending on established airport policy. Police and security personnel are best equipped to assist the individual and decide whether he actually was lost or should be regarded with suspicion. It is never a good practice to ask such a person leading questions (unless, perhaps, they are asked with a specific intent to catch the person in a lie). For example, "Are you supposed to be on Flight 395 to Chicago?" provides the intruder with an easy "yes" answer, unless there is no Flight 395 to Chicago. Questions should be phrased so that the individual must provide the authenticating information. If someone claims to be scheduled on a specific flight, he or she should be asked to show the ticket. However reasonable the explanation, though, the individual should be turned over to a law enforcement or security officer for a final decision.

Passengers deplaning onto the ramp should proceed directly into the terminal. Carrier employees should be present to make certain no passengers stray toward restricted areas and to assist passengers as they leave the airplane.

U.S. CUSTOMS BADGING

The U.S. Customs Service issued interim regulations on February 3, 1986, requiring that airline and airport employees who have access to Customs security areas at U.S. airports apply to Customs for an approved identification strip, seal, or badge to apply to their identification cards. I recall being issued my sticker which was a hologram of the Statue of Liberty. The regulation became effective on March 5, 1986, and the final rule was published on September 12, 1986 (19 CFR Part 6).

Under the regulations, employers must attest that they have conducted background checks of the employment history of employees hired since November 1, 1985. When proposed, the cutoff date was November 1, 1984, but an FAA program had previously established November 1, 1985 as the trigger date for background checks. The carriers protested the disparity and Customs agreed the 1985 date. For employees hired before that date, the carrier only need attest to the fact that the employee was hired before that date and was considered trustworthy. The background check consists of references and employment history for the preceding five years.

The Air Transport Association also protested that the FAA had been assigned the role of security overseer of the U.S. airline industry and argued that only one federal agency should exercise that responsibility. They protested that the issuance of the interim regulations without prior notice or opportunity to comment was unjustified.

VEHICLE IDENTIFICATION AND MARKING

The requirement for airport vehicle identification also stems from Part 107 of the FARs. Section 107.13(a) provides that the airport operator will have a security program that:

> (1) Controls access to each air operations area, including methods for preventing the entry of unauthorized persons and ground vehicles.
>
> (2) Controls movement of persons and ground vehicles within each air operations area, including, when appropriate, requirements for the display of identification.
>
> (3) Promptly detects and takes action to control each penetration, or attempted penetration, of an air operations area by a person whose entry is not authorized in accordance with the security program.

In some cases, vehicle control is accomplished by requiring vehicles that are authorized access to any air operations area to display visual identification (such as a large decal or sign) while operating in that area. There are exceptions. For example, an emergency vehicle responding to an emergency situation may be permitted in the air operations area without such identification, if it is under escort by an authorized vehicle that has two-way contact with the airport's control tower or emergency control unit.

Vehicles requiring identification fall into two categories. The first category covers vehicles regularly assigned to the airport. These must be identifiably painted either with the carrier's colors and design and logo or with appropriate colors and design showing that the vehicles are the property of the city or airport authority. Under FAA guidelines, these vehicles need no further identification.

This is acceptable guidance in general practice, though the security officer should be alert to the possibility of subterfuge. These vehicles pose the same "Trojan horse" problem as identifying individuals by their uniform. Because a vehicle purchased or painted with appropriate colors is not readily distinguished from the genuine vehicles, good security practice would require additional identifying items, such as a highly visible window or bumper decal, for access to air operations areas past guard posts.

The second category of vehicles requiring identification includes those needing occasional access, but which are not painted or marked in such a way as to make them readily identifiable as authorized for restricted-area access. These vehicles should have identification that is visible from a distance of at least 300 feet. This can be a large card placed in the window or a magnetic sign attached to a door, for example. Naturally, a small window decal is not sufficient. Because the means of identification suggested for this occasional-use category are removable, and hence temporary in nature, the card or sign must be protected or secured when outside the restricted area. For example, to steal the magnetic sign from the door of a parked vehicle, or to remove the access card from an unlocked vehicle would be appallingly easy. At a minimum, when such access identification signs are outside the restricted area, they should be within the locked vehicle. They should also be completely out of sight, since locked cars can be broken into with relative ease. The ultimate solution is probably to not allow these means of identification to leave airport property. They can be picked up as the vehicle exits through the security checkpoint.

Whatever the category, all vehicles authorized for access to restricted areas should be registered with the airport operator, and the same type of accountability given to identification badges must be maintained for vehicle identification media.

The failure to properly identify vehicles can have disastrous results. On September 5, 1986, four Arab men dressed as security people and driving a vehicle similar to a security vehicle with a flashing amber light, were allowed to drive to a 747 that was loading passengers at Karachi, Pakistan. They stormed the aircraft with weapons, firing numerous shots in the process. The climax of the 16 hour hijacking occurred when the planes APU (auxiliary power unit), which powers the lights and air conditioning, failed. The hijackers panicked, not realizing what had caused the lights to go out. They began spraying the aircraft with automatic weapons fire and detonating grenades into the center area of the aircraft where all the hostages had been gathered. At least 22 were killed and over 100 injured.

Chapter 13

TERMINAL AND RAMP SECURITY

The terminal building presents special access security problems because, within it, there is a dividing line between the public areas and the air operations areas. How this division is established and actually enforced presents some problems. On the surface, locking all the doors between the terminal's public areas and the air operations area might seem a simple solution. Actually, this would bring carrier and terminal operations to an immediate halt, as both airline and airport employees must pass to and from the ramp areas. A quick reminder that authorized persons must have access might come from a pilot who has preflighted the plane in subzero weather and then found that the door from the mechanics' stairs into the jetway--and warmth--is locked for security reasons. The ultimate solution, whatever it may be, must allow access for authorized people while reliably halting unauthorized individuals outside the air operations area.

A concrete example of this type of access control problem is the sterile concourse. Due to the practical realities of airport operations, the sterility of an area is hard to maintain. The greatest problem in keeping the restricted area under control stems from the doors to and from the ramp. If a concourse or similarly controlled area is to be truly sterile, it must be impossible for unauthorized people to enter the controlled area from the ramp area. The presumption that unauthorized persons will not be on the ramp has been proven wrong on a number of occasions.

At airports without jetways where passenger access to the aircraft is from the ramp, it is often necessary to post a security guard to prevent people meeting incoming flights from venturing out on the ramp to greet a loved one.

In theory, these people have been screened, but nonetheless, they should not be out on the ramp unless they are passengers.

A similar problem occurs where jetways are in use. At the aircraft end of the jetway, a set of stairs provides access from the ground to the jetway; these are generally referred to as "mechanics stairs." If the service door at the top of the mechanics' stairs is left unlocked on the ramp side (and many are), anyone on the ramp can gain access through the jetway to the sterile concourse. At some stations, a push-button lock has been installed to secure the ramp-side jetway service door, and the combination has been given to all authorized personnel. Although the door from the concourse to the jetway may not be locked from the jetway side (as it may be needed as an emergency exit from the aircraft), it must of course be kept locked from the concourse side. Jetway doors could also be equipped with push-button locks on the concourse side so that personnel who know the combination could gain access to the jetway from the concourse. How this ramp-to-concourse access problem is handled differs from airport to airport, and the solution may depend upon the preferences of the carrier and the rule interpretation of the local FAA Civil Aviation Security Field Office (CASFO).

A related problem that also appears to be handled differently at various airports is the control of doors to the ramp from within the concourse. Here the security officer must consult with the local fire department because local fire codes generally incorporate definite rules on the subject of exits from places of public assembly. At some airports, these doors are simply not locked. At others, the doors have exit alarm locks. Doors so secured are normally locked, and authorized exit and entry is by key. In an emergency, however, these doors can be opened from the concourse side by merely pushing a panic bar or a similar paddle actuator, which sounds an alarm. The alarm usually sounds at the door, but it can also be readily wired to signal at a central station or guard post. Computerized access control (FAR 107.14), once again, will resolve many of these problems.

The United Airlines terminal at Kennedy Airport has an excellent solution to this problem of locked exits that was designed and installed by United's own people. The doors leading off the concourse are kept locked, but they are automatically unlocked in case of an alarm or fire. In addition, any door may be released individually from the central location.

Whatever system is installed, the safety of passengers must be considered primary, and a means of easy egress from the terminal must be provided for use in case of fire, explosion, or other emergency.

Once the doors from the public area are secured, the question arises as to who should be given a key or the push-button combination to the locks installed on these doors. The purpose of the antihijacking procedures is to prevent unauthorized persons from gaining access to air operations areas or to aircraft. Employees of the airline or the airport are usually authorized persons. However, not all employees need access to air operations areas, though a surprising number do. Because of the intense activity surrounding the arrival or departure of a flight, many people must have rapid access to the aircraft and the adjacent area. Cabin cleaners, fuelers, mechanics, technicians, food handlers, crew members, baggage handlers, passenger agents, and a seemingly endless number of others must have ready access to various locations on the ramp in servicing the aircraft. It would seem reasonable that all of these people should have the key or lock combination for these ramp doors. Frankly, if the combination is not given to them, they will either have it soon or the door will repeatedly be found propped open.

Key locks can be installed on these doors, but key operation presents all sorts of problems, including key accountability, the cost of key blanks and making keys, lost keys, duplicates made by employees, changing locks, and so forth. The use of a combination identification and computerized access control card would seem a reasonably practical solution, though a costly one. Card reader installation can run from $3000 to $5000 per door. The primary difficulty with this method, however, is that the badge, which is worn, must be unclipped from the uniform, inserted in the receptacle, and be refastened to the uniform on each occasion, which becomes a considerable nuisance. The alternative of wearing the badge on a chain around the neck, as is done in some offices, is an unsafe practice when working around airplanes. A proximity card can alleviate some of these problems. I have always had a desire to take hidden videos of people attempting to get the badge they are wearing in close proximity to the card reader. Depending on where the badge is worn, the gyrations can be quite humorous.

There are also two security-related factors that should be taken into consideration with the use of card-or badge-actuated access control systems. On some systems, an optional feature is required if the door actuation system is to continue to operate when power or computer control is lost. Second, each door can be monitored by some signaling or alarm circuit to determine whether it is open or closed. A door not under observation may be opened legitimately and then propped open without detection. Patrolling officers should test these doors to make sure they are is secured when closed; that is, that the electronic strike, which opens to release the door, has not been jammed open or has not failed to reclose after an entry. Computerized access control systems often have a function that sounds an alarm if the door remains propped open beyond a preset time such as 30 or 45 seconds.

KEYS

To determine what type of access system the airport will adopt, it is first necessary to decide what level of security is sought. On doors leading from the terminal to air operations areas, the people you want to stop are the potential hijackers, the mentally deranged, the curious, and those who might inadvertently stray onto the ramp while trying to find their way. Most often, it is those individuals suffering from mental problems who gain access to the aircraft and inform you that it is their airplane because the President gave it to them. Although these people are usually harmless, the ease with which they gain access to the aircraft is alarming to those charged with the security of the area.

I feel that any airport access control system is acceptable if it achieves the exclusion or significant delay of those individuals who should be stopped, while permitting the access needed by airport and carrier employees. This standard is not, admittedly, the most secure system that can be devised, but it achieves an appropriate level of security without hampering airport operations. In such a system, all the doors from the terminal to the air operations area can be keyed alike and a key issued to each authorized individual. If each carrier is to have its own keys for those doors that are under its control, a submaster keying system may be used.

FAR 107.14, which is discussed in detail in Chapter 12, requires airport operators to install an access control system that can be zoned for time and area and will be specific to the employee using the card. Where such a system is installed, to the extent that readers are installed on individual doors, it will eliminate the need for keys, except for an extremely limited number of bypass keys.

Nothing can eliminate the need for good key control, of course. Keys should be numbered sequentially, issued and recorded to specific employees, and stamped, "Do Not Duplicate," although this is just a request of the locksmith and has no legal sanction. Keys should be picked up whenever a key-holding employee leaves or is transferred, and they should never be passed from one employee to another.

The easiest and least expensive method of rekeying is accomplished through the uniform use of core-removable lock cylinders. With the use of a special core-removing key, which must be held under the tightest security, in a matter of two or three minutes unskilled maintenance personnel can rekey a door by removing and replacing the cylinder core. To capitalize on the advantages of this system, security or maintenance people can hold one or more sets of cores sufficient in number to replace each core-removable lock cylinder in use, whether keyed alike or differently, and submastered and master-keyed, in accordance with the keying of the original cylinders. The removed cores can be reused elsewhere, recombinated to a new keying, or held for use at another time. Rekeying by this method can be accomplished rapidly. As an example, the security personnel of a telephone company with key-holding employees on strike were able to change the keying of all exterior locks at all facilities in a 4000-square-mile area in a little over four hours.

Key-cutting equipment must be kept in a secure area, and access to the machine must be restricted and controlled. Key blanks and unissued keys should be kept locked up, preferably within a key-control cabinet specifically designed for this purpose.

A continuing problem for key control is that it often becomes a status symbol to have a master or grandmaster key. The issuance, if any, of such keys should be highly restricted and tightly controlled. Some companies solve the problem by not issuing master keys to individuals; when needed, the keys are checked out of the security center on a temporary basis.

For obvious reasons, keys must not be numbered or coded in such a way as to reveal which door they open. The lock of Door E5 should not be operable by a key marked E5 or 5E. Master keys should never be marked as such. In fact, to an expert, such keys are identifiable on sight, another reason why they are best not carried by key-holders.

LOCKS

If combination padlocks are used, the changeable-combination models are well worth the additional original cost. When an employee familiar with the combination leaves, particularly if discharged for cause, the combination can be changed, eliminating the necessity of buying new padlocks. Most combination padlock models are not recommended for outdoor use, and if used outside, they must be protected from the weather as much as possible. Like all locks, they should be kept well lubricated with a graphite spray lubricant. It would be best, however, to check with a locksmith before lubricating locks because some lubricants, such as WD 40™, may free up the lock more than is necessary and render it useless.

Push-button combination locks are generally and properly used for controlling access to specific areas, but they should never be regarded as being designed to resist forced entry.

Lock Installation

Where and how the lock is installed is most important. Often, a very durable and expensive lock is mounted on a relatively fragile door and can be ripped from its mounting by no more than a heavy screwdriver. This is particularly true in the case of padlock hardware. If the screws holding a hasp to a surface are exposed when the hasp is in its locked position, padlocking becomes all but useless when conventional screws are used. Hinge pins on doors and gates leading to critical areas should be spot-welded in place to prevent hinge-side entry, unless security hinges are used.

The importance of the lock's environment was shown in a situation where--despite good key control--unknown persons were gaining access to a wire high-value cage. There were no signs of forcible entry. It was eventually discovered that the wire cage door, which rolled open and closed on a set of overhead rollers, could be opened when locked by pushing down on the metal runner that guided the bottom of the door and springing open the door. This created a gap wide enough for a man to squeeze through, though the front edge of the door remained securely locked. This weakness was countered by padlocking both the front and rear edge of the cage door.

In exterior gates, there is usually a receptacle in the pavement into which drops a bar or rod mounted at the leading edge of the gate. Even in ordinary use in clean environments, these receptacles tend to fill with dirt. They must be cleaned out regularly because the security of the gate will be weakened, if not nullified, if the bar does not drop all the way into the receptacle. A parallel situation exists with a lock in the bottom of a glass door that bolts into the threshold. Here, the filled strike results in frequent and erroneous reports that the lock is out of order because the bolt cannot be thrown with the key.

Latches and bolts of door locks should be protected against prying by the securely attaching a steel flange to cover the opening between the door and the jamb at the lock. Excessive space between the door and the jamb should be corrected, as it permits prying attacks. A less common but effective attack can be made on a door that is securely locked on the outside but has a decorative glass panel next to it that can be broken to reach the lock lever on the inside.

ELECTRIC LOCKS

With the advent of the computerized access control system, it is important to know something about electric locks because the access card normally activates some type of electric lock to permit or deny access. Electric locks are used in three basic situations. They are used to hold a door secure until released by someone presenting an access card or successfully completing a biometric access test or both. Generally, the amount of time the door can be open without setting off an alarm can be regulated. This is important to combat the problem of employees who want to prop a door open for ventilation or some other purpose.

Second, an electric lock is used when it is desirable to release the latch from a remote location. For example, the latch may be released by a security officer who has verified the person's identity.

Third, computerized access systems can be time controlled, making it possible to unlock a door only during specified time periods. For instance, it may be desirable to lock the front door of a building from 5:00 p.m. until 8:00 a.m. the next day and all day on Saturdays, Sundays and holidays. This information can all be programmed into the computer so that access is controlled automatically. Because access cards can be programmed by area or zone and by time, this system has the advantage of selectively allowing authorized individuals into the building. Thus, a company can allow some employees into the building after hours while denying access to others, or it can allow only some employees access to the computer room.

If the time, date, and employee identification number are registered each time the door is activated from the outside, how is the person who exits from the inside of the building handled? When the door is opened without benefit of the access card, an alarm condition should result. To take care of this situation, a sensor is placed nearby, generally in the ceiling over the door. When an individual approaches the door, the sensor detects the presence and shunts the alarm for a specified time period, allowing the person to exit without activating the alarm. These sensors are generally of the passive infrared (PIR) type. They may be referred to as a "REX" (request to exit).

Of course, a card reader could be put on the inside of the door to record exits in the same manner that the reader on the outside records entries, and this is proper in a high-security area such as a computer room. This situation is known as "card in/card out." The problem that arises is that, because of fire codes, the individual doesn't have to use the card reader to get out of the room as required when entering. A person can simply hit the panic bar and exit. An alarm will record in a central station, but on a large campus of buildings, a security officer may not be able to get there to investigate for several minutes, and eventually, security personnel will tend to disregard these alarms.

Psychologically, one of the best means of enforcing the "card out" rule is to put a "screamer" or local alarm on the door. If the card is used to exit, the alarm is deactivated or shunted for a matter of seconds until the person exits. If the person exits without using the card, the local alarm sounds, and if others are around, peer pressure will tend to enforce this requirement.

The four basic types of electric locks are the electric strike, the electric deadbolt, the solenoid lock, and the electromagnetic lock. Because each has different applications, a fundamental knowledge of each lock is helpful to the security practitioner. During construction or modification of buildings, a part of the planning will concern how to get power to these units and the interface with the access control system.

> *Electric Strike.* This lock is placed in the door frame rather than in the door. It operates by rotating the striker plate out of the way of the bolt. It is generally sold without power supply, control logic, or electronics.
>
> *Electric Deadbolt.* This lock is the opposite of the electric strike. The door strike remains fixed in the frame and the bolt moves. Most operate by means of a solenoid although they can be motor driven. They also are sold without power supply, control logic, or electronics.
>
> *Solenoid Lock.* Although it operates similarly to the electric deadbolt, the solenoid lock does not fit into the door like normal door hardware but is mounted externally onto the door and the door frame.
>
> *Electromagnetic Lock.* This lock utilizes an electromagnet attached to the door frame, often at the top, and a metal plate mounted onto the door opposite the magnet. When the magnet is energized, the door is held closed with a force of from 500 to 1500 pounds. The locks are excellent and commonly used, but may be difficult to fit to some types of glass doors, especially the Herculite™ doors. Once again, security needs may come in conflict with the architect.

INTRUSION ALARM SYSTEMS

The criteria for using of intrusion alarms in the airport environment are much the same as those applicable to alarm systems for other types of business installations. The purpose of these alarms is to protect premises where economics preclude posting a guard at each point to be protected. By using intrusion alarms, a security officer at a single post can monitor alarms from an entire airport and dispatch security or police response by radio. Although alarms are generally used to protect closed premises after people have gone home, they are also used to signal the use of a normally locked door that is only opened to remove or replace valuable stores. The alarm announces the use of the door and, if an authorization code has not been phoned to the guard monitor, signals the need for a guard response.

Three basic types of systems are available, and the airport operator or his designee must determine which type will best suit the specific needs of the airport.

A proprietary alarm system is one that is owned and operated by the airport. It is usually monitored and responded to by airport police.Proprietary alarm systems are selected and purchased by the user and, usually, installed by the manufacturer. The airport either may contract with an alarm company for maintenance of alarm sensors and circuits, or may train airport electrical maintenance employees to service the alarms themselves. Proprietary alarm systems should be chosen only after visiting other proprietary installations by the same manufacturer for frank discussions with the users at each installation.

A central station alarm system is owned, operated, monitored, serviced, and maintained by a commercial alarm company that charges an installation fee and a monthly rental. The service includes monitoring all alarms at the central station and notifying police and the customer in the event of an alarm. Also, the alarm companies usually respond to the alarm themselves, dispatching at night an armed uniformed serviceman to investigate the cause of the alarm and to restore the alarm protection, if possible.

In some communities, police departments permit firms to connect their alarm system to an alarm display panel at the police station, where it is monitored (but not maintained) by police communications officers. Generally, however, police departments are reluctant to have alarms wired into their stations because of the excessive demands on police personnel created by false alarms. Across the United States, the average rate of false alarms ranges between 90 and 95 percent of all alarms signalled. However, what the police statistic identify as "false alarms" are any alarms that are not caused by an actual criminal intrusion and include those instances when the system is (correctly) responding to a sign falling down within the store, a sonic boom, or an interruption in the connecting telephone line. It is understandable why the central station alarm company employees who respond to alarms are servicemen and troubleshooters.

Unfortunately, the proprietary alarm station cannot be profitably used by those concessionaires whose insurance coverage or insurance premium costs are contingent on alarm protection. This results from insurance industry standards for the recognition of alarm protection, which virtually require the use of commercial alarm services. However, carriers may choose to contract with the airport's proprietary alarm center for alarm protection and response. Most major airlines are self-insured and thus independent of these requirements. Another reason why proprietary alarm service and response is not usually extended to tenants is the potential for problems of liability. While commercial central station contracts are written for a relatively small liability limit, upon occasion courts have overruled the contract provisions and awarded substantial compensation where the alarm service has been held to have been at fault in a major loss. In addition, in order to equal commercial alarm service, the proprietary center would have to have an alarm serviceman available 24 hours a day, seven days a week. Also, the security officer investigating a tenant's alarm is in greater danger of being mistaken by responding police for a cornered thief than is the alarm company serviceman, whose highly identifiable service vehicle will be parked outside for police to see as they arrive.

A local alarm system is one that no connection remote from the protected location. An alarm merely sounds a siren or a bell at the premises. The theory is that someone nearby or passing will hear the alarm and call the police, but this does not occur reliably or promptly in practice. On the other hand, the local alarm is more likely to frighten intruders away before they can steal much property, and it is sometimes chosen for this purpose.

ALARM SYSTEM SENSORS

Local, commercial, and proprietary alarm systems use the same alarm sensors. A variety of sensors and detectors are available on the market.

Contacts are the most common and elementary alarm sensor. These act by breaking (i.e., interrupting) a normally-closed electrical circuit. Examples are contact alarms on doors and windows that break the circuit when separated by the opening of a door or window. They are used in conjunction with foil tape on plate glass, which is severed when the glass is broken, and with grilles incorporating a thin wire that is broken by forced entry. Grilles are used to cover openings (96 square inches or more), such as ventilator shafts and ducts.

In its simplest form, this basic perimeter alarm system is rather easily compromised, particularly at doors and windows, and cannot protect against the individual who hides inside before the alarm is set or who enters through a ceiling or wall. In such cases, the alarm is triggered only when the thieves break out of the premises, and they will be long gone by the time the police arrive. One simple variation that offers protection against thieves who hide inside or enter through the roof is to alarm interior doors the burglar would use in searching for valuables.

Photoelectric cells use a light transmitter and light-sensitive receiver to create a beam of light that, if interrupted, will cause an alarm. Because even a relatively inexperienced burglar will avoid visible beams, infrared or infrared-filtered beams are usually used. Light-emitting diodes (LEDs) are pulsed, and their light is invisible, making it virtually impossible to deceive the beam receiver with a counterfeit source.

Beam systems are used to detect the passage of a vehicle through a gate or into a protected area, and they are sometimes used along fences. However, they are liable to being set off by animals, birds, and fog interrupting the light beam. This is true of laser transmissions, also.

Microwave (or, less commonly, *radar*) *systems* are popular, portable, and relatively inexpensive, but they have interfered with avionics equipment near airports and are particularly subject to spurious signals. Before investing in such a system, its effect on navigation equipment, both on the ground and in aircraft, should be thoroughly explored. Because microwaves penetrate even brick walls, the system must be tuned to restrict it to the protected area, and watched for signs that the alignment has drifted. Because the system penetrates ceilings, walls, and glass, it will detect motion outside the building or protected area. In some cases, this is intended, but in many cases, it is not. It can be exasperating to set up a microwave system to protect a room, and have the alarm triggered by every passerby in an adjacent public corridor. On the other hand, reducing (tuning down) the microwave area to prevent penetration will leave the perimeter without protection.

Ultrasonic sensors are triggered by the disturbance of high-frequency sound waves that emanate constantly from a transmitter. Like microwave sensors, ultrasonic systems fall into the category of space alarms. Ultrasonic waves are contained within the protected space and are a good choice for the protection of a room where no movement of any kind is anticipated. They are sensitive to wind currents, which may include air conditioning or heating devices turning on in response to a thermostat, and are inappropriate for locations where strong air currents might be expected, such as a stairwell. Incidentally, ultrasonics can be left on during the day with the alarm turned off. However, users should know that some people can, in effect, "hear" the ultrasonics and find it uncomfortable.

Passive infrared detectors are one of the most popular types of detectors, accounting for about 70 percent of the market. Passive infrared (PIR) detects the difference between body and reference temperature. No energy is transmitted by the sensor. It merely receives and analyzes radiant energy.

The best system for most applications is a dual system that involves a combination of technologies. It provides for lower false alarm rates and lower sensitivity. The system is generally set up to detect one intrusion twice before alarming.

Although there are many other kinds of intrusion alarm sensors on the market, these are the most common. Alarm systems should be designed, selected, tailored, and tested for a specific need. Wise security specialist will first determine what they are trying to protect and then set about finding the best way to protect it. One source of information about which sensors to use is the local commercial central station alarm company. Also, security consulting firms specialize in this type of work. And last but certainly not least, listen to what the vendors have to say.

FIRE PROTECTION

No one contemplating an alarm system is likely to omit the inclusion of fire sensors, nor should their use be limited to those installations that have or will have proprietary alarm reception centers. They are a first-line addition to all measures for fire protection and life safety and should be considered as part of fire protection as a whole.

Certain basics of fire protection and prevention will almost certainly be incorporated in every facility if only because of local, state, or federal regulations. Insurance considerations may prompt the inclusion of certain minimum measures where local or state ordinances are not particularly strict. It is important that the security operation be familiar with all the fire protection features that have been incorporated in the alarm system, and security management should be constantly aware of, and campaign for, appropriate improvements in protection, particularly when new construction is anticipated. The security manager should keep in mind that all fire protection planning begins with a careful and imaginative study of one basic question: What is the specific risk of fire and its possible consequences presented by this particular area, its traffic (or lack of it), its contents, and its normal occupancy?

Fire Sensors

Today's improved equipment for the early detection of fire even before ignition has had significant effect in reducing the number of fires and the extent of fire loss where the equipment has been effectively used.

Of the three forms of automatic fire detection--heat, smoke, and flame--fire-sensing planning is chiefly concerned with the first two. Flame detection is employed in the protection of such risks as refineries and volatile fuel storage facilities where flame is often the first detectable sign.

Smoke detectors, a category that includes units that sense presmoke products of combustion, probably offer the earliest warning of potential fires. These *ionization-type* detectors detect the almost invisible products of combustion that occur at the earliest stages of a fire. As a consequence, depending upon circumstances, location, and installation planning, they may be productive of numerous petty alarms that might be undesirable in some instances and preferred in others. Certain locations such as kitchens may be more appropriately protected by either rate-of-rise or fixed-temperature heat detectors, or by heat detectors that incorporate both features.

Ultraviolet detectors react to a narrow band of ultraviolet radiation given off by actual combustion. They are intended to protect where a fire will develop almost instantaneously with little or no incipient or smoldering stage.

Smoke detectors that use a *photoelectric* beam are triggered when smoke passes between a transmitter and a receiver, interrupting a beam of light. They are not the best for such areas as data processing centers where the air circulates rapidly. *Thermal* detectors are triggered by the rise in heat from the fire site.

Considerable independent and continuing research is likely to necessary to assist the security practitioner in choosing between fire sensors for the different areas and occupancies of airport facilities, because there are few genuinely experienced and entirely disinterested experts in this specialized field.

The best detectors are worthless if not responded to immediately. In the Illinois Bell fire at Hinsdale, Illinois, which wiped out phone service to a large area for weeks, alarms were ignored until one hour and five minutes after the fire started. Finally, a security officer was sent over in a car to check it out.

Sprinklers

The areas of a facility that are required to have sprinklers may have been determined by ordinance or fire insurance standards, but a good understanding of sprinkler operation, applications, and maintenance will not only maximize the protection these systems offer, but will also make it possible to suggest other areas in which sprinkler protection would be advantageous.

Where there are sprinkler systems, security personnel in particular should be cognizant of the difference between a water-flow alarm from a sprinkler system and other alarms. Knowledgeable response to water-flow alarms is necessary to ensure that sprinklers opened by fire and broken pipes will receive the prompt attention necessary for limiting water damage. Also, the use of cabinet hoses--to wash down a spill, in horseplay, or by vandals, for example--will also activate waterflow sensors. For these reasons, responding security personnel need to know that they are seeking a flow of water and not necessarily a fire. It should be noted that in colder climates, a dry-pipe system may be in use in which the pipes are filled with air to prevent freezing. When the air escapes through a valve, it allows the water to enter the pipes and extinguish the fire.

At airports where maintenance engineering personnel are not on duty around the clock, knowledgeable instruction should be obtained--and developed in the form of procedures--from such sources as the local fire department and fire insurance surveyors about the correct procedures to follow during and after sprinkler operation or damage, and these responsibilities must be clearly assigned. Closing a valve too soon, locking a valve closed, and failing to restore sprinkler protection after a system has been opened can lead to catastrophic fires--and has.

Having survived the Loma Prieta earthquake on October 17, 1989, and having seen the resulting damage and problems on a campus of about sixty buildings, I now anticipate that sprinkler pipes will break loose and flood buildings after major earthquakes. Naturally, the water has to be turned off immediately to prevent further damage. At the same time, however, gas pipes will be ruptured, and alarms may or may not be operative. Therefore, at a time when fire danger is perhaps the greatest, the sprinklers will have been turned off and there probably is no way around this. Security managers must, however, be aware of the increased risk and put guards in each building with the specific duty of patrolling for fires.

Special Applications

Early-warning fire sensors are often used in conjunction with sprinklers so that incipient fires can be smothered before they actuate the system. This eliminates water damage, mop-up operations, and the need to refit new sprinkler heads in order to restore fire protection to the area.

Cut-off areas, such as janitorial closets, rubbish and mail chutes, and storerooms, particularly benefit from the protection offered by a smoke or fire sensor. Occupancies such as offices that are locked off from view for 12 to 16 hours a day (let alone over two-to four-day weekends) are frequently protected in the same manner in modern buildings. Fire insurance premiums can sometimes be reduced if there is an hourly guard tour (properly called a "fire watch"), but the goal of fire loss reduction is better served by constant monitoring.

Preplanning

New construction planning presents a unique opportunity to incorporate innovations that increase fire protection and life safety. Many modern facilities have incorporated some of the following applications for fire protection. Although conventional fire doors and fire walls have long been in use to close off fire-involved areas when heat build-up reaches a certain point, provision can now be made to actuate their closing before such temperatures can develop. At the design stage of construction planning, arrangements can be made to employ fire or smoke detection signals to turn off air conditioning and to actuate special air-venting arrangements in the affected areas to prevent smoke spread and build-up. Zoned emergency public address systems can be employed for the reliable communication of recorded or live instructions in the event of fires, explosions, disturbances, or other emergencies in which life safety may depend upon being able to reach all those persons in the affected or endangered areas with prompt and appropriate directions. Corridor and stair doors can be equipped with magnetic hold-back devices that automatically release to close them when a nearby smoke or fire sensor is activated, and so one.

In existing facilities, on the other hand, such applications can be prohibitively expensive and even impossible to install after the fact. This makes it doubly important that today's potential for better protection be fully canvassed in the preliminary stages of construction planning. Now that such innovative equipment and applications for efficiently achieving higher levels of fire safety are available and tested in use, those new facilities that fail to include such features are becoming out-of-date before they are built.

SECURITY OF PARKED AIRCRAFT

The security of parked aircraft does not represent a new problem. Many years ago, I was a member of a U.S. Air Force Office of Special Investigations, "penetration team." Our assignment was to try to penetrate U.S. military installations. At this time, the Strategic Air Command B-47 bombers were protected by barbed wire, guards with automatic rifles, and dogs. Still, some team members managed from time to time to get through and plant their wood blocks, which simulated bombs. (Unfortunately, my fate was usually to end up spread-eagled on the ground at the wrong end of an automatic weapon while a snarling dog was being held at bay.)

Now the industry is concerned about the security of domestic aircraft because, in some instances, the bombs have not been simulations. Each carrier must estimate the degree of security its aircraft require at which points. Any commercial airliner represents a formidable investment, yet the estimate must be realistically weighed for the probability of sabotage or vandalism at that place under normal conditions. This varies with the circumstances; a parked aircraft at Cedar Rapids, Iowa, will not require the same measures as an aircraft on the ground in a Middle East country.

Factors other than international tensions must be considered, and conditions are not always normal, as evidenced by the two Pacific Southwest Airlines B-727s that were seriously damaged November 23, 1973. The cockpit of one jet was almost completely destroyed by hammering and acid, while the second aircraft had extensive fire damage. Mechanics and maintenance workers had gone on strike against PSA a week before. When a strike is in progress, increased security protection must be given both aircraft and facilities to prevent violence directed against the company by the most knowledgeable of all attackers who are familiar with every detail of the carrier's operations.

Juvenile vandalism, on the other hand, is a constant risk, and it seems obvious to say that aircraft should not be parked within a stone's throw of the perimeter. Nevertheless, two juveniles throwing rocks at a parked aircraft broke a cockpit window. They were apprehended, and the juvenile court judge asked the airline to advise him of the cost. The carrier reported that the repair of the window at the site (the plane could not be flown without it) cost $3000 and that the cost of having the plane out of service was $11,000. The further indirect costs of bringing in other aircraft were not calculated.

Part 107 of the FARs provides that the airport operator shall adopt and put into use facilities and procedures designed to prevent or deter persons and vehicles from unauthorized access to air operations areas. Part 108.13 requires the carrier to prohibit unauthorized access to its aircraft and to conduct a security inspection of the airplane before placing it in service and after it has been left unattended. In some locations, aircraft that remain overnight must be sealed or otherwise secured to prevent tampering, and in some cases, they must be under guard. In high-risk locations, even employees servicing the aircraft, such as caterers, cleaners, and mechanics, must have the items they are bringing on board searched every time they board the aircraft and must be under observation while on board.

It is rather frustrating, to say the least, for a carrier to implement a thorough predeparture screening program only to have a hijacker ride a bicycle out to the plane (with a rifle balanced on his handlebars) through a point where there should have been a fence. That feeling is compounded when a hijacker gains access to an aircraft twice at the same airport, as happened in Reno, Nevada, many years ago.

The first line of defense for parked aircraft must be the airport perimeter fencing and access controls. This is discussed in more detail in Chapter 11, but the fact is relevant here. No one will seriously contend, however, that fences are or can be the total answer to the problem.

It should be basic to the carrier's security program that unattended aircraft be parked close to occupied company premises and in well-lighted areas. The aircraft should be completely secured, and all jetways, and ladders removed when not being used for servicing. The ventral stairs should be up and locked on aircraft so equipped, and these stairs should not be used as a tail stand. This may present a problem on some aircraft but not on most. The only Boeing aircraft with any problem in this regard is the B-727-200 series and by properly shifting fuel loads when taxiing in, this problem can often be alleviated. If not, the plane can be refueled to 23,000 pounds (normal distribution throughout the three tanks). This will move the center of gravity far enough forward to make the airplane stable, and the ventral stairs can then be retracted to provide security while parked.

Attempted or actual intrusions have been made under the cover of darkness. A high level of illumination not only serves as a deterrent to intrusion but acts as an aid in observing parked aircraft. This can be accomplished by making maximum use of existing light sources and augmenting these with portable flood lights. The far side of the aircraft should also be well illuminated and not left in a dark shadow.

One effective lighting solution that has been employed to enhance the security of a number of parked aircraft is to place lights around the aircraft at ground level so that a guard can see under all of them at one time. People walking around the aircraft can be readily detected under these conditions, both by sight and by the conspicuous moving shadows they project.

If company guards are present, their duties should include regular inspection of areas where aircraft are parked. In other stations, arrangements can sometimes be made for periodic police patrols on an unscheduled basis, private security patrols, or night watchmen. In some cases, airport police are available to oversee these aircraft. At other airports at which the tower is open all night, tower operators have sometimes agreed to look after planes parked nearby in well-lighted areas. Except in a situation of unusually high risk assessment, it may not be necessary to provide a full-time guard to maintain a constant watch on each aircraft, provided the area is well-lighted and not readily accessible to outsiders.

One last line of defense is the preflight inspection conducted by the pilot prior to each flight. A thorough preflight inspection should uncover explosives in wheel wells or other obvious places on the exterior of the aircraft.

If a jetway is attached to the aircraft during predeparture servicing or cleanup after arrival, the door from the terminal building to the jetway must be locked; this is an FAA requirement. To allow for evacuation of the aircraft in case of fire or other emergency, the jetway side of the door should be operable by panic hardware. Consideration might also be given to locking the door into the jetway from the outside mechanics' stairs to prevent unauthorized access to the aircraft from the ramp. Of course, with so many locked doors and so many people who have a legitimate need for access to the aircraft, the control of keys (or the combination of push-button locks) are used will be an important factor. Some system must be devised to furnish the combination to every flight crew that taxies into the dock because the crew member who conducts the preflight inspection has to get back onto the plane via these mechanics' stairs.

If the purpose of locking doors is to keep out the general public, the mentally ill, and potential hijackers, there is probably no reason why the entire concourse or even the entire airport could not have all locks keyed alike and the same combination used for every push-button lock. At this point, trained security professionals may shudder, because such a plan results in a low level of security with a high risk of compromise. Yet, such an arrangement would probably accomplish the stated purpose, which was to keep outsiders locked out of the working area of the airport and, particularly, the aircraft without hindering operations. The goal of effective security management is not always to have the tightest security possible. If security is so strict that the airline has difficulty operating, then changes need to be considered. The implementation of the computerized access control system will greatly enhance security in many of these restricted areas.

A number of manufacturers market alarm systems designed to protect parked aircraft. Several types of alarms are available that operate on a variety of scientific principles. Some use the principle of radar, including the microwave Doppler radar and the range-gated Doppler radar. Infrared intrusion alarms include monostatic infrared and bistatic infrared. Closed-circuit television, particularly if motion detection is incorporated, can be effective as long as monitoring is maintained. Capacitance alarms are probably among the best of the available options.

Unfortunately, a number of problems are associated with just about all of the intrusion devices designed for protecting aircraft, although the state of the art has advanced significantly in recent years. The alarm problems are understandable under the conditions: Systems need to be immune to false alarms caused by birds or small animals, blowing leaves, or weather phenomena. Setting up the alarm is another problem. If setting up an alarm system becomes an every-night task and particularly if it includes lining up transmitter/receiver units with one another, it becomes too impractical an operation for the carrier who is parking aircraft at a small station.

Cost is certainly a factor. Probably the greatest drawback to all of these systems, however, is the problem of the rate of response. A local alarm sounding a bell or siren at a deserted airport is of little value. The alarm must be wired into a central station or to a police department, both of which are normally several miles from the airport. If it takes 10 to 15 minutes for a police car to arrive, the alarm system loses most of its value. Where the airport is large enough to have its own police department, the alarm should of course be wired into its 24-hour watch station or perhaps to the airport's crash, fire, and rescue (CFR) station. A remote alarm that transmits a signal to crew members as far as 10 miles away is discussed in Chapter 15.

If an alarm system is considered, the capacitive system has a number of advantages. It covers the entire aircraft structure, yet is insensitive to the movement of other aircraft, people, ground equipment, and animals in its vicinity. The equipment is self-contained and simple to mount and has a low power consumption.

The cost of alarm systems, when balanced against the cost of providing the other means of security coverage already discussed, may augur against purchasing such a system, but that decision can only be made after a thorough investigation and evaluation of all the factors.

Chapter 14

LAW ENFORCEMENT, COMMUNICATIONS, AND CONTINGENCY PLANNING

The indiscriminate use of the term "guard" or "security officer" has confused many because of its occasional use to designate the person assigned to the screening point as the required law enforcement presence. This confusion increased when carriers contracted with guard agencies for personnel to perform predeparture screening functions. As a result, both contract security guards and law enforcement officers were stationed at some screening checkpoints. Actually, the men and women hired from guard agencies were not performing a guard function; they were baggage inspectors. Although the carriers contracted for these personnel in part because the function was related to security, they primarily did so because these agencies offered a ready pool of appropriate labor from which to draw.

The official law enforcement officer, according to FAR 107, must be authorized to carry and use firearms, must be vested with the powers of arrest, and must be identifiable by some indicia of authority. There are about as many approaches to satisfying these requirements as there are U.S. airports. As previously noted, the requirement to provide the law enforcement officer is placed upon the airport operator, not on the carrier. Conversely, the carrier has no power to determine whether the airport will employ a city police officer or a specially empowered guard for the required law enforcement function--though the widely varying cost ($30,000 to $50,000 a year for a police officer versus $10 an hour for a guard) is borne by the carriers indirectly.

Some airports have contracted with guard agencies for personnel to perform the law enforcement officer function with restricted powers of arrest. Larger municipalities often employ port police who have responsibility for the city's harbors and as airline terminals. Where airports are governed by semi-independent commissions, the airport protective function may be assigned to a security force responsible to the airport management executive. This force may be deputized (airport police) or nondeputized (airport security).

LAW ENFORCEMENT FUNCTION AT AIRPORTS

Although the federal requirement is that the law enforcement officer be present during predeparture screening, whenever and wherever that may be--at the boarding gate or at the entrance to a concourse--many other areas of major airports require the attention of law enforcement agencies, such as traffic control, vagrants, intoxicated individuals, persons attempting to commit crimes associated with ticketing, credit card fraud, missing persons, pickpockets, stolen cars, and a long list of other enforcement activities.

Most major cities now have an airport police department or an airport detail of the city police department that is made up of officers specially trained for that function. The Department of Transportation sponsors an excellent Transportation Safety Institute at Oklahoma City, Oklahoma, where police officers can be sent for airport security training.

The law enforcement officer assigned to an airport must know the laws and ruling precedents covering crimes aboard aircraft, interference with flight crew members, search and seizure, and narcotics; into which jurisdiction various crimes fall; and the relevant Federal Aviation Regulations. The officer should be proficient in the use of firearms, have a knowledge of first aid and public relations, be trained in the security control measures appropriate to civil disturbances, natural disasters, fire, and bombs and explosives, and have a good knowledge of air carrier handling procedures for baggage and air freight.

Terrorist activities have with increasing frequency moved the point of attack from the aircraft to within the terminal. This has been recognized in at least one U.S. airport by the creation of a bulletproof station for the law enforcement officer on duty that faces the entry area and screening point. The need for some such protection has, unfortunately, already been shown. Thus, for many reasons, it has become essential that airport terminals have a law enforcement presence made up of competent, well-trained police officers.

JURISDICTIONAL PROBLEMS

Local jurisdictional problems occasionally arise. They can be result from a variety of causes, such as geographic location or overlapping legal jurisdictions. A noteworthy problem of geographic jurisdiction is San Francisco International Airport (SFO), which, despite its name, is not located in the (coterminous) City and County of San Francisco, but in San Mateo County. This caused problems when the federal requirement for law enforcement presence was established. Greater Cincinnati Airport (CVG), serving Cincinnati, Ohio, is located across the Ohio River in Covington, Kentucky. Such differences must be worked out by the local authorities and governments concerned. The Port Authority of New York and New Jersey is one example of cooperation between two states which is working well. While no one rule can be generally applied to their solution of these problems, it is certain that the airport operator must know who does has jurisdiction over crimes occurring at the airport well before a need arises.

In addition, federal agencies have primary jurisdiction over certain airport crimes. The Federal Aviation Administration can levy civil penalties or take other punitive measures against airport operators, air carriers and in some cases, passengers. The Federal Bureau of Investigation has jurisdiction over certain crimes aboard aircraft, interference with flight crews, and thefts from interstate shipments. At some larger U.S. airports, FBI Special Agents are assigned full-time to an FBI Resident Agency. Immigration and Naturalization Service personnel are stationed at official points of entry, as are the U.S. Customs Service men and women who inspect incoming baggage and freight for contraband and articles on which duty must be paid. United States Customs agents also investigate thefts of in-bond cargo. The Drug Enforcement Administrationquite often becomes involved in matters related to illegal drugs and attempts to smuggle narcotics into the United States. Bombs and explosives are the purview of the Treasury's Bureau of Alcohol, Tobacco and Firearms.

This multiplicity of agencies sometimes results in jurisdictional and other types of disputes. One of the more serious of these confrontations became public in December 1989. Federal Aviation Administration security was called upon to mediate a dispute between airport security officers and government agents, including U.S. Customs, the Immigration and Naturalization Service, the Drug Enforcement Administration, and the Agriculture Department. Some of the confrontations have led to arrests of federal agents and threats against local security officials. There have even been cases of federal agents displaying their weapons when challenged by airport security personnel.

The issue revolved around agents who refused to wear identification or present it upon request. Customs sent the FAA a letter stating, "We cannot permit a situation to exist where private citizens employed [at an airport] are permitted or encouraged to challenge a Customs officer not wearing airport identification."[1] Airport managers nationwide experienced difficulty as these confrontations increased. They were reluctant to complain officially. They claimed that when such issues were raised with local Customs officials, virtually all special handling requests were denied for months afterward and passenger flow-through and processing time slowed down as much as 30 percent for lengthy periods.[2] Airport operators interpreted these actions as retaliatory in nature.

Another type of jurisdictional problem occurred at a large U.S. airport. The mayor instructed the airport police to test the screening operations by pretending to be passengers and attempting to breach security with a weapon in carry-on luggage. The city then published the results of those tests, sometimes to the embarrassment of the carriers. The carriers, supported by the FAA, talked not only with the mayor, but took their case all the

---

[1] *Aviation Daily*, December 6, 1989, (Washington, D.C.: Ziff-Davis Co.): 434.

[2] Ibid, December 28, 1989: 579.

way to the Secretary of Transportation. They protested that the FAA, and the FAA only, had the responsibility for conducting tests of the checkpoints. No one really won. The problem simply went away.

Currently, another problem has surfaced in the same city. The city has indicated a desire to take over screening in the international terminal from the carriers. Although the responsibility for screening is currently with the carrier and not with the airport, the carriers might go along with this proposal except that the city proposes to charge the carriers for the screening and want to start screeners at a minimum of about $7.50 to $8.00 per hour, which is significantly more than they are making now. Therefore, the carriers are opposed to the city's offer. These types of jurisdictional problems continue to surface from time to time across the country and require the utmost in diplomacy to resolve.

One of the changes that may occur is that the responsibility for screening may be given to a governmental agency, such as the municipality or even the federal government, as has been proposed. The air carriers have long opposed being placed in the position of acting in a quasi-law enforcement capacity. The Aviation Security Improvement Act of 1990[3] created the position of Federal Security Manager (FSM) to coordinate the efforts of all these agencies and to avoid the rivalry that sometimes develops when jurisdictions overlap.

AIRPORT COMMUNICATIONS

Communications networks are not only an integral part of airport operations, but they are so vital to the operation that an interruption of air traffic communications, for example, would require most airports to stop all landings and takeoffs. Therefore, the security and protection of such systems is a legitimate and acute security responsibility.

Telephone Communications

Telephones are so much a part of U.S. life that their availability and reliability can easily be taken for granted. Actually, protection of telephone communications must be part of the security plan of each airport.

Frame rooms and rooms containing private branch exchange (PBX) equipment, must be kept locked. Where a PBX operator is working, the switchboard room door should be lockable from the inside and, if near public traffic, kept locked. Voice Mail Systems (VMX) and PBX systems have both become targets for hackers in recent years. Obviously, locked doors will not totally protect one from hackers, but they are an important deterrent. One hacker arrested in April 1990, has been charged by the Justice Department with masterminding a series of break-ins into Pacific Bell offices in San Ramon, San Francisco and Palo Alto, California. The government alleges that in most of those cases, he dressed as a workman and used stolen or man-

---

[3] Public Law 101-604, November 16, 1990

ufactured identification.[4] In his apartment, investigators found blank I.D. cards, a laminating machine, a stack of mug shots of himself and blank birth certificates. They also found business cards indicating he had passed himself off as a locksmith and as a magazine staff writer and photographer.

Local telephone company representatives should be consulted to determine the effect an electrical power failure would have on the telephone system. Will incoming calls be received and, if so, on all extensions? If not on all extensions, on how many and which ones? On incoming calls, will the bells ring? Will the visual indicator lamps light? Will it be possible to make outgoing calls and if so, from all extensions? Will extension users be able to dial other extensions within the internal system? If not, can a PBX operator manually ring and connect extensions? How rapidly? Is there space for an emergency operator to assist?

Electrical power failures usually do not affect public pay telephones because power for these telephones is furnished from the telephone company's main office and is generally backed up by the telephone company's own emergency stand-by power.

Listings of emergency telephone numbers and alternates should be provided to all key airport and carrier management people in the form of wallet-sized cards or directories, so that key personnel can communicate in an emergency. If the loss of power will affect telephone operations, special instructions for such occasions should be printed on the cards. I have effectively used portable cellular phones on such occasions, and they should be assigned to individuals with emergency responsibilities. Certain car phone models can be detached from the car mount and used as a portable, while still enjoying the range received with the car antenna while used in the car.

Radio Communications

Much of the day-to-day work of an airport is tied to radio communications. An aircraft receiving landing instructions from the control tower, a truck wanting to cross a runway, and an aircraft mechanic talking to the pilot of the aircraft at the gate, are all using radio communications. Such a volume and variety of traffic, of course, requires a variety of bands, frequencies, and types of transmitters and receivers.

All U.S. radio communication is licensed and governed by the Federal Communications Commission (FCC). Even the student pilot who merely wants to communicate with the tower must have a special FCC permit to do so. Volume V of the FCC Rules contains the regulations for airport radio services.

Ground radio communications are usually in the VHF-FM or UHF-FM frequencies, and such systems consist of one or more base stations, a number of mobile units, and the number of hand-held portables needed for the operations served. The FCC has assigned to airlines the exclusive use of frequencies within the 450 to 470 MHz range, while airport operators are authorized to use frequencies within the 150 to 162 MHz band. Because many

---

4 San Jose Mercury, May 27, 1991.

airport vehicles must have both ground control and ramp communications, combination AM-FM two-way radios that meet this need have been developed and are in use at some major airports. With these sets, both channels can be heard simultaneously, ensuring that a traffic warning is not missed due to a ramp transmission.

Because of the importance of ground radio communications to airport security, professional guidance should be obtained when planning for any new system. With good advance planning, an integrated system can be developed that not only is less expensive in the long run but will provide greater versatility than a hodgepodge system created by adding one limited system to another. Here again, similar users of equipment under consideration can provide invaluable information based on their experience.

EMERGENCY POWER

One or more auxiliary sources of emergency power should be available that are independent of the main power system; otherwise, a power failure or successful attack on the airport's power line or source would black out communications and operations. This need is most properly taken up early in planning as it is not economically feasible to support all communications, lighting, and power at normal levels with auxiliary power supplies. Choices must be made as to which requirements will be uninterrupted.

One fully automatic emergency power source might support the essentials of the air traffic operation. A secondary system might maintain safety and security functions such as security radio communications, fire-sensing alarm systems, and key lighting, while omitting closed-circuit television systems.

Automatically operating emergency lights can provide enough light for safety in interior areas of the terminal without auxiliary power supply support, but fall-back security must be provided to any operation or area which will be without power for the duration of the interruption. If electronically operated gates will not have emergency power, are they designed to lock or to unlock in a power outage? If electronically locked exit doors must be fail safe[5] in a power outage, who will be responsible for these exits while they are unsecurable? These are only some of the questions involved in a thorough consideration of emergency power planning.

BOMB THREAT CONTINGENCY PLANNING

Response to threats to aircraft and the receipt of threat calls are discussed in Chapter 10. The threat procedures discussed here relate primarily to threats against facilities. Although threats against aircraft have been more frequent

---

[5] Fail Safe means that when the power supply is disrupted the door unlocks. Fail secure means that the door locks when power is interrupted. Fail soft means the door will be locked but will still operate in a normal fashion. There generally will be no audit trail or record of entrance in such cases.

than those against facilities, bombs have been exploded in terminals, terminals have been swept by gunfire, and threats against airport facilities have been received.

As discussed in Chapter 9, bombs have exploded in rental lockers with lethal effect. As a deterrent against such attacks, rental lockers should be moved behind the security checkpoint. Where the screened area is secured around the clock, this should effectively rule out the most common explosive devices.

From thousands of threats received, very few actual bombs have materialized. In those cases where an actual explosion has occurred, it was rarely preceded by a threat. Hence, an explosion following a bomb threat may be considered improbable, but it must be considered possible. In evaluating the risk represented by a specific bomb threat, the probability indicated by actual experience must be considered. Federal agencies such as the FBI and the Bureau of Alcohol, Tobacco and Firearms can advise on the details of a threat that make it more probable and threat assessment advice is available from several sources.

Once a threat is received, decisions must be made, for example, whether to evacuate the terminal. Although the principle of "play it safe" would suggest evacuating the terminal every time such a threat is received, doing so would be self-defeating. Such evacuations, experience indicates, only increase the incidence of threats, and they raise havoc with airport and airline operations. Unless factors indicate that a relatively high degree of credibility should be given to the specific threat, it is not only more practical to proceed without evacuation but also less conducive to further nuisance threats.

Threat assessment has become a science in itself in the field of psycholinguistics. Dr. Murray S. Miron of Syracuse University, has lectured and written numerous articles on threat assessment.[6] He evaluates the seriousness of threats for a number of clients. Closely related to the field of threat assessment is the area of potentially violent employees. James E. Turner, Ph.D. operates Behavioral Litigation Support Services, Inc., in Savannah, Georgia, which deals with the handling of such situations. In the San Francisco area, Work Trauma Services works with corporations to handle potentially volatile situations and the post-traumatic stress disorder that follows such situations.

---

[6] Murray S. Miron, Ph.D. and John E. Douglas, Threat Analysis, "The Psycholinguistic Approach," *FBI Law Enforcement Journal*. September, 1979: 5-9.

Murray S. Miron, "Psycholinguistics in the Courtroom, "The Language of Scientist as Expert in the Legal Setting," Vol 606:55, *The Annals of the New York Academy of Sciences*. November 19, 1990.

Murray S. Miron and Thomas A. Pasquale, "Psycholinguistic Analyses of Coercive Communications," *Journal of Psycholinguistic Research*, Vol. 7, No. 2, 1978.

Murray S. Miron, "The Resolution of Disputed Communication Origins," Department of Psychology, Syracuse University, Syracuse, New York.

Planned Response

Every airport should prepare a written plan for handling bomb threats, whether these are threats to aircraft or threats to facilities. Unless such a plan exists and is followed, bomb threat response will only result in frustration and chaos. Although the plan will probably be written by the individual responsible for security at the airport, the plan also needs to be coordinated with all air carriers who operate out of the terminal to make certain that the airport plan is compatible with the response required by each carrier. For example, a carrier's bomb threat response plan may require the pilot to taxi the plane back to the gate and discharge the passengers, while the airport plan may state that a threatened aircraft may not discharge passengers into the terminal. It is essential that any incompatibility between airport and carrier plans be identified and resolved well in advance of the need, rather than disputed at the time an aircraft is threatened, as happened in an early case. One way of ensuring a coordinated effort is to hold mock drills, followed by a critique that includes all parties involved. Problem areas, if any, will show up then.

Before the airport plan is finalized, every agency that will be involved in responding to the threat should have a copy of the proposed plan and an opportunity to comment on it. The FBI and the Bureau of Alcohol, Tobacco and Firearms have jurisdiction in bomb threat cases. The FBI should be notified of bomb threats related to aircraft and airports. The FAA, police, Postal Service, Customs, fire department, and emergency medical services should also be involved. Tenants and concessionaires should also be aware of the plan, which will include how they are to be advised in the event of an evacuation or of the need to search their areas. Again, testing the proposed plan by simulating emergency situations is valuable for revealing shortcomings and unforeseen problems.

When a suspected bomb is found, a law enforcement bomb squad or military Explosive Ordnance Disposal (EOD) unit will be called upon to disarm or detonate it. Certainly, it should never be touched or moved by untrained personnel. The FAA suggests that formal support agreements be established with such agencies, rather than have cooperation rest on oral agreements between individuals.

Prior to 1971, the burden of responding to and neutralizing improvised explosive devices, except for a few major metropolitan areas was carried out by United States Armed Forces EOD units. Based on the increase of radical and terrorist bombings in the late sixties and early seventies, the federal government [in the form of the Law Enforcement Assistance Administration (LEAA)] and public safety agencies [in the form of the International Association of Chiefs of Police (IACP) entered into agreements to respond to this urgent situation. The National Bomb Data Center (BDC) was established and the Hazardous Devices School (HDS) was organized at Redstone Arsenal to provide bomb technician training to the civilian public safety community. A professional organization, the International Association of Bomb Technicians and Investigators (IABTI) was founded in 1973 with a charter mandate to maintain and enhance the professionalism of bomb technicians and bomb crime scene investigators. The operation of the BDC and the administration of the HDS have been designated as FBI responsibilities.[7]

The FBI Bomb Data Center provides bomb scene cards upon request that contain, in part, the following information:

BOMB THREAT OPERATION

T Target assessment; accurate information on site and occupants, i.e., controversial, political, prominent, etc.

H How to search; coordinate a systematic search.

R Remote command center established at a safe perimeter.

E Evacuate if necessary based on threat analysis.

A Awareness of other targets; beware of secondary devices.

T Tape and man appropriate cordons; keep people out at all costs.

BOMB CRIME SCENE OPERATION

S Seal off area; save crime scene.

C Command center established in safe area; mandatory control point for personnel and evidence.

E Evacuate injured; leave obviously dead.

N No unnecessary disruption of bombing crime scene by noninvestigative personnel.

E Essential personnel only within cordon area; full awareness of secondary devices.

Critical airport areas should be identified in the plan. These should include fuel farms; fire, crash, and rescue equipment; cargo facilities; emergency power generators; the control tower; critical communications facilities; and any other areas deemed critical to airport operations. Responsibility for searching these areas in the event of a threat should be clearly assigned in advance to specific, knowledgeable individuals and crews, preferably those personnel most readily able to identify objects that are foreign to the particular search area. Depending on the amount of time indicated in the warning and the credence attached to the threat, the evacuation decision must be made by the airport manager or a designated representative; therefore, someone so authorized must be available every hour of every day.

---

[7] Federal Bureau of Investigation. *Bomb Summary 1988*, (Washington, D.C.: U.S. Department of Justice, 1988), 25.

Search Procedures

When time allows, a covert search of the threatened area might be completed before the evacuation decision must be made if the area of search is not too large or undefined. If no suspect object or package is found and the search has been relatively thorough, the probability that the threat is valid is reduced. In order for this procedure to be reliable, however, the caller must have given some indication of where the bomb is supposedly located, even if only by narrowing it down to the terminal, a ramp, an airplane, or some other specific area.

The effectiveness and thus the reliability of a search also depends on whether the search must cover a public area or one restricted to airport employees. Within an airport facility area to which public access is restricted, employee are the best searchers. An employee who refuses should not--and probably cannot--be required to participate in the search. This question needs to be resolved ahead of time. It may be necessary to negotiate with the union and provide added life insurance or some other concession for those who will be engaged in searching aircraft. These employees, who are familiar with the area, its surroundings, accessible hiding places, and the objects to be expected in the search area, are much better prepared to make a rapid, effective search. A stranger, even a law enforcement or EOD officer, is far less likely to recognize that a package doesn't belong where it is found or that a certain object is out of place. What the searcher should be looking for in a search subsequent to a threat is usually an object foreign to its environment. When any such object or package is found, it should be left undisturbed and bomb disposal experts should be called to investigate it.

Once a bomb or something suspected of being a bomb is found, then the evacuation of the immediate area becomes a necessary subject for prompt decision. It is rarely (if ever) necessary to evacuate an entire building, and the extent of the evacuation forms part of the evacuation decision. Clearing the terminal area for a distance of about 100 yards from the suspect bomb is usually sufficient, provided people are also moved away from large windows that might shatter if the bomb were to detonate. In multistory steel and concrete buildings, a smaller area should be cleared on the floors immediately above and below. If the suspect package is near an elevator or emergency stairs, these should be secured from use until the threat is past.

Clearly, any bomb search must be done systematically; if it is not systematic and thorough, areas will be missed. Actually, a planned and thorough search takes less time than an unplanned one, particularly if the plan has assigned responsibilities for specific areas, rather than depending upon last-minute direction. To eliminate overlapped or missing areas, search areas must be specified in detail.

Unless the caller has given some clue as to the location of the bomb (the caller should always be asked where the bomb is), a search of the entire terminal area will be necessary. Public access areas should be searched first, followed by areas that are considered restricted but to which a visitor could gain access without much difficulty. Because the bomber usually needs some time without observation in which to place and perhaps arm the bomb, the location is more likely to be away from heavy traffic areas. The bomb's placement and possibly its appearance will have been chosen by the bomber to reduce the chance of its accidental discovery.

The Alcohol, Tobacco and Firearms Criminal Enforcement Section suggests[8] that the first step in a room search should be to stand quietly with eyes shut and listen for a clock ticking. If there is little background noise in the area, an explosive that has a clockwork timing mechanism for detonation can sometimes be detected by the sound of the clock.

The next step is to divide the room into areas, making specific individuals responsible for each specific area. Then, the room should be searched by levels or heights, starting with objects resting on the floor. The next searching height might be from the hip to the chin, then to the ceiling, and finally the ceiling itself, including the light fixtures and the space above dropped ceilings.

Waste receptacles in the rest rooms are likely objects for checking, as are wall openings in restrooms. Actually, every waste receptacle in the search area should be checked because of its accessibility and the ease with which a bomb can be concealed in one. If possible, care should be taken not to tilt or jar the receptacles while making the search.

Where dogs trained in detecting explosives are available, their use is by far the best method. When these dogs are employed, the search can be conducted with speed because the dog is able to detect sizable amounts of explosives immediately upon entering a room, before checking individual suitcases. The dog can be walked alongside rental lockers, through rest rooms, into stairwells, and through other obvious areas.

Some airports with dogs also have large mobile X-ray units that may be employed at remote locations at the airport to inspect suspect packages. Caution must be exercised in the use of airport carry-on inspection units, however; before a suspect bomb can be taken into a passenger terminal to be inspected by a screening X-ray unit, the area should be evacuated or the X-ray moved to another area. Again, moving the suspect package should be left to the bomb squad or EOD personnel.

---

8 *Bomb Threats and Physical Security Planning*, Bureau of Alcohol, Tobacco and Firearms, ATF P7550.2 (7/87) (Washington, D.C., GPO).

Chapter 15

GENERAL AVIATION SECURITY

General aviation traffic is big business at many airports. Airport security, already a problem, becomes more vital in direct proportion to increases in traffic. The ten busiest airports in the United States in 1989 and the percentage of the landings that were general aviation are as follows:

Airport	Total Operations	General Aviation Operations	%
Chicago O'Hare	789,384	28,266	3.6
Dallas/Ft. Worth	693,614	18,501	2.7
Atlanta International	669,543	24,730	3.7
Los Angeles Int'l	632,237	47,981	7.6
Santa Ana	533,237	438,161	82.2
Van Nuys	499,087	496,473	99.5
Ft. Worth/Meacham	492,743	490,512	99.5
Phoenix Sky Harbor	479,790	119,977	25.0
Denver Stapleton	468,490	38,911	8.3
Long Beach	462,177	431,683	93.4

It may be surprising to some that Van Nuys Airport (VNY) in California, which has no commercial scheduled operations, is the sixth busiest airport in the United States and that Kennedy Airport in New York (JFK) didn't even make the top ten. Following close on the heels of Van Nuys Airport is Meacham Airport (FTW) in Ft. Worth, another airport that is not well known outside the general aviation community, but which is the seventh busiest airport in the United States. The fifth busiest airport is Santa Ana (SNA), which is also known as John Wayne Airport or by its former name of Orange County Airport. Even though SNA has considerable service from certificated air carriers, over 82 percent of its operations is from general aviation. Similarly, Long Beach (LGB) is served by several air carriers, but 93.4 percent of its operations is general aviation aircraft.

I have a commercial pilot's license and, between flying and my work in aviation security, I have a keen interest in and awareness of the security problems that confront both the general aviation pilot who flies for business or pleasure, and the fixed base operator (FBO) who provides the services on which the general aviation pilot must depend. The airplane and its equipment represent a considerable investment for the private pilot, who can rarely afford either the theft of the aircraft or the loss from the aircraft of avionics gear worth thousands of dollars.

The theft of aircraft or avionics equipment remains the primary security problem facing general aviation pilots today. However, the hijacking of general aviation aircraft still remains a risk, along with that of vandalism and malicious destruction of aircraft. Since 1978, when there was a high of five general aviation aircraft hijackings, there has not been more than one or two in any year; there were none in 1988, but two during 1989.[1]

With so much general aviation activity, how is the airport operator to protect the ramp, the parked aircraft, and their theft-attracting avionics equipment? Indeed, the question might be whether the effort can be made. The National Business Aircraft Association has observed, "At airports where there is no airline service, security may be virtually nonexistent or dependent on the financial capability of the airport ownership or airport service organization."

Airport management and the FBO must be genuinely concerned about security. The service to which the tie-down or hangar fee entitles the aircraft owner legitimately includes protection of the aircraft and its equipment. The presence of security-minded employees of security-conscious operators is far more important than any locks, alarms, or ingenious devices available to the aircraft owner.

Although it might appear to be an elementary precaution no one would overlook, the aircraft should be locked whenever it is unattended. Obviously a locked hangar adds even more protection. Self-contained alarm systems for parked aircraft are available, but these are effective only if someone will hear and respond to the signals. One company, Shepherd Intelligence System, Inc.,[2] an EG&G affiliate, has come up with an aircraft alarm system coupled with a pocket pager. If an alarm occurs, the person carrying the pager is alerted immediately if he or she is within a range of 10 miles. This is particularly handy if the pilot is staying in a nearby motel. The system can monitor up to four aircraft and seems to be particularly well suited for corporate aircraft. At $125,000 to $150,000, it may be a bit expensive for the private pilot who wants to protect a Cessna 172. Defense Technologies, Inc. (DTI), of

---

[1] Federal Aviation Administration, *Semiannual Report to Congress on the Effectiveness of the Civil Aviation Security Program, July 1, 1989--December 31, 1989.* (Springfield, VA: National Technical Information Service, February 1991):10

[2] 12 Hartwell Avenue, Lexington, MA 02173.

Bedford, Massachusetts, which used to market an aircraft alarm known as System 72, was acquired by Shepherd Intelligence Systems in January, 1989.

Another available system is marketed by Securaplane USA,[3] a subsidiary of Civil Flying Service Pty. Ltd. of Cheltenham, Victoria, Australia, a company which specializes in aircraft maintenance. I have talked with the flight crew of one U.S. Corporation who praise this system for its simplicity and reliability.

Two systems are marketed out of England. Shorrock Security Systems Ltd.[4] manufactures a portable microwave fence aircraft security unit. The system produces a sensing zone that covers an area up to 120 square meters. RDS (Electronics) Ltd.[5] sells a portable infra-red intruder detection system called Sentry. Sentry uses established passive infra-red detection technology coupled with radio transmission of an alarm signal to a central location, alerting a security guard on his normal communications frequency. The system is also marketed by IAL Security Products under the designations Sentinel 600 and 610.

Less expensive approaches are available. One is a custom-fitted theft prevention plate that simultaneously prevents operation of the aircraft and blocks access to avionics equipment. Another is an aircraft throttle lock constructed of heavy gauge cadmium-plated steel. It fits all push-pull type throttles or mixture controls. It is simply slipped over the throttle or mixture control and locked in place with a padlock. Another type of throttle lock fits push-pull throttles and vernier type (push, pull and twist) throttles. Similarly, a device called a "prop lock" is available. It is an adjustable series of links constructed of hard steel encasing a seven-strand braid of hardened steel cable. The manufacturer claims the links are impervious to hacksaw blades and bolt cutters. It is attached by fitting the device around the propellor and spinner. Nonetheless, even if there were an affordable equipment solution for every security need, an aircraft equipped with all of them will still be at risk as long as it is kept where security is not provided.

CORPORATE AIRCRAFT

Aviation Data Services, a Wichita, Kansas based firm, reports that of Fortune 500 companies, 332 companies operate nearly 1,300 aircraft. Those companies which operate corporate aircraft generally have a larger employee base and a larger net profit than those who do not. Companies which operate aircraft have an average of 33,666 employees and a net income in 1988 of $315 million. This compares with 9,333 employees and a 63.3 million profit for companies without corporate aircraft.[6] FAA statistic reflect that in 1988,

---

[3] 3710 East 43rd Place, Suite 101, Tucson, AZ 85713.

[4] Shadworth Road, Blackburn, Lancashire BB1 2PR.

[5] Unit 12/13 Halcyon Court, Stukely Meadows Industrial Estate, Huntingdon, Cambridgeshire PE18, 6DG, England.

[6] "Palm Beach (Florida) Post," November 5, 1989.

45,800 small planes flew 8.7 million hours on business trips. During the first 9 months of 1990, deliveries of business jets increased from 99 to 109, a 10.9% increase over the previous year.

Security would seem to be particularly important for corporate air fleets. A recent survey of Fortune 1000 companies revealed that 37 percent of the CEOs use private planes. Less than 2 percent have their aircraft inspected for sabotage or espionage. Terrorist acts and threats and the corporation's attraction for irrational attack suggest that company aircraft security should incorporate special precautions against sabotage, explosives, and well-planned hijacking attempts aimed at seizing top executives for purposes of extortion.

For example, the company should re-examine the desirability of displaying the company logo on the exterior of the plane. Does the advertising benefit exceed the risk of identifying the aircraft as belonging to a corporation marked for attack? Indeed, an overall review of policy and the development of appropriate rules is probably in order. How shall passengers be identified? Which executives of the company can authorize the boarding of persons who are not employees?

Physical security of the hangar is always important, and it becomes paramount if the hangar is shared with other corporations. In such cases all tenants must participate in a cooperative effort and a specific security plan. Preflighting the aircraft should include a check of baggage compartments and the cabin, to ensure that nothing has been put aboard, and special attention to the points pilots know to be vulnerable. These and other measures, however, must be only additional to the security precautions that apply to all general aviation aircraft and operations.

IN-TRANSIT RAMP SECURITY

One of the most frustrating problems for general aviation pilots at some airports seems to be their exclusion from the scheduled-carrier ramp areas. If passenger terminal ramp areas are to be secure, general aviation pilots cannot be permitted to enter the terminal from the ramp along with screened passengers from a scheduled carrier. At some airports, however, FBOs are located across the field with their own terminals and lounges, and the problem does not occur. The ground controller can and does refuse any request from a general aviation pilot to discharge passengers at the main terminal.

At other airports, the problem is not so easily resolved. The FBO may be immediately adjacent to the main terminal. Unavoidably, the private pilot will want to enter the terminal--to conduct business, to rent a car, to eat lunch, or to file or close out a flight plan at the FAA Flight Service Station in the terminal. Erecting a large fence across the middle of the ramp and requiring air carriers to taxi to one side and general aviation to the other is not a feasible solution.

The normal approach in this situation is to post signs at the general aviation parking area, advising private pilots that they must enter the terminal only through certain specified gates and that in no case may they enter the main ramp area on foot. Gates providing access to the terminal from the air carrier ramp area should be kept locked so that private pilots disregarding the signs cannot gain entrance. If they elect to jump over the fence, as some brave and hardy aviators might well do, the incident should be reported to both local law enforcement and the FAA. Although compliance with airport security rules has become a matter of habit for most pilots, the rules will not be followed by everyone unless they are enforced.

At most airports, general aviation pilots normally enter the office or lounge of the FBO, where they can close out their flight plans by telephone, arrange for fueling their planes, and perhaps get some lunch, in addition to any other business they need to take care of. If pilots wish to go into the main passenger terminal, they must enter from the public side and not from the air operations area; once they have parked their aircraft, as far as airport security is concerned, they are a member of the public and must obey the same rules as the general public. After all, the rules protect the pilots as well as the flying public. They add to the security of the airport, and most pilots welcome the added protection.

AIRCRAFT AND AVIONICS THEFTS

Every aircraft owner should record the serial numbers of all the avionics equipment, keeping the list in a separate relatively secure location. When a theft is discovered, the local law enforcement agency should be given a list of the equipment and the serial number for each part. (The FAA should not be notified; they have no jurisdiction over such thefts.) Ask the law enforcement officer if a description of the stolen items will be routinely entered in the Stolen Article File of the National Crime Information Center (NCIC) at the Federal Bureau of Investigation. If the officer doesn't know about this computer data bank, the owner of the aircraft should courteously press for this procedure, as it greatly increases the possibility that the items will be returned. The NCIC computers receive information from and reply to more than 6000 law enforcement agencies. The NCIC contains files on wanted persons; stolen vehicles, license plates, and guns; stolen identifiable articles and securities; stolen aircraft, snowmobiles, and boats; and offenders' criminal histories.

In January 1989, the Aviation Crime Prevention Institute (ACPI) reported a continued decline in the theft of aircraft. Fifty-eight aircraft were stolen in 1988 at a total value of $14.5 million; however, $10 million of the total is represented by just seven aircraft, all turboprops. Although the number of aircraft thefts declined, there was a 55 percent increase in aircraft burglaries compared to 1987 figures. The ACPI estimates the value of the equipment stolen in 1988 at $2.8 million.[7]

The International Aviation Theft Bureau has been set up by the Aircraft Owners and Pilots Association (AOPA) to help combat aircraft theft problem. It is funded by annual membership subscriptions from numerous insurance companies. Valuable advice is available from this source.

---

[7] Aviation Crime Prevention Institute, P. O. Box 3443, Frederick, MD 21701. 1-800-234-5444. Sponsored by Associated Aviation Underwriters.

DRUG SMUGGLING

General aviation has been involved to a considerable extent in drug smuggling. In some cases, aircraft are stolen by smugglers, but more often, they can afford to buy new aircraft for their flights. Because of the high profits involved, some general aviation pilots have become involved in the smuggling of drugs into the U.S. Preliminary figures for 1988 show that the smuggling of cocaine by general aviation aircraft is a mushrooming problem. The total amount of cocaine seized on such aircraft exceeded 55 tons in 1988; that's more than 55 percent of all cocaine seized in the U.S. that year. That volume is up from 38 tons or 36 percent in 1987.[8] Sixty-six aircraft accidents related to illegal drug activity in 1988, according to the FAA.

As a part of its plan to more strictly enforce drug smuggling regulations, the FAA plans to totally renovate of the Airmen and Aircraft Registry in Oklahoma City. Actually, what is planned is the computerization of a manual microfiche-microfilm system, replacing it with optical disk technology. This information, as well as photographs, can be transmitted over telephone lines. The registry will provided a dedicated mainframe computer capable of providing daily, 24-hour service to law enforcement, which will be able to access the registry's database for information on airmen and aircraft. The FAA expects to begin implementing the plan in 1991 and said it will take about four years to complete it.

HIJACKING

United States private aircraft owners, charter pilots, corporate pilots, airfield operators, and FBOs should be employing adaptations of those preventive security measures proven effective in scheduled flight operations.

Baggage inspection has obvious application. A quick hand search of all baggage should be a basic precaution never omitted when unknown passengers are to be transported. In one hijacking the passenger said he was transporting musical instruments. The "instruments," which turned out to be weapons, and a rifle was taken from a trunk while the plane was in flight.

Passenger screening, so successful a part of commercial aviation security, can have some application in general aviation. Although no measures appropriate to general aviation can be as foolproof, some deterrent effect can result from requiring both adequate personal identification and individual signatures from all unknown passengers, and promptly comparing the signatures provided against the exemplars on the identification.

Aviation hitchhikers are a familiar sight, and their success in obtaining free air travel has been publicized in at least one national news magazine. Although no hitchhiker has as yet turned hijacker, caution suggests that strangers should not be readily accommodated. If a pilot chooses to do so,

---

[8] FAA report to Congress on its plans for enforcement actions related to the Drug Enforcement Assistance Act of 1988. *Aviation Daily* (October 15, 1989):163

however, the same precautions should be employed as for unknown passengers.

Publicizing security may be the most effective antihijacking method of all. Deterring the potential hijacker or hijackers from the attempt is not only more effective but far safer than any precautions a pilot can take. This deterrent effect can be created by prominently posting signs in passenger arrival areas that list all the precautions that are in effect on all flights originating at that field or FBO terminal.

Field access at airports without scheduled flight operations is rarely controllable, a fact reflected in general aviation security problems, from vandalism and theft to hijacking. However, the circumstances under which these occur differ markedly. Vandalism and theft are most apt to take place under cover of darkness and involve unattended aircraft. Hijacking attempts, on the other hand, are most probable during the hours in which the field is in operation, at a time when pilots and field operating personnel are present.

This difference provides an opportunity because it is possible that the risks of insecure perimeters and the ease of access to operations areas may be countered, at least in part, through careful planning and the cooperation of both personnel and those pilots who frequent the station.

At first glance, cooperative effort may seem to be of little utility as an antihijacking measure. However, official reports of hijacking attempts and incidents (the latter being those threats and demands not officially classified as hijacking attempts) indicate that a large proportion, if not a majority, of one-person tries involve deranged persons, and that their condition is frequently apparent in their first response to any challenge.

At small fields, many persons will be aware of a stranger; at larger fields, the task is necessarily more difficult. Nonetheless, a general awareness of and an agreed approach toward unfamiliar persons can save lives. A polite welcome can include an inquiry. and When both personnel and pilots make a consistent effort to take the moment or two required, irrational persons can sometimes be detected well before a crisis situation develops. Like all cooperative efforts, of course, such a program will require consistent use and encouragement, and a general recognition of the common danger.

Section IV

AIRLINE SECURITY

Chapter 16

INVESTIGATION MANAGEMENT

In the preceding chapters, our discussion is chiefly concerned with he protection of assets through the application of preventive measures developed to reduce risk of loss. These measures can and will reduce loss, but they cannot totally eliminate it. Preventive measures make it possible for honest people to remain honest, but they cannot reform character; neither deterrence nor difficulty will make dishonest employees honest. In any situation, there are always those who subscribe to the creed "What's yours is mine if I can take it."

It is this irreducible minority of larceny-oriented persons that creates the need for an established, well organized investigative department within every large corporation. It is this department's purpose to identify losses due to criminal conduct and to obtain the kind of evidence management must have to decide what action should be taken against the perpetrators and to prevent future depredations.

OBJECTIVES OF AN INVESTIGATION

Investigations should serve both aspects of management decision; identification of the culprit and correcting procedures so that the same situation is less likely to occur again. Understandably, investigators can become so intent on obtaining evidence of an individual's dishonesty that they neglect to develop the information needed to establish the circumstances and practices that allowed the loss to occur. Given competent internal investigation, each assignment should yield practical information regarding procedures. Every final investigative report should detail not only the evidence gathered on the misconduct of particular suspects but also those procedural defects and problems that led to the loss. To leave these holes unplugged is to invite almost certain recurrence; to omit these weaknesses from the investigative report, whatever the attitude of management at that time, is indefensible.

The investigative report should include recommendations for correcting both the demonstrated defects in current procedures and those vulnerabilities that were evident to the investigator but have not as yet been capitalized on by thieves. The procedures officially in effect may be entirely adequate, but a shortage of supervisory personnel may be encouraging their neglect.

The investigative report that omits recommendations for preventing further losses of a similar nature is particularly incomplete. It is in this respect that the internal investigative report differs from the traditional investigative report of law enforcement. The latter confines itself to facts and excludes opinion; to make recommendations requires opinions. To ensure that recommendations are not omitted, the standard format of the internal investigation report should include a point at which the investigator's recommendations and conclusions are to appear. This can be an administrative page or cover sheet or a special section within the body of the report.

Investigation of suspected dishonesty by nonemployee outsiders, however, is usually the province of local, state, or federal agencies. Airline investigators may conduct preliminary inquiries into a matter in which a member of the public may be criminally involved, or they may discover such involvement in the course of a loss investigation. Once their investigation has established this involvement, the information developed should be referred to the appropriate law enforcement agency. Airline investigators deal primarily in internal investigations involving the airline's own employees. Certain kinds of investigative work or portions of an investigation may occasionally be contracted for with an outside security agency.

CREDIT AND CRIMINAL CHECKS

A "credit and criminal" has long been considered a minimum check for any background investigation. However, in recent years, right-to-privacy laws have made such information more difficult for corporations to obtain. In some states, these investigations are illegal.

Federal Restrictions on Criminal Checks

The U.S. Department of Justice issued an order in May 1975[1] prohibiting law enforcement agencies from disseminating criminal history information to outsiders and from confirming the existence or nonexistence of such information for employment purposes, except where provided for by statute. This order, backed by fines and the loss of funding in some cases, ensured that applicants for most positions could readily conceal past criminal records and that applicants with clean records were unable to prove it. However, in March 1976,

---

[1] Department of Justice Order 601-75.

the Department of Justice modified the order to permit the distribution of "conviction data" and "criminal history information contained in court records of public judicial proceedings." The modified order also allows dissemination of information on arrests for which prosecution is pending.

The Civil Rights Act of 1964 prohibits discrimination; discharge, failure, or refusal to hire; limitation, classification, or segregation of employees or applicants in any way adversely affecting their status or employment opportunities on the grounds of race, color, religion, sex, or national origin. Applicants cannot be asked where they were born or what country they are from. They cannot be asked what their religion is or whether they have ever been arrested for a crime, but they can be asked if they have ever been convicted of a crime. They cannot be asked their age or the year they were born.

Credit Checks

The Fair Credit Reporting Act[2] places a number of constraints on third-party access to credit records. The law states that the credit reporting agency may furnish information only under the following circumstances and no other:

1. Court order.
2. Written instruction of the consumer.
3. Upon request to a person who the agency has reason to believe intends to use the information for:
   a. A credit transaction.
   b. Employment purposes (within certain limits).
   c. Underwriting of insurance.
   d. Eligibility for license or other benefits.
   e. A legitimate business need in connection with a business transaction involving that customer.

Significantly, the statute defines "employment purposes" as including evaluating the consumer for employment, promotion, reassignment, or retention as an employee. However, the reporting of information dating back more than seven years is prohibited with some exceptions.

The Federal Consumer Credit Protection Act of 1976[3] requires that any organization that procures or intends to procure an *investigative consumer report* (a report made by a consumer reporting agency containing information developed through *personal interviews* with those who have a personal knowledge of the subject of the investigation) on a person who has applied for employment, retention, or reassignment must give the applicant written notice of that intention. The employer must advise the applicant or employee of the areas the investigation will cover. If adverse action is taken as a result of the information in a consumer report, the applicant must be informed of that fact and of the name and location of the consumer reporting agency that provided the data. The employer does not have to tell the subject the nature of

---

2 15 U.S.C. 1681, et seq.).

3 Title 6 U.S.C. Public Law 91-508.

the adverse information. Using internal security staff to conduct an inquiry on an applicant does not make the company a consumer reporting agency. The act incorporates both civil remedies and criminal penalties for violations of its provisions, and every investigator should be completely familiar with the provisions of the act and its restrictions.

In May 1991, Electronic Data Systems Corp. (EDS), who had been screening applicants for General Motors Corp., without admitting it did anything wrong, agreed to overhaul its procedures,in response to charges by the Federal Trade Commission that they had violated federal laws. Job applicants had been denied employment on the basis of credit reports, but had not been advised that their rejection was based on credit information as required by the Fair Credit Reporting Act.[4]

Upon request, credit agencies must disclose to the consumer the nature and substance of information in its files, the sources of that information, and--under certain guidelines--the recipients of any reports of this information on the consumer. It should be noted that some credit agencies routinely advise consumers of the identity of anyone requesting a report on them. TRW Inc., a major credit reporting agency, sells a service to consumers for an annual fee of $35, that includes automatic notification when anyone receives a copy of the consumer's TRW credit file.

SURVEILLANCE

Company policy on surveillance should be well established and the guidelines provided to those responsible for instituting and supervising surveillances. If special company approval is required for a surveillance, clear instructions must exist as to when this approval is needed and how it is to be obtained.

Policy will vary from company to company. A carrier may permit surveillance of its employees by its own investigators on company property, while forbidding surveillance off company property. The approval of top management may be required before an outside agency can be engaged to conduct any type of surveillance activity. Some companies in the business community go a step further and directly employ undercover operatives, while others make undercover the first assignment of new members of the corporate security department.

A surveillance conducted for the purpose of observing and identifying those employees engaged in criminal activity is an entirely legal and moral undertaking. Of the numerous surveillances in which I have participated during my career, I can recall no instance of an individual being troubled who was only incidentally observed and who was not criminally involved. Although cries of "Big Brother is watching" and "invasion of privacy" occasionally arise from foes of investigation, it is generally those who are observed in some type of felonious conduct who raise objections to the surveillance as if it were constant, unprovoked, or arbitrary. I--with the rest of his profession and the preponderance of judicial opinion--believe that where evi-

---

4 San Jose Mercury News, May 29, 1991: 3D.

dence exists of mysterious disappearance, theft, or other criminal conduct, surveillance is not only justified but required.

A surveillance can involve one person sitting in a parked car or several individuals equipped with binoculars, two-way radios, and even video cameras and video recorders. For reasons of safety and to ensure effectiveness, any undercover or other surveillance activity must be approached with extreme caution. Knowledge of the specific operation must be restricted to the smallest possible number of people. Certainly, undercover operations should only be begun with the knowledge and approval of the president of the company. A surveillance or undercover operation should only be used where criminal activity is suspected. Because of federal law, such measures cannot be used, for example, to gather information regarding union activities, and the undercover investigator reporting on the progress of a criminal investigation should not include observed union activities in the reports.

If the surveillance involves sitting in a parked car for an extended period of time or other suspicious activity, particularly within a residential area, the local police should be made aware of the activity. People often mistake an investigator sitting in their car for a drug dealer or police officer and deaths have resulted.

CELLULAR AND CORDLESS TELEPHONES

Federal law[5] extends the provisions of the 1986 Omnibus Crime Control and Safe Streets Act to the more sophisticated types of communications devices, such as cordless and cellular phones. However, courts have recently ruled that users of wireless devices have no reasonable expectation of privacy and therefore knowingly forfeit their privacy. Thus, it is perfectly legal to monitor a household cordless phone, and many people do just that. *Monitoring Times*, a publication with a monthly circulation of 30,000, lists technical information and frequencies of cordless devices. In addition to the telephone, other frequencies monitored by people with scanners include McDonald's takeout windows, the local highway patrol, hospital paging systems, taxi dispatchers, Alcoholic Beverage Control field agents, sheriff's departments, and, perhaps the most vulnerable of all, baby monitors, a device many parents use to transmit sounds from the crib to a speaker elsewhere in their home. As useful as these are, parents must be aware that they have just planted a highly effective bug in their house and anyone in the neighborhood listening on that frequency can hear what is going on in the entire house.

I have already been involved in one situation wherein an informant produced a tape recording of a 45-minute phone call between two senior executives of a firm. The person monitored the call by parking outside the home of one of the executives with a scanner and waiting for a phone call to be made or answered on a cordless phone. Cellular phones are just as vulnerable. They are theoretically off-limits according to a 1986 statute[6] that is

---

5 Electronic Communications Privacy Act (ECPA) (October, 1986).

6 Ibid.

commonly flouted and never enforced.[7] They are routinely monitored by an ever increasing number of eavesdroppers and recreational snoopers, as well as those who are looking for an inside bit of information for their publication or for the competition.

EMPLOYEE PRIVACY IN THE INFORMATION AGE

With the advent of the information age, new privacy issues have arisen. For example, a new software program allows network administrators to read files stored on hard disks in users' computers. Managers now have the ability to search user files at will, allowing them to maintain surveillance on their employees. Central to the issue is establishing what belongs to the company and what is covered under employees' privacy rights.

Similarly, with regard to fax machines, automated phones, and electronic mail (E-mail), significant new issues remain to be resolved. I have already encountered a situation in a case of information loss, in which a suspect's electronic mail, which he thought had been destroyed, could actually be retrieved. Again, the question of whether that communication belongs to the company or the employee is paramount. A case is currently under consideration that will test the privacy rights of employees with regard to electronic mail. It involves a company where managers routinely read employee's E-mail correspondence. Employers need to develop policies in this area that enables them to exercise certain rights without making employees feel criminalized. Courts have found that though employees cannot be guaranteed the same privacy rights as in their homes, they should enjoy some "reasonable expectation of privacy."

SEARCHES

Whether, when, and how to search an employee's locker, lunch box, toolbox, or his car are questions that trouble many managers. Some feel that searches are disruptive of employee morale and their possible benefit is not worth the probable consequences. Other managers make inspection a regular practice, and their employees habitually open their lunch boxes or packages for inspection as they pass the guard on leaving the plant or location. Others restrict the searching of employees to a spot check or random check system, or they search employees only when there has been a specific theft.

It seem reasonably well established that evidence obtained as a result of searches "by private persons" is admissible in court proceedings, as discussed in Chapter 6. Certainly, it seems to be unquestioned that evidence so obtained is admissible in a grievance hearing.

Some problems can be traced to the sudden imposition of searches without warning or precedent. For this reason, most companies have established--and provide to employees--company regulations stating that employees' packages, briefcases, lunch boxes, toolboxes, lockers, clothing, vehi-

---

7 The Wall Street Journal, (October 9, 1990).

cles, etc., are subject to inspection at the request of security or supervisory personnel. Employees must be put on notice that such inspections will take place, or they may indeed be deemed illegal.

Under some circumstances, it may be advisable to have a union representative present if, for example, the company plans to have all lunch boxes or all lockers inspected on a certain day. Should an employee refuse to allow an inspection, the search generally is not conducted, but the employee may be terminated or disciplined on the grounds of insubordination. As a fine point, the employee's supervisor should be present or be called to the scene to issue the order to the employee, and the employee should be reminded of the probable consequences of refusal. The language of one Board of Adjustment Arbitration is typical:[8]

> Here the Company's supervisor, faced with highly suspicious circumstances, was attempting to get at the root of the trouble; and he had every right to expect that [the employee] would comply with his request to turn out his coat pockets or have them searched. Indeed, under the circumstances, it would seem to be insubordination on [the employee's] part to refuse to cooperate or comply.

The absolute right of a company to search under any and all circumstances, however, is questionable. One arbitrator held that a distinction must be made between a "random search" and a search of personal items resulting from an informant's tip or management observation. The arbitrator ruled that a random search is "indiscriminate, admittedly serves no purpose other than to have a deterrent effect, and consequently its sole motivation is one of intimidation." The arbitrator felt that the company should be prepared to demonstrate a justification or basis for search.[9] Indeed, random searches appear to be the most difficult to justify.

This investigative tool should be used with discretion, of course, but not abandoned. Some managers believe that any search is accusatory in nature, and they do not use it on those grounds. I find it difficult to accept this viewpoint. After all, positions involving trust in the handling or custody of valuable items and negotiable instruments regularly involve some form of audit. These measures are accepted as sound management practices developed to keep honest people honest, to detect misconduct in a timely manner, and to protect the innocent from the consequences of others' dishonesty. Audits are not accusations; on the contrary, they create confidence in the honesty and integrity of the employees and the company.

When locker and lunch box searches are conducted only when a problem is believed to exist, however, then the searches must be seen to relate to the specific problem; if the freight terminal is missing television consoles or large boxes of clothing, a search of lockers is clearly irrelevant to the prob-

---

[8] UAL--IAMAW System Board of Adjustment, Arbitration between Airline Employees District 141, IAMAW and United Air Lines, Inc., Grievance No. 78516.

[9] TWA--ALSSA System Board of Adjustment, Arbitration between Airline Stewards and Stewardesses Association, Local 550, TWU, AFL-CIO and Trans World Airlines, Inc. (April, 1973)

lem. In many cases, companies provide the locks used on employee lockers and retain a master key. Where employees are allowed to use their own locks, however, management cannot search a locker unless it has the employee's consent or is prepared to break the lock or the employees have furnished management with a copy of their keys to keep on file in a secure location. Unassigned lockers, of course, should be kept locked with a company lock at all times to avoid their being used as a convenient, nonidentifying place to stash contraband or pilfered items.

The flat tops of employee lockers can also be used for hiding articles. If the lockers are set against a wall rather than recessed into it, a board should be affixed slanting from the front edge of the locker top to the wall. Many lockers come with such a fixture already installed. On lockers in the middle of a room, the same effect can be obtained by topping the lockers with a tent-shaped form closed at each end. New lockers can be purchased so constructed. Where there is space under lockers, this area should also be closed off to prevent concealment of items.

The company regulations of some air carriers allow the search of employee automobiles on company property. Restricting the search of employee's vehicles to those parked on company property can seriously hamper an investigator. One of the prime methods of air freight theft is for employees to park their cars at those airport locations they will pass with a tractor during their shift, probably under cover of darkness. If such a clause is currently in company regulations, it should be removed.

Whatever the company policy in regard to employee searches, that policy must be spelled out in detail in company regulations and in employee information booklets so that all employees are aware of it. Failure to publish the company rules on searches can have adverse effects, particularly in grievance cases. The validity of the search will most certainly hinge on the employee's expectation of privacy, and that turns on whether he or she has been put on notice.

INTERVIEWS AND INTERROGATIONS

The primary legal questions that arise in the area of interview and interrogations have to do with the admissibility of the resulting statements in court. The rules regarding statements made to law enforcement officers, and the warnings and advice that these officers must give a suspect prior to the taking of an admissible statement, are well established in law.[10] However, case law generally clearly holds that these warnings are not required in an interview being conducted by an individual who is not a law enforcement officer. Investigators and security guards who are company employees have repeatedly been held to be "private persons" in law and are not required to give these warnings before interrogating a suspect or before taking a confession.

On the other hand, the admissibility of a statement may legitimately be threatened by any indication that the statement obtained was not wholly voluntary. Statements made because the employee was threatened with loss of

---

[10] *Miranda v. Arizona*, 384 U.S. 436 (1966).

job, promised immunity, or told that cooperation would make things easier can be challenged and may be held inadmissible on the grounds of intimidation or coercion.

Some question exists as to the employee's right to remain silent when the company wishes questions answered about company matters. Wisely, many companies have regulations requiring employees to cooperate with company investigators in the course of an official investigation as a condition of employment. Threatening an employee with loss of his job if he or she refuses to furnish a statement may invalidate any admission so obtained, but terminating an employee for refusing to cooperate is generally accepted as a discharge for insubordination. Some union contracts include a work rule that employees must answer questions about company business during working hours.

Some union contracts also allow the employee to request the presence of a union representative during any interview. A landmark court decision has extended this to a legal right if the employee reasonably assumes the interview may lead to disciplinary action. The decision, *National Labor Relations Board vs. Weingarten, Inc.,*[11] should be read closely by security professionals and investigators. A purely investigatory interview seems to be exempt, while an interview that might lead to disciplinary action clearly comes under *Weingarten.* Although the investigator cannot require the employee to have the interview in the absence of the requested union representative, the investigator is not required to interview the employee:

> The employer is free to carry on his inquiry without interviewing the employee, and thus leaving to the employee the choice between having an interview unaccompanied by his representative or having no interview and foregoing any benefits which might be derived from one.

The decision also explicitly states that the employee can choose to have an interview unaccompanied by his union representative. There is no requirement stating that the employee must be asked if he wants to have a union representative present, unless it is specifically spelled out in the contract.

Weingarten was decided in the absence of any specific contract provisions relating to union representation during an interview. Had there been contractual agreements, those agreements would have been controlling. The language of the contract will have a significant impact upon the employee's right to representation. The union contract may contain rules of its own--for example, that the employee must be advised of the right to representation. For this reason, the investigator must be familiar with the specific language of each union contract, as the contract covering mechanics may differ in this regard from the agreement covering pilots or flight attendants. In each investigation, it is important that the investigator know the specific contract involved, his or her obligations under it, and the rights and privileges of employees covered by it.

If the employee insists upon having a union representative present, it may be better to terminate the internal investigation and refer all the informa-

---

11 95 S.Ct. 959 (1975).

tion gathered to the law enforcement agency that has jurisdiction. This move could result in a worsened position for the employee, because the chance of prosecution is increased, but there is limited utility in questioning an employee in the presence of another worker, let alone a union representative. In some cases, the union representative is only an observer and may not participate in the interview or give advice to the employee. In appropriate cases, the investigator may legitimately offer the employee the choice of talking with the company or with the law enforcement agency having jurisdiction. In the absence of consent to a company interview, the investigator may turn the case over to the law enforcement agency. The police officer is not bound by *Weingarten*, but is bound by *Miranda*.

Signed admissions taken from employees and other suspects should include a preamble stating that the investigator was identified by name and job title to the employee and has advised the employee of the matter under investigation. There also must be a positive declaration to the effect that the statement is made voluntarily, of the employee's own free will, and that no duress or promises of immunity were used to obtain the statement.

Company employees are occasionally asked by outside investigators to make written statements that can affect company liability on a given matter. One example is when the FAA or another governmental agency takes statements from company employees in the course of an investigation of an alleged violation of Federal Aviation Regulations. The company may wish to consider instructing employees not to discuss any case with government investigators without the advice and consent of the company's law department. Should employee statements become necessary, it is perhaps better that they be taken by company investigators, so that the statements may be reviewed by the carrier's law department before they go to the government. This, of course, assumes that the government agency is willing to go along with the procedure.

Recording Interviews

Some record of the entire course of each interview is essential. Although tape recording provides the most complete reproduction, several considerations are involved in its use. Recording without the knowledge of the person being interviewed (provided it is done with the knowledge of the interviewer) is legal under federal law, but many states, such as California and Illinois, have enacted statutes prohibiting this activity unless agreed to by all parties of a conversation. Wiretapping or electronic surveillance by persons other than those authorized law enforcement officials acting under court order is totally prohibited by the Omnibus Crime Control and Safe Streets Act of 1968. In the absence of state legislation prohibiting it, the law may permit managers or private security officers to electronically intercept a conversation over the telephone, provided the purpose is a legitimate one. This is an area in which security practitioners should make certain they are acting on the advice of their legal department, lest they inadvertently expose the company and themselves to civil and criminal liability.

Many interviewers routinely record with the knowledge of the person being interviewed, and some make a practice of always placing the recorder in full sight. With the tape running, the interviewer states that the interview is being recorded and establishes that the person does not object. If the situation is such that the *Miranda*[12] or *Escobedo*[13] rule applies, the advising of rights should also be on tape.

It is true that recording, particularly with the machine in full sight, often has an inhibiting effect, and obtaining admissions under such circumstances may require more expertise on the part of the interviewer. Nevertheless, a tape recording is the most positive evidence against later claims of duress or coercion. In my opinion, however, no form of recording can be substituted for having another person present when a male investigator interviews a woman.

Many law enforcement agencies are utilizing video cameras for taping interviews and interrogations and stops of intoxicated drivers. There is a place for the video camera in private security interviews. Although in its infancy, this use deserves some innovative exploration.

Whether the interview is being taped or not, a log of the interview should be openly maintained. It should include such information as who is present, the exact times the interview began and ended, who advised the subject of his or her rights and at what time, together with anything unusual that may occur. The subject's requests to smoke, to go to the bathroom, to have a drink of water, and to see an attorney or a union representative should also be logged, together with the action taken on those requests. This log can be used to counter charges of duress and such claims as that the subject was interrogated six hours nonstop rather than the actual 60 minutes the interview lasted.

Currently, one firm is the target of allegations by several hundred former employees. They claim that they were forced to confess to thefts they did not commit and pay back money they did not steal, through coercion and extortion by the security personnel of the firm. One of the investigators has given sworn affidavits that he chose suspects at random and falsely told them he'd seen them steal or had their crime on videotape. Aside from the damage done to those who were forced to confess to crimes they now say they did not commit, this type of coercion has no place in the security profession. It subjects both the investigator and the firm to serious civil liability and seriously questions the professionalism of the investigator.

TRUTH VERIFICATION DEVICES

The two primary truth verification devices in use today are the polygraph, known colloquially as "the lie detector" and among investigators as "the lie box," and the psychological stress evaluator, or PSE.

Most investigators are reasonably familiar with the polygraph and its operation. Certain relevant questions and many irrelevant questions are asked

---

12 *Miranda vs. Arizona*, 384 U.S. 436 (1966).

13 *Escobedo vs. Illinois*, 387 U.S. 476 (1964). "Focus of the investigation" test.

of the interviewee as he or she sits in a chair being monitored by attachments which register galvanic skin response, blood pressure, respiration, and heart rate. The amount and kind of changes or the lack of change in these functions are recorded on chart paper, providing the examiner with visible data with which to evaluate significant answers as truthful, untruthful, or inconclusive, the latter means that a significant distinction between reactions is lacking, and the subject has neither passed the test nor demonstrably lied. Most investigators feel they receive more than their fair share of reports marked "inconclusive."

The PSE is a more portable device that analyzes recorded muscular microtremors in the voice. The degree of stress in the voice is read as indicating whether a specific statement is false or not. A recording of the interviewees yes and no responses to perhaps 12 to 15 structured questions is played through the PSE, producing a graphic display of the voice pattern on a continuous roll of graph paper. This vocal response can be analyzed for stress in 32 different ways. The striking advantage of the PSE is that investigators trained to structure questions properly need no more than a tape recorder to complete field interviews that can be analyzed elsewhere by a trained evaluator using the PSE. The tape can be mailed to the evaluator or transmitted over the telephone.

The PSE has come under criticism from the American Polygraph Association, whose members question its reliability. As widespread use of the PSE, which requires only a few days training, would cut deeply into the business of polygraph operators, their opposition is not regarded as entirely disinterested. Polygraph operators, who fight a continuing war against laws banning the polygraph, have succeeded in getting legislation in some states to limit the use of the PSE. In Illinois, for example, a law forbids the use of any device as a truth verification device if it measures less than two body functions. Because no other single-function equipment is presently available, the legislation appears to be directed at the PSE.

Generally, the results of polygraph examinations are not admissible in court proceedings unless their introduction is agreed to by both defense and prosecution, a notably rare event. However, of the 11 U.S. Circuit Courts and District Courts, including the District of Columbia's court, nine have admitted polygraph evidence. However, the chief value of a truth verification device to the investigator does not lie in its admissibility. As a practical matter, most confessions obtained by a polygraph examiner are made during the pre-examination interrogation before the individual is tested. Hardened thieves will not confess, even though the device indicates they are not telling the truth, but the results can still provide useful indications to the investigator.

The Employee Polygraph Protection Act of 1988 (Public Law 100-347), which became effective December 27, 1988, was praised as a "holiday gift to working people" by the American Civil Liberties Union. It has had a significant effect on when, where and how the polygraph is used in the United States.

The law prohibits an employer engaged in interstate commerce from requiring, requesting, suggesting, or causing an employee or prospective employee to take or to submit to a lie detector test. Nor can the employer discharge, discipline, discriminate against, or deny employment or promotion to an employee or prospective employee for exercising any right under the law. The act encompasses not only the polygraph but the psychological stress evaluator, the voice stress analyzer (VSA), and any other electrical or mechanical device used for the purpose of attempting to determine whether a person is telling the truth.

In addition to a fine up to $10,000 against the employer, the employee or prospective employee can file a civil suit for damages against the employer. There are some exemptions to the act. It does not apply to federal, state, or local governments. Nor does it apply to testing by the Federal Government of certain private individuals engaged in national security related activities. It can be administered in the private sector, subject to restrictions, to certain prospective employees of security service firms (armored car, alarm, and guard companies) and to pharmaceutical manufacturers, distributors, and dispensers. The act also permits polygraph testing, subject to restrictions, of employees of private firms who are reasonably suspected of involvement in a workplace incident (theft, embezzlement, etc.) resulting in economic loss to the employer.

The conditions under which an examination may be given are relatively strict; they are so strict, in fact, that they may totally discourage some companies from administering these tests. More probable, however, is that the law departments of may corporations may prohibit security personnel from administering the tests because they are unwilling to assume the risk of a civil suit.

Before a test can be given the employer must execute a statement describing the particulars of the incident and the basis for testing the employee, the specific economic loss or injury, the employee's access to the property, and the basis for the employer's suspicions. The employee must be furnished a copy of the statement, and the company should retain a copy for a minimum of three years.

The company must also agree that:

- The employee can terminate the examination at any time.
- The employee will not be questioned in a degrading or needlessly intrusive manner.
- No questions will be asked about religious beliefs or affiliation, beliefs or opinions about racial matters, political beliefs or affiliations, sexual behavior, or labor union beliefs, affiliations, opinions, or lawful activities.
- No test will be given if written evidence by a physician indicates that some condition or treatment might cause abnormal test responses.
- The examinee will be given a written description stating the nature of the examination; whether there are two-way mirrors, cameras, or other devices in the test area; whether any of the devices will be used during the examination, and that either the examinee or the interviewer may (with mutual knowledge) record the examination.

- Before the test, the examinee must read and sign a written notice that states that the examination cannot be required as a condition of employment and that any statement made during the examination may constitute additional supporting evidence for adverse action and that enumerates the individual's specific rights under the act.

- Before the test, the examinee will be provided the opportunity to review all of the questions to be asked and informed that he or she has the right to terminate the examination at any time.

- During the test, no questions will be asked that were not reviewed by the examinee before the examination.

- After the test, no adverse action will be taken unless the employer interviews the examinee on the basis of the test results and provides the examinee with a written copy of the polygraph examiner's opinion or conclusion, a copy of the questions asked, and a copy of the charted responses.

- The polygraph examiner cannot conduct more than five polygraph examinations in one day, and each test must be not less than 90 minutes in duration.

- The examiner must be licensed in the state where the test is conducted if a license is required; must have professional liability coverage of not less than $50,000; must provide any opinions in writing, based exclusively on the polygraph charts, and free of any information other than case facts, examinee admissions, chart interpretations; and must make no employment recommendations in the report.

- The polygraph examiner must maintain all records related to the examination for at least three years after the examination.

One of the chief dangers that the polygraph presents to the investigator is the temptation to substitute its use for genuine, admissible investigation. In simple terms, the investigator can get lazy and use the polygraph as an investigative shortcut, rather than as a last-resort resource for use when the investigation cannot clearly allocate responsibility or when it is essential for insurance recovery to have admissions of amounts stolen.

Although some polygraph operators advocate wider and more frequent use of the polygraph, I believe that its usage by business should not be too widespread. Corporate controls should establish that specific approvals will be required before an employee can take a polygraph examination to prevent indiscriminate usage. Employees normally fear that the operator will ask questions about areas of their lives that they properly consider off limits, or that the machine may indicate that they are not telling the truth when in fact they are. Although such fears can be reduced by a skilled examiner, the truth verification device has the potential of being an invasion of the individual's privacy and should not be used unless the totality of the circumstances justifies it. Nevertheless, the major benefit of the truth verification device is that it offers independent evidence of a person's innocence, and indeed the most frequent use of the polygraph is for this purpose.

INTEGRITY TESTS

A number of paper and pencil tests designed to test employees' honesty, attitudes, or drug involvement are on the market. I feel that they have some validity, but I'm not an expert on those tests. My limited experience in utilizing them was in the screening of security guard applicants for a guard force of about 400 people. The reject rate was so high, we were having difficulty staffing our posts. Then, in some instances, people who had passed the test with flying colors were caught in acts of thievery. Similarly, other applicants who had worked for us under a previous guard contract and whom we felt were excellent employees were rejected by the new contractor because they did not pass the test. Our skepticism continued to increase until we ultimately abandoned the test instrument. In this case, we were using two tests; one designed to test for honesty and the other for drug usage.

Two Chicago area companies, London House and Reid Psychological Systems publish approximately three-fourths of the integrity tests used in the U.S., and about 50 other companies publish the tests. Significant controversy has existed over the accuracy of the tests and the fairness of using them to hire, and especially, to turn away job applicants. Massachusetts and Rhode Island have made it illegal to deny someone a job because of the results of an integrity test and other states are considering such legislation. The U.S. House Committee on Education and Labor requested a study of the tests that likely will serve as a basis for hearings. The two-year study by the Office of Technology Assessment (OTA), was one of two major reports released in 1990. The second report was an extensive study conducted by the American Psychological Association.

If I were to use integrity tests again, I would explore the results obtained by other companies and closely monitor the effectiveness of these test instruments during their usage. Critics of these tests cite a high yield of false negatives--scores that label honest applicants as risky. Even the strongest advocates of the polygraph generally only claimed a 90 to 92 percent accuracy. If the written tests yield similar accuracy results, then they're worth a try, yet I confess I would be angry and feel I had been dealt with unjustly if I were rejected for a job based on the results of an integrity test.

PROSECUTION

When an employee is discovered to have stolen from the company, the first step toward preventing further and similar crimes--prosecution--must be taken. If the company is willing to accept restitution or to merely terminate the employee with or without restitution, further theft problems can be expected. Word of company discipline--or the lack of it--travels fast. An employee who knows from past reports that dismissal is the worst possible consequence of theft is more likely to steal than if the employee knows he will be prosecuted. Few criminals, amateur or professional, believe they will be caught. Where the rewards of dishonesty appear great, detection improbable, and the consequences of being caught no more than minor, few employees will resist temptation for long.. On the other hand, the sight of uniformed police officers escorting an employee in handcuffs from the place of business to jail works wonders in reducing the incidence of theft.

In any case, standards of conduct must apply equally to all employees. It is not only wrong but dangerous to get tough with blue-collar workers while allowing white-collar workers to resign quietly, particularly when their thefts total as much or more than those of others more severely punished.

RESTITUTION

A company may elect to accept restitution in isolated cases, but this must be done with care. If an arrest has been made or prosecution is contemplated, the company should consult the relevant law enforcement agency or prosecutor's office before any agreement to accept restitution is approached. Many good prosecutions have been dropped because the employer agreed to accept a few hundred dollars in restitution before the criminal case could be brought to trial.

Restitution agreements can also impair recovery of otherwise insured theft losses. Once an agreement to accept restitution has been made, the employee merely owes a monetary debt for which there is no more recovery than for any other debt. The employee who fails to pay cannot be proceeded against unless he or she was prosecuted and convicted and restitution was made a condition of probation or parole.

DRUGS IN THE WORKPLACE

The impact of drugs and alcohol on a company is far-reaching. General Motors, for example, pays health benefits for 1.2 million people that amount to $2 billion per year. Nearly $1.5 billion of that cost is related to drugs or alcohol. In a study of 70,000 employees at GM[14], it was found that 10 percent of the workforce was using seven times the amount of health benefits they should have. One of three groups in the study was identified as having particularly poor work records, and 75 percent of the people in that group had a drug or alcohol addiction. This group used an average of 90 sick days per year. When you consider that there are only about 220 work days available each year, it becomes clear that these employees were missing nearly 41 percent of the work year. A control group of average employees used only about 4 days per year, or less than 2 percent. Not only is drug and alcohol abuse costly in the area of medical claims and absenteeism, but in the GM study, it was found that employees who were drug or alcohol dependent were involved in four times as many accidents as their fellow employees. Workers

---

[14] "How GM Fights Drugs," *Advisor*, Team Building systems, Inc. 1988.

with a drinking or drug problem are absent 16 times more often, have four times more accidents, use a third more sickness benefits and file five times more compensation claims than an average worker, according to Vice President Dan Quayle, then chairman of a subcommittee that issued a report on the subject.[15] Additional costs relate to many other areas such as poor and reduced work performance, low employee morale, theft, and embezzlement.

The National Institute on Drug Abuse (NIDA) estimates that as many as 23 percent of American workers abuse drugs on the job[16] and according to the *Journal of the American Medical Association* as many as one of every 18 employees is a drug dealer.[17] The threat in the transportation industry has been identified as being particularly severe. Mechanics who overhaul and repair aircraft and the crews who fly them must be free from drugs and alcohol or the safety of the traveling public is jeopardized.

Drug usage is particularly dangerous in the transportation industry. The National Transportation Safety Board (NTSB) attributed a fatal 1983 air accident to illegal drug abuse. Autopsies showed that the pilot of a cargo plane had been smoking marijuana, possibly while flying, just before he crashed at Newark Airport, killing two. In March 1985, an air traffic controller who had been injecting three grams of cocaine daily at work put a DC-10 on a collision course with a private plane. At the last moment, the smaller aircraft made an emergency landing. According to Dr. Robert Wick, corporate medical director for American Airlines, a computer operator who was high on marijuana failed to load a crucial tape into a major airline's computer reservation system. The system was out of service for around eight hours, costing the company about $19 million.[18] On January 19, 1988, a Metroliner crashed during its approach to Durango, Colorado. The captain and eight passengers died in the crash. The National Transportation Safety Board has advised that there was evidence of cocaine use by the captain

The argument that what employee does on their own time is not the employer's business is losing credibility. It is not easy to turn off the effects of drugs and alcohol between 8 a.m. and 5 p.m. In November 1985, researchers at Stanford University School of Medicine and the Palo Alto Veterans Administration Medical Center published the results of a study on how marijuana use affects the ability of pilots to land planes.[19] The pilots tested their flying skills in a simulator after smoking marijuana. A full day after taking the drug, long after any sensation of being high had passed, the pilots were still swerving dangerously upon landing. One "crashed" his plane beside the runway. The study concluded that marijuana users have difficulty performing complex tasks or doing work that demands quick reaction time for as long as 24 hours after smoking the drug.

In November 1988, President Reagan signed into law the Drug Free Workplace Act of 1988. The law, which became effective on March 8, 1990,

---

15 Wall Street Journal, February 27, 1986.

16 Michael Marrone, "Drugs in the Workplace: No Quick Fix" *Security* (September 1988):62

17 Ibid.

18 "Battling the Enemy Within," *Time* (March 17, 1986):55

19 *American School of Psychiatry* 142:11. *Psychology Today* (February 1986):68

applies to all companies in the United States that have government contracts greater than $25,000. Therefore, any air carrier that carries mail is affected. The act requires that employees of federal government contractors be forbidden to manufacture, distribute, dispense, possess and use illegal drugs and controlled substances in the workplace, and it imposes stiff penalties--including the loss of government contracts--upon companies that do not meet certain requirements. The new law also requires affected companies to:

- Notify their employees of their drug policy and the penalties for violating this policy.
- Notify the government of an employee's conviction involving drugs in the workplace.
- Establish a drug free awareness program so that employees can understand the dangers of using drugs in the workplace.
- Notify employees of any available counseling or rehabilitation services and employee assistance programs.

Under the law, compliance with the drug free policy is generally regarded as a condition of employment, although termination is not mandatory. Many companies offer drug rehabilitation programs, but they would be in a position to require participation on the part of the employee. The employee is required to notify the company within five days following their conviction of a criminal drug violation for making, using, possessing, dispensing, or distributing drugs in the workplace. The company is then under legal obligation to notify the government.

DRUG TESTING

Drug testing has been in the newspapers and courts constantly over the past few years. Regardless of the industry, drug testing has generally been introduced by management and fought by the unions. It would take another book to document historically all of the arguments and court cases that have helped develop the law on this subject. Questions still need to be resolved relating to privacy, education, counseling, and treatment, but in the meantime, the courts have said that affected airline employees will be subject to random drug testing.

In March 1988, the FAA published a proposed rule that would required airlines and other large aircraft operators to test employees involved in safety- and security-related work for drug use. The directive, applicable to all Part 121 carriers and all Part 135 carriers with 50 or more covered employees, was ultimately issued. (A Part 121 carrier conducts operations with aircraft having a passenger seating configuration of more than 60 seats. A Part 135 carrier operates aircraft having from 31 to 60 seats.) The Drug Abatement Program required the carriers to submit plans by August 18, 1989, on how they would test safety-related employees for drug use. The tests were to be implemented by December 18, 1989.

In July 1990, the 9th U.S. Circuit Court of Appeals in San Francisco, in a 3 to 0 decision, upheld the government's random drug testing of an estimated 538,000 commercial airline employees and air traffic controllers. In 1989, the U.S. Supreme Court upheld drug testing of Customs Service employees who sought promotions to jobs involving drug enforcement or guns. The Appeals Court, relying on the earlier precedent, analogized "The government interest in preventing drug use by persons holding safety-sensitive positions in the aviation industry is at least as compelling as the interest in preventing drug use by Customs officers." Additionally, the court noted that the FAA had provided evidence of some drug use by airline employees, including treatment of a number of flight crew members for cocaine overdoses or addiction and the presence of drugs in the bodies of two pilots after crashes.

Interestingly, the court decided that the FAA's "interest in promoting air safety through random drug testing outweighs aviation industry employee's Fourth Amendment right to be free from unreasonable searches and seizures." That language is almost identical to that used by the courts to uphold the predeparture screening of passengers. For example, in *Barrett v. Kunzig*,[20] the court stated:

> When the interest in protection of government property and personnel from destruction is *balanced* against any invasion to the entrant's personal dignity, privacy and constitutional rights, the government's substantial interest in conducting this cursory inspection *outweighs* the personal inconvenience suffered by the individual [emphasis added].

Random testing means that employees who are not suspected of drug use can be forced to submit to urinalysis without notice. One interesting innovation in drug testing is being utilized by San Diego Old Town Trolley and by the oil tankers operating out of Valdez, Alaska. Instead of a urinalysis, workers are asked to sit down and play a video game designed to test hand-eye coordination. If they pass the test they go to work. If not, they don't.

Customs pays close attention to airline employees when checking for drugs at points of entry. Because of their ease of travel, a number of airline employees have become involved as couriers or in the purchase, sale, and distribution of drugs.

ALCOHOL

In March 1990, three Northwest Airlines pilots operated a flight from Fargo, North Dakota to Minneapolis/St. Paul while under the influence of alcohol. The captain and first and second officers were arrested and subsequently discharged by Northwest for violating a company rule prohibiting Northwest crewmembers from drinking within 12 hours of a flight departure. Federal

---

[20] 331 F. Supp 266 (1971).

Aviation Administration regulations require that the blood alcohol level not exceed 0.04 percent. The blood alcohol level of the captain was 0.13 percent, according to tests taken two hours after the flight landed.

News reports state that the captain was reported to have drunk from fifteen to twenty rum and colas the evening before the flight and to have left the Speak Easy Bar at about 11:30 p.m. The first officer and flight engineer were reported to have shared six or seven pitchers of beer. The captain returned to the bar to get directions to his hotel, which was about three blocks away. None of the crew got more than 5 1/2 hours of sleep. Blood tests showed the captain still had the equivalent of eight drinks in his system after the plane departed at about 6:25 a.m. The tests also showed that the first officer and the flight engineer had the equivalent of three beers each still in their systems.

The captain's defense at the trial was that he was an alcoholic, and therefore, he had built up a tolerance to alcohol. The crew was found guilty of violating a 1986 federal law that prohibits transporting commercial passengers across state lines while under the influence of alcohol or drugs. Conviction carries a maximum penalty of 15 years in prison and a $250,000 fine; however, federal guidelines call for sentences of 12 to 18 months on a first offense.

The incident heightened public awareness over a long-standing problem in the industry and was responsible for several new programs proposed by the government and industry. Certainly, the problem of alcohol abuse is no worse in the airline industry than in any other industry, but the thought of the pilot of a plane carrying several hundred passengers being under the influence of alcohol is frightening. The Air Line Pilots Association (ALPA) has said that its 42,000 members have no tolerance whatsoever toward the use of drugs or alcohol in the cockpit.

Since 1984, sixty-one professional pilots have had their licenses suspended or revoked for flying while under the influence of alcohol.[21] A 1987 check of the National Driver Registry and pilot records found that more than 1000 professional pilots had been convicted of driving while under the influence of alcohol and had failed to report the convictions to the government as required.[22]

Also in March 1990, the Department of Transportation proposed a federal alcohol abuse prevention program that was opposed by the Air Transport Association (ATA) and the Airline Industrial Relations Conference (AIRC). The ATA argued that no known accidents were attributable to alcohol consumption and that the industry had long ago confronted alcohol misuse with policies and programs designed to identify individuals with drinking problems. The carriers were opposed to random, periodic, preperformance and pre-employment testing as encompassed by the federal program.

One week after the Northwest incident, the FAA announced a six-point plan for handling such matters in the future. The plan focused on noti-

---

21 San Jose Mercury (August 21, 1990).

22 Ibid.

fication to airline management and more education and training for FAA inspectors rather than on an alcohol abuse prevention program. Virtually every major airline has a rehabilitation program through which pilots can seek confidential help for alcoholism and return to the cockpit after passing a battery of tests.

THE FUTURE

I considered including a section in this book on computer security, something of relative unimportance when the first edition of this book was published in 1976. Because the subject has been thoroughly covered in other publications by those far more qualified than I, I'll continue to rely on their expertise. Computer security is now of such vital importance that the security professional must have at least a basic understanding of computers to do THE job properly. I have just completed the investigation of a theft-of-trade-secrets case where it is alleged that computer programs written by one company were stolen from that company and taken to another where they were marketed to a third. Although security professionals will necessarily rely on those with degrees in computer science or electrical engineering for assistance with cases like this, a general familiarity with computer operations is essential.

Additionally, it is rare now that an investigation is conducted without some type of computer interface. The ease with which spread sheets can be prepared for analysis of patterns and trends; the clarity which graphics or a chart can bring to a complicated investigation; the ease with which reports can be written on a word processing program, all make computer literacy important.

In an article on the future of airline security[23] I suggested "Moore's Law," a new law of security and loss prevention. Slightly expanded it reads as follows: "As new marketing and production methods are developed through new technology, former security vulnerabilities are remedied by the new technology. However, commensurate new opportunities for fraud occur, which are more difficult to prevent and investigate than the former." Our best efforts will be required to stay current in our profession.

---

23 Kenneth C. Moore, "Airline Security in the Year 2000," in *Security in the Year 2000 and Beyond*, ed. Louis A. Tyska and Lawrence J. Fennelly, (Palm Springs: ETC Publications, 1987), 192.

Chapter 17

TICKET FRAUD

Significant progress is being made against the problem of stolen tickets. Because most ticket thefts occur at travel agency offices, the adoption of the Air Traffic Conference's mandatory security procedures for travel agents' ticket stock has contributed to reducing the number of stolen tickets on the market.

A great deal of confusion exists for the average consumer in purchasing airline tickets. Even for legitimate tickets purchased through reputable travel agencies, the public is aware that a multitude of fares exist to get from point A to point B. The fares may differ not only by carrier and class of service but by advance notice, which days of the week the passenger is traveling, whether they are staying over a Saturday night, and whether the moon is full or not. It is quite possible for a person who paid $150 for a ticket to be sitting next to someone who paid $500 for the same flight.

Because everyone is looking for the lowest possible fare, passengers have become extremely resourceful in buying tickets. There is a "gray market" in airline tickets. Some of these tickets are great buys, and some will create problems for the purchaser. Particularly risky from a financial standpoint is purchasing a ticket from a personal ad in the newspaper. Some are stolen tickets, some are frequent flyer tickets, and some are being sold by consolidators; some are O.K., and on some you will get hurt.

There is an old rule that says, "If it sounds too good to be true, it probably is." One example of that is a Chicago-based travel agency that advertised $29 round-trip fares to Hawaii, a good deal even in 1987. The agency agreed to pay the California Attorney General a $50,000 fine and to stop its misleading advertising, which had noted only that "restrictions apply." The "restrictions" turned out to be be inflated rates at hotels in Hawaii that had to be booked through this agency, at prices significantly higher than travelers would have paid if they had booked themselves. The $29 deal usually ended up costing from $225 to $1200. Parenthetically, in such cases the laws of California and several other states give telephone buyers a right to cancel their purchases within three days. Those who pay by bank credit card should have 60 days to notify banks to cancel the payments.

An important contribution has certainly been made by the use of available computer memory to identify stolen tickets. For example, United Airlines enters into its Apollo Reservations System the identifying numbers of all stolen tickets, stolen or recalled credit cards, and the numbers of checking accounts on which bad checks have been written. To check the validity of a ticket, ticket agents enter the ticket number into their VDT[1] and receive a virtually immediate response from the computer. If the ticket is listed as stolen, the employee can qualify for a cash reward by retaining the ticket to prevent its use. Other air carriers have similar systems, and many subscribe to TRW's Validata system, which provides its subscribers with the same service. Validata is said to average between 50 and 100 detections per day of forged or stolen tickets, voided credit cards, and bogus checks.

British Airways was the first carrier to have used a system known as "TicketCheck," manufactured by Rand McNally & Co., the major manufacturer of airline tickets. TicketCheck is merely a bar-code reader wand that scans the ticket number, air fare box, and other key sections. It takes two seconds to check the ticket number against a hot list in the database to determine if the ticket has been stolen or whether alterations have been made.

Tickets may also be computer checked after the flight has departed, an excellent and often productive practice. Postdeparture checks, which need only be done on a spot-check or random basis and at larger stations, are particularly effective when done for flights of two hours or more. If a stolen ticket is discovered, a phone call to the destination station can arrange for airline management or security personnel to meet the flight. This affords an opportunity to interview the passenger, to determine where the ticket was obtained, and to make arrangements for payment. The disadvantage of postdeparture checks is that the probability of collecting the fare due for the travel is much less when the passenger is confronted after the flight is completed; the passenger challenged before boarding will usually pay the fare and make the flight.

TICKET THEFTS

Because stolen tickets are so easy to sell, travel agencies have long been plagued with burglaries and armed robberies. In November 1989, a Joint Task Force on Crime Against Travel Agents was formed to study methods by which carriers can track stolen tickets and agents can lessen the risk of ticket theft. The task force was made up of members of the Air Transport

---

[1] Video display terminal. Also referred to as CRT (Cathode Ray Terminal).

AssociationAir Transport Association (ATA) and Airlines Reporting Corporation (ARC) and it was formed because of a marked increase of robberies, one of which resulted in the death of a travel agent. The task force was asked to focus on four major areas: (1) industry procedures for communicating stolen ticket serial numbers to airlines; (2) technology that could improve detection of stolen tickets, such as magnetic stripes or bar codes; (3) antifraud programs conducted by individual airlines; and (4) security at travel agencies.

In the first 11 months of 1989, twenty-three armed robberies of travel agencies were reported, in contrast to only six in 1988. All but two of the 1989 robberies occurred in southern California. There were 62 burglaries of travel agencies in November 1989 alone, the same number as for the first 11 months of 1988. There were 70 travel agency burglaries in 1988. Not only is ticket stock stolen, but validators and validator plates are taken also, so that the tickets will be easier to pass. During the first 11 months of 1989, 4800 stolen tickets were recovered, compared to 7000 in 1988. It has been estimated that each stolen ticket represents an average loss of $800 to the carrier.

USE OF STOLEN TICKETS

The black-market sale of stolen tickets remains highly profitable. The tickets are usually sold to individuals in such places as bars or beauty parlors by fellow patrons who profess to want to do the person a favor. Stolen tickets have actually been advertised as bargains in underground newspapers and have been sold through telephone information centers set up to provide referral services for drug counseling, abortion clinics, free venereal disease checkups and legal aid. In such cases, the exchange gave a referral number for "inexpensive travel." Tickets sold this way might be accompanied by detailed instructions on how to fill the tickets out and how to use them. One such set of instructions advised, in part:

> You cannot be prosecuted for the use of this ticket, if you are reasonably cool. The airline is not losing cash; you are filling an otherwise empty seat. You cannot be prosecuted unless it can be proven that you know that the ticket was stolen. It is not worth their time and money to prosecute, unless they think you can finger someone. It is advised that you go clean (i.e. without drugs, narcotics, marijuana, etc.), as contraband can complicate an already touchy situation. The use of this ticket involves an element of risk.

The stolen ticket may also be sold to someone other than the would-be traveler. In 1974, at least one air carrier bought back 2000 tickets from underworld representatives for a sum said to be less than $70,000. The carrier stated that the payment for recovery was a business decision; the ransom paid was $10 per coupon versus the average loss on a stolen coupon, which the Air Traffic Conference then put at $300. Now, the figure is closer to $800. The carrier felt the potential loss to the airline, had the tickets been used for travel, could have been $2 million or more, and that a ransom payment of the much lower price was justified. Others disagreed, believing that the purchase of the carrier's tickets from the underworld was paying blackmail. The *New York Times* called the transaction "a shocking breach of corporate responsibility."[2]

Indeed, such a decision entails other considerations than just the prevention of a potential monetary loss to the carrier. Some people in the industry believed that such a practice, by in effect creating a ready resale market, might motivate persons to steal tickets who would not otherwise have seen the possibilities of profit in doing so.

The active pursuit of stolen or fraudulent tickets is often hampered by a mistaken opinion that pervades the airline industry itself. Many believe that if a stolen ticket is used on a flight, the only cost to the the airline is the meal served to the freeloading passenger; the plane was flying to the destination anyway, and the seat would probably have been empty otherwise.

This is a highly questionable and potentially dangerous thesis. Besides creating a feeling of general apathy about the stolen ticket problem, the theory ignores several key facts. In the first place, stolen tickets are most attractive at peak travel periods, and a revenue-paying passenger would often have occupied that seat. Also, stolen tickets can cost the carrier their full face value in cash; many stolen tickets are resold to professional ticket refund fraud operators.

Also, when the stolen ticket of one airline is used for travel on another, if there is no loss-sharing agreement, the issuing airline must reimburse the actual carrier. Where tickets from travel agencies are involved, however, the airline honoring the stolen ticket usually must absorb the loss, on the theory that, by failing to check the stolen ticket bulletin, the airline has chosen to assume the monetary risk. Where loss sharing agreements are in effect, all parties suffer loss. Obviously, any payment that is made against nonexistent revenue must necessarily be deducted directly from profits.

TICKET CONSOLIDATORS

An airline seat has a unique value. Before the plane leaves the ground, a seat may be worth several hundred dollars. Once the plane takes off, an empty seat is absolutely worthless. Therefore, if it is possible to sell that seat, even at a greatly reduced price, the airline is money ahead to do so. To fill these empty seats, airlines will sell blocks of tickets to consolidators, who discretely sell the tickets for less than the published fares that may be printed on the ticket. They are usually stamped "nonrefundable." This enables the airline to maintain its posture of not selling below the lowest published fares. It was not possible for airlines to sell tickets this way prior to deregulation.

On many trips, particularly overseas flights, the airlines know in advance how many tickets they can make available to consolidators. The forecasting of loads is one of the functions of the new science of "yield management," which has generated significant new revenues for many carriers.

---

[2] New York Times, May 31, 1974

The consolidators advertise in major metropolitan newspapers and operate under such innovative names as Cost Less Travel, Cheap Tickets, Inc., and other names that can't be missed. (I still remember one travel agency of questionable reputation, which ultimately went out of business, with the name Fly by Night Travel. Who would buy an airline ticket from an agency with a name like that?) One of the disadvantages is that if the flight is delayed or canceled, the gray market ticket can't be used to switch airlines. The airlines may do this for the passenger, but it's not required. The reason, of course, is that airline B will bill airline A full price for the ticket. In summary, the tickets purchased from consolidators are generally good tickets, and passengers should have no problem using them.

FREQUENT FLYER TICKETS

In about 1981, airlines began rewarding brand loyalty by issuing free flights for mileage traveled. The programs were extremely popular and worked for the airlines as nothing ever had before. They also worked for a number of people who did not work for the airline, but saw frequent flyer tickets as a way to make money. Initially, the members of the airline security departments saw misuse of these programs and made senior management aware of it. The programs were so successful at generating new revenue and brand loyalty, that the marketing departments did not want the security departments upsetting the apple cart, and they were willing to absorb some fraudulent losses.

A number of things happened to change that. Coupon brokers got into the business, and this distorted the original purpose of the frequent flyer programs. Travel agents were upset because the coupon brokers were taking legitimate business from them. There are at least 18 cases of airlines suing award brokers. Additionally, the accountants began to look at the liability of unused travel the airlines were amassing with the programs -- estimated by the *Chicago Tribune* in June 1990 at $1.24 billion.[3] The bonus miles, redeemable for more flights or better seats, were debits that could decrease revenue.

As a result, the programs were modified so that the mileage expired after a certain date if it was still unused. Seats allocated for the program were reduced, so that passengers now find out they need to reserve months ahead of time in many cases in order to be able to use their mileage.

The coupon brokerage business created additional security problems for the air carriers. The brokers have established prices that they will pay for specific mileage awards. For example, they may pay $1700 for a 100,000 mile award or $1200 for fewer miles. In several cases, employees of the firms handling the frequent flyer accounting for the carriers were caught setting up fictitious accounts, pumping fictitious mileage into them, and sending the certificates to a post office box. Then, they sold the certificates to a coupon broker. There have also been several recent federal prosecutions of travel agents for frequent flyer fraud.

---

[3] San Francisco Chronicle, July 22, 1990.

Passengers soon learned that it could be profitable to fly, particularly after they reached "premier" levels where the mileage accumulated faster. They could then sell their certificates, still realizing a profit after paying for their air travel or after their company paid for their travel. It was interesting to note the progression of this.

First came a few individuals who simply flew for a living. They might start out in New York, fly to Honolulu, and return just to accumulate mileage. Then, they either became lazy or were not accumulating mileage fast enough, and they began paying others to fly in their name for them. The final straw came when they began buying tickets, checking in at the gate, giving their ticket to the flight attendant at the head of the jetway, then returning to the gate area because they forgot something, and never flying. The airline had one less person on board the aircraft than it had tickets. The first time we caught this, we thought we had a sabotage threat. It was possible that someone had checked a bag on board the flight and not flown: a certain threat situation. The flight had to take a delay while the bags were unloaded for a passenger-bag match.

Another type of problem occurred when instead of accumulating a specific number of miles, the passengers had only to fly a certain number of flight segments. The more enterprising began buying up all the tickets for short haul flights such as Phoenix to Tucson or San Diego to Los Angeles, so that regular passengers had no chance of getting a seat.

Carriers finally tightened their rules for the use of these tickets. Current United Airlines Mileage Plus literature states,

> Award tickets acquired by redeeming Mileage Plus AwardCheques can be designated to a friend or relative, subject to restrictions set forth in the Program Rules. You must designate your award tickets in person at a United ticketing location or city ticket office, and of course, you may not barter or sell your AwardCheques, certificates or award tickets. Be sure to bring official picture identification with you such as a driver's license or passport. Your ticket will be issued in the name of the person traveling. When traveling, I.D. will also be required.

American Airlines has a clear, but stern, warning that states:

> At no time may AAdvantage mileage credit or travel awards be purchased, sold or bartered. Any such mileage or awards are void if transferred for cash or other consideration. Violators may be liable for damages, litigation, and transaction costs.
>
> Fraud or abuse concerning AAdvantage mileage credit or award usage is subject to appropriate administrative and/or legal action by American Airlines including, without limitation, the forfeiture of all award certificates, tickets issued against award certificates and any accrued mileage in a member's account, as well as cancellation of the account.
>
> Use of award tickets that have been acquired by purchase or for any other consideration may result in the tickets being confiscated. If a trip has commenced, any continuing travel will be at the passenger's expense on a full-fare basis.

The cancellation of frequent flyer membership was not taken lightly by passengers. Several obtained attorneys and waged a strong campaign in an attempt to retain their membership.

To combat the coupon brokerage business, gate agents began asking for identification on all frequent flyer tickets to make certain the frequent flyer member had not sold the ticket to a coupon broker who had then resold it to a different individual. The tickets were easy to detect because the fare was shown as "Free." In such cases, the tickets were confiscated and the passengers had to purchase new tickets if they wanted to fly. In many cases, the tickets were first observed on the return flight. Passengers who wanted to return home had to come up with a credit card or cash to buy a new ticket. The airlines reasoned that even though the passenger may have paid several hundred dollars to a coupon broker for the ticket, the carrier had not received the revenue it was entitled to. Also, mileage certificates could no longer be exchanged through travel agencies, a small number of whom were circumventing the rules, to earn additional profits. The mileage certificates had to be exchanged at an airline ticket office. Coupon brokers continue to act as "brokers." After they are contacted by the seller, they locate a buyer, and the seller then arranges with the airline for the ticket to be issued in the name of the purchaser .

These dealings are risky for the program member. A notice appearing in United's inflight magazine states:

> Don't buy or sell coupons. Be warned. Any persons selling, purchasing or bartering United Airlinesd Mileage Plus AwardCheques, certificates, award tickets and mileage credit may be liabile for payment of full fare, damages, litigation and transaction costs. All such items are void and may be confiscated.[4]

United requires the member sign the following statement on the back of their AwardCheques:

> I hereby certify that I have not received and will not receive any compensation for this AwardCheque or the award for which it is being redeemed. If United Airlines determines this AwardCheque has been sold or bartered, I agree to be responsible for the full fare value of the ticket, the forfeiture of corresponding amounts of mileage and other penalties set forth in the Program Rules.

Currently, air carriers are adding alternative premiums and awards for frequent flyers. American offers savings or discounts on a variety of services. For a 75,000 mile award, they offer a $1500 discount on a new Toyota Cressida or Supra. Their catalogue offers discounts on everything from pianos to grandfather clocks to fur coats to speedboats. Perhaps the most luxurious plan is that offered by MGM Grand Air, the all-first-class airline that flies between Los Angeles and New York. Among the top awards offered are two round-trip tickets to Paris on the Air France Concorde, worth

---

[4] *Vis a Vis* (June 1991):112

$10,616, and the use of a Porsche 944 S2 Cabriolet for one year, which is valued at $15,000.

Another question that has arisen in relation to frequent flyer programs is who owns the mileage earned through these programs. The programs are not available to corporations, only to individuals. Some companies have attempted to get employees to turn over the awards they have earned. One real estate syndicator, Southmark/Envicon Capital Corporation, sued seven of the major carriers, charging them with commercial bribery in the operation of their frequent flyer programs and saying that the inability of their employees to transfer mileage and other awards to their companies induced workers to breach their fiduciary trust with employers. The New York Supreme Court dismissed the case. The court also dismissed a separate charge that American had failed to provide employee account information of mileage accrued and benefits conferred to employers.

The policy of most carriers is that the mileage information is personal and private and cannot be released to an employer, even though the employer may have paid for the ticket. Law enforcement agencies have also requested records of frequent flyer mileage because it provides an accurate itinerary of dates, flight numbers, city pairs,[5] and mileage flown; a handy tool for anyone trying to keep track of a drug trafficker. These records should not be released without a *subpoena duces tecum* or a signed release from the passenger. Each carrier has an established policy on the matter.

OVERBOOKING AND DENIED BOARDING COMPENSATION

Air carriers face a problem in accepting reservations for a flight. If they have 250 seats on a plane and stop accepting reservations at 250, there is a predictable percentage who will not appear ("no show") for the flight. That percentage varies according to the flight, time of day, day of the week, and a number of other circumstances. Carriers generally come fairly close to their prediction because of new yield management techniques. However, some passengers book reservations on two or three carriers to make certain they get a seat. Others change their plans and fail to notify the carrier. There have even been a few isolated cases of airline employees, who fly space available (SA), making phantom reservations on a flight, to ensure that there will be empty seats for them. When a flight inexplicably falls apart at the gate and there are 30 to 50 no shows, the first direction to look is at the SAs who are on the flight.

To avoid that empty seat with its lost revenue, carriers have considered a number of possible solutions, including nonrefundable deposits or tickets. The most workable solution seems to be the practice of overbooking a flight. When the percentages don't work out, 270 passengers with confirmed reservations may show up for a flight where the plane that only holds

---

[5] An airline term referring to the origin and destination. In a flight from San Francisco to Los Angeles, the city pairs are SFO - LAX. An air carrier may have a market manager for specific city pairs, such as San Francisco - Chicago. That manager would be responsible for advertising, promotion and sales on that route.

250 passengers. In cases like this, it is necessary to deny boarding to some of the passengers.

Initially, denied boarding compensation was governed by the Civil Aeronautics Board (CAB). With the demise of the CAB on December 31, 1984, carriers were free to implement their own rules, but for the most part, they adhere fairly closely to those that were imposed by the CAB. Those rules[6] provided that:

> [A] carrier...shall pay compensation to passengers denied boarding involuntarily from an oversold flight at the rate of 200 percent of the sum of the value of the passenger's next stopover, or if none, to the passenger's final destination, with a maximum of $400. However, the compensation shall be one-half the amount described above, with a $200 maximum, if the carrier arranges for other transportation used by the passenger that, at the time either such arrangement is made, is planned to arrive at the airport of the passenger's next stopover or if not, at the airport of the passenger's destination, not later than 2 hours after the time the direct or connecting flight on which confirmed space is held is planned to arrive in the case of interstate and overseas air transportation, or 4 hours after such time in the case of foreign air transportation.
>
> Carriers may offer free or reduced rate air transportation in lieu of the cash due...if (1) the value of the transportation benefit offered is equal to or greater than the cash payment otherwise required, and (2) the carrier informs the passenger of the amount of cash compensation that would otherwise be due and the passenger may decline the transportation benefit and receive the cash payment.

A passenger with confirmed reservations was not entitled to denied boarding compensation in three situations: (1) if the carrier was unable to accommodate the passenger because of substitution of equipment of lesser capacity when required by operation or safety reasons; (2) if the passenger was offered accommodations or was seated in a section of the aircraft other than that specified on the ticket at no extra charge, except that a passenger seated in a section for which a lower fare is charged would be entitled to an appropriate refund; and (3) the carrier arranged comparable air transportation or other transportation used by the passenger at no extra cost to the passenger, that at the time such arrangements were made was planned to arrive at the airport of the passenger's next stopover or, if none, at the airport of the passenger's final destination not later than 1 hour after the planned arrival time of the passenger's original flight or flights[7].

Each carrier now has its own denied boarding procedures and rules. Generally, they offer a space-available round-trip ticket to anywhere they fly domestically, a cash payment, or a combination of both. The tickets are bearer tickets and can be used or resold. The carriers are also required to get put the passenger on another flight to the same destination, whether on their flight or that of another carrier, within two hours, or the denied boarding benefits increase. Because some of the benefits of a denied boarding (DB) are

---

6 Part 250 of the Economic Regulations, §250.5.

7 Part 250 of the Economic Regulations, §250.6.

quite attractive, most carriers first offer passengers the opportunity to volunteer for denied boarding. This was also required under the CAB. Often, if it appears the flight is overbooked, an announcement will be made in the boarding area that the flight is overbooked and that volunteers are sought. Volunteer passengers then put their names on a list in the event that the DB situation occurs.

The cost to an airline of a bonus ticket is partially offset by "take-alongs." Take-alongs are those customers traveling with a denied boarding ticket holder who would not otherwise have flown on a specific carrier. "Breakage" also reduces the cost of a denied boarding ticket. Breakage occurs when a denied boarding ticket is not used within the one year validity period. One airline calculated the system average cost of a voluntary denied boarding at $77.00. The cost to the airline of an involuntary denied boarding averaged $413.00

On September 11, 1990, the 9th U.S. Circuit Court of Appeals ruled that airline passengers bumped from overbooked flights can turn down the free tickets and refunds they're offered and sue for damages. The case involved a passenger with a nonrefundable ticket. The airline switched from a 146-passenger plane to a 78-passenger plane without warning ticket agents. The court did not decide whether the passenger had a valid claim under the law, only that he had a right to sue.

Individuals looking to get something for nothing were quick to take advantage of the denied boarding situation. A number of people began making reservations and purchasing tickets on flights that frequently encountered overbooking. They the tickets on their credit card with no intent of traveling. They stood by for the flights, and if the denied boarding situation did not occur and they were told to board the flight, they would make some excuse about leaving important papers in the car or being ill and state that they were going to have to miss the flight. They then went to the ticket counter and turned in their tickets, obtaining a credit on their charge accounts. When the airlines became aware of this and started recognizing the same people pulling the same scam, they refused to refund the tickets.

On one occasion, after having caught an individual working this scam, the reservation computers were checked. It was discovered that the person had reservations on seven or eight different flights leaving from the same concourse in 15 to 30 minute intervals to a variety of destinations. When the fraud is that blatant, the possibility of criminal prosecution exists.

Employees can become involved in denied boarding frauds, also. Because an individual was repeatedly observed in the concourse attempting to obtain denied boarding compensation, another computer analysis was made. It was discovered that the same employee had made all the reservations for this individual, regardless of when the reservations had been made, a highly unusual situation. It also became evident that the employee had checked loads to find out which flights had a greater possibility of being oversold. When it was discovered that the employee and the passenger were former roommates, the conspiracy was obvious.

Because the area of denied boarding compensation has such a potential for fraud, strict controls and frequent audits are needed.

UNREPORTED TICKET SALES

In the opinion of experienced observers, unreported ticket sales fraud more often involves ticket agencies than airline personnel. In such cases, the detected discrepancies are handled as an accounting matter, and agencies are invoiced for the value of the unreported tickets. It is worth noting that such unreported sales have often proved to be a prelude to a prime default on the part of the travel agency, and thus should be regarded as a warning signal by the airline.

Investigation of this kind of ticket fraud by airline employees continues to occupy a major portion of the investigative time of most airline security specialists. The most common method used is for a ticket agent to sell a ticket for cash, pocket the money, and destroy the auditor's coupon, usually believing the discrepancy will not be caught. Indeed, the mills of accounting operations often do grind slowly, and the thief may enjoy a run of two to three months before accounting establishes that there is no auditor's coupon to matches the used ticket.

By this time, the employee has succumbed to the delusion of having discovered a unique method of undetectable crime. Naturally, the employee proceeds to sell more tickets and pocket more money. By the time the carrier's accounting department has evidence of the first unreported sale, many other such tickets will have already been written, because this kind of thief never stops with pocketing the payment for a single ticket sale. Success in theft breeds greed, and the process is usually accelerated by stupidity. Just to make the investigator's lot an even more difficult one, tickets are now sold aboard aircraft. Now, besides all the potential for ticket fraud on the ground, the same crime can now occur several miles above ground.

Once a theft has been detected by the accounting process, investigation focuses on identifying the seller, beginning with the agent's sine[8] on the ticket. However, the theft-bent employee of average intelligence can foresee this possibility and will either omit his or her agent's sine from the ticket or will enter the sine of another agent, real or fictitious.

The introduction of computer-linked automatic ticket-printing machines has added another difficulty to the task of identifying the agent who prepared the ticket. Each agent has a confidential code that he or she enters into the computer at the beginning of the shift. Indeed, the computer will not accept a sale until the confidential code has been entered. Agents sign out each time they leave their assigned position by removing their code from the computer. Should an agent forget to remove the code (and agents do), any other agent present can cause tickets to be printed with the absent agent's sine. This system's other shortcoming in accountability is that the confidential sines are, all too often, not confidential. Too many employees know each other's sines. When a dishonest employee uses this information, the investigative trail becomes even more difficult to follow.

The best evidence in these cases is an eyewitness--the customer who can and will testify about purchasing the ticket for cash from a specific agent and is prepared to identify that agent in court, if necessary. This evidence is

---

[8] Usually a two digit identification code assigned to an individual employee.

and is prepared to identify that agent in court, if necessary. This evidence is so desirable that, though it involves considerable work for the investigators, the passengers who purchased unreported tickets should routinely be located and interviewed.

Furthermore, it is important to identify the counter position at which the agent who sold the ticket was working, and--as nearly as possible--at which time of day the ticket was purchased. This information can prove invaluable when checking the schedules that show which agents were working at that time on that date. The investigator should also obtain as good a physical description of the ticket agent as possible. The passenger may recall no more than that the agent was male rather than female or white rather than black, but this information, coupled with other, more complete descriptions obtained from other purchasers, will eventually lead to the guilty party.

Another form of "pocketing" occurs at small stations where the standard accounting precaution of double accountability is not always enforced. Where the agent who writes and receives cash for a one-way ticket is the same person who receives or can gain access to the ticket lift[9] for the flight, it is possible to destroy all evidence of the sale. This can be countered by "shopping" the small stations where this can occur in much the same way test shopping is done in retail stores. An investigator not known at the station makes a cash purchase of a one-way ticket and departs. A later check is made to ensure that the ticket is not missing from the lift. For prevention, besides the procedural control of allowing no employee to have responsibility for two portions of any chain of custody, the ticket lift must be protected by sealing it and keeping it locked up prior to mailing.

COUNTERFEIT TICKETS

The first counterfeit airline tickets appeared in 1970, mostly in the Pacific and Southeast Asia regions. The first counterfeits were of poor quality and consequently easy to detect. In 1973, prompt action by carriers led to the arrest of four counterfeiting suspects in London. Thus far, the emergence and availability of color copiers has not created counterfeiting problems for the industry, in part because of the alertness of carrier personnel in spotting and picking up the counterfeit tickets as they are presented for travel and refund.

ALTERED TICKETS

Tickets are altered in a number of ways in order to obtain an illegal profit or benefit for the perpetrator. Usually, alteration involves the purchase of a legitimate ticket for a destination a short distance away. Then the ticket is altered to a transcontinental or international flight. This necessarily includes altering the indicated fare to reflect an increased amount. The next step for the experienced criminal is to get rid of the altered ticket as quickly as possible.

---

[9] The ticket lift is that group of tickets for a specific flight which are picked up from the passengers as they enter the aircraft or the jetway.

This is usually done by taking it to another carrier's ticket office or ticket counter, where it is exchanged for a new ticket with a different routing, a process called "laundering." On occasion, the transaction may also result in a partial cash refund to the individual, but the primary purpose of the scheme is to provide the perpetrator with a legitimate ticket that he can sell, use for travel, or turn in to obtain a refund.

For this reason, tickets presented for reissuance should be carefully inspected by the ticket agent for any indication of tampering or of raising the fare to a higher amount. Further, refunds and reissued tickets should not be made out for anyone other than the person named on the original ticket, and confirming identification should be required. It is not unusual for a ticket to be found in an airport or to be stolen from a passenger waiting for a flight. This is another reason why ticket agents should always ask for identification when they are reissuing a ticket and, in addition, make a notation of the identification obtained. Tickets showing obvious erasures should not be accepted for reissuance, and the number of the ticket presented should be checked against the carrier's ticket blacklist. Tickets can be altered with chemicals and mechanical erasure and by changing ticket numbers and raising the fare .

MISCELLANEOUS CHARGES ORDERS

Miscellaneous charges orders (MCOs) are issued for air or surface transportation, excess baggage charges, land arrangements for tours, deposits or partial payments for tickets, and for refundable balances. They may be issued to any carrier and, as such, are negotiable documents. Although they cannot be directly used for transportation, they are exchangeable for a ticket or cash refund. Blank MCO forms must be protected with the same care in storage and usage as ticket stock, because anyone who knows the proper way in which to fill them out can employ them for illegal gain.

TICKET-BY-MAIL FRAUD

Tickets by mail (TBMs) or Ticket arrangement mailout (TAMs) are a valuable marketing aid. Their sale is handled in the following manner. A passenger calls the airline and makes reservations for a flight and the airline sends the tickets by mail. Previously, only about 40 percent of TBMs were paid for prior to the flight; the balance provided a fertile field for individuals with larceny in their heart.

Fortunately, the implementation of a few precautionary measures has helped airlines limit the losses that previously had occurred. Callers who request a one-way ticket or a TBM with an open-origin schedule are offered the choice of a "will call" ticket (paid for when picked up) rather than a TBM. Requests to mail the ticket to somewhere other than the billing address of the credit card are refused, and the caller is offered the will call prepaid option. If tickets are to be mailed to a private residence, a check should at least be made, through a regular or a reverse telephone directory or a city directory, to determine if an individual of that name lives at that address. Airlines can request advance payment on all TBMs, or on all TBMs having a value in excess of a specific dollar amount.

To reduce losses, carriers occasionally use a "pink envelope" system that requires that a cash or check payment be sent to the carrier before the ticket is mailed. Even then, the carriers have experienced unusually high losses in some of these cases due to bounced checks. When one carrier attempted to use pink envelopes for certain geographical areas by zip code where losses had been unusually high or for callers under age 22, the Civil Aeronautics Board's Bureau of Consumer Protection charged them with violating the Equal Opportunity Credit Act and the Federal Aviation Act.

One study of just over 400 tickets, admittedly a small sample, showed that unpaid TBMs have a number of factors in common. When the passenger did not give a telephone number, 75 percent of the tickets were unpaid. On invoices on which more than two tickets were ordered, 66 percent were unpaid. Seventy-two percent of tickets over $500 were unpaid, and 76 percent of first class tickets were not paid. Among indirect routing tickets, 63 percent were unpaid.

Persons who illegally obtain TBMs will either use them for flight, sell them on the street, attempt to launder them by a process of exchange and refund through other carriers, or attempt to refund them at the ticket counter of the carrier that issued them. They may also have been contracted by individuals to provide tickets to a specific destination.

It has been difficult to prosecute TBM fraud cases, for a number of reasons. Often, the evidence needed to prove mail fraud is lacking. Although this problem can be resolved by using registered or certified mail to send TBMs, the additional cost this service would add must be compared with the losses occurring due to lack of registration. In addition, many U.S. Attorneys feel that the airline has exposed itself to this liability by mailing out tickets on the basis of a phone call, and therefore should be prepared to absorb these losses. They are reluctant, as they put it, to be "used as a collection agency" for the airlines. Investigations are often coordinated with the Postal Inspectors and may involved controlled deliveries of the tickets.

Because of these problems, most carriers now require advance payment, by either credit card or check, before mailing out a TBM.

PREPAID TICKET ADVICE FRAUD

Another form of telephonic ticket sale is subject to credit card fraud. This is the prepaid ticket advice (PTA), when a passenger calls to order a ticket, gives his or her name, credit card number, and other authorizing information, and asks that the ticket be held for pickup at the airport. Fortunately for the prevention of loss, fraud involving the PTA will invariably involve a ticket that is to be made out to someone other than the credit card holder. The perpetrator of this kind of fraud obtains the legitimate credit card information of another individual; this information can be obtained from a number of sources without theft of the card. This fraud can be prevented and the perpetrator arrested, if the agent will take the precaution of calling the person who supposedly ordered the ticket on the pretext of asking to verify the account number or the expiration date of the card. This is a worthwhile precaution to take routinely, because the frequency of someone ordering a ticket in another person's name is not great. The best precaution, exercised by some carriers, is to require the individual ordering the PTA to come in to a ticket office or airline ticket counter and complete the transaction, usually paying an additional fee for the PTA service. This provides an airline employee the opportunity to check identification and verify the transaction.

Prepaid tickets are also ordered by travel agencies. Fraudulent orders from this source are usually called in after business hours, and the caller gives a fraudulent PTA number. Ideally, the carrier receiving the after-hours order should make a return phone call to the agency owner or manager or to an agency employee known to the carrier for verification. The Air Transport Association now secures home addresses and telephone numbers from new agents applying for accreditation so that calls can be made.

In March 1985, the Area Settlement Plan developed a special computer program to help airlines detect fraudulent use of PTAs. With the program, a subscribing airline receives a record of all of the PTA numbers that have been shipped to each travel agency over the past two years but not yet reported by the agency.

REFUND FRAUD

Not all stolen ticket blanks are used for travel. The individual who holds the ticket may fill it out for substantial travel and attempt to obtain a refund. For this reason, it is essential that all tickets presented for refund be checked through the computer or on other blacklists to make certain that they are not stolen tickets. If a ticket was purchased with a credit card, the card number should also be checked.

A scheme sometimes used in refund fraud is to purchase a ticket with a bad check. When tickets paid for by check are presented for refund, the passengers should be advised that the refund will be mailed to them. If, during banking hours, a ticket holder insists on an immediate refund, the bank should be called to see if the check has cleared. Outside of banking hours, the refund should be delayed unless a supervisor is confident that the individual's identity has been positively established through photographic identification such as a driver's license, through signature comparisons with the license, credit cards, or other identification cards, and by major credit cards in the possession of the passenger.

Legislation that became effective in California on January 1, 1991, forbids merchants from asking for credit card numbers as identification when cashing checks. The justification for the law was the illegal use of credit card numbers. Checks already have a name, address, phone number, and sometimes even a driver's license number. The addition of a credit card number on the check gives an individual all that's needed to start ordering merchandise over the phone. Merchants can be fined from $250 to $1000 for each offense.

In the case of refunds, floor limits should be established to ensure that refunds over a specific dollar amount will be processed through the airline accounting office and paid by check. If a check or draft issued in payment of a refund provides space for the signature and address of the payee, the agent--without fail--should witness the filling-in of this information by the payee. This requirement is in addition to the agent's responsibility for positively identifying the individual presenting the ticket for refund and for determining that a refund is legally due. If refund drafts or checks are mailed, they should at least be sent by certified mail. The best method is to send all refund requests to the home office refund department, where receipt of payment for the ticket or MCO can be confirmed prior to issuing the refund.

On occasion, the auditing department will become aware that an employee is falsely claiming to have refunded a ticket for a passenger. In some cases, employees have taken tickets from passengers as they boarded or from the ticket lift, and made refunds to themselves on tickets that belonged to passengers who actually flew. This version of pocketing is best controlled by mailing sampling letters to passengers shown to have received a refund. If the passenger denies having received a refund and claims to have flown as planned, investigation should begin. The use of these sampling letters on a continuing basis will usually isolate dishonest employees before they can do a great amount of damage.

PROSECUTION AND LEGISLATION

The probability that persons responsible for crimes against air carriers will be prosecuted is frequently directly proportional to the interest and cooperation of the law enforcement agency into whose jurisdiction the offense falls. The prosecution of those responsible for the theft of airline tickets is no exception. Local prosecution is difficult to obtain because of the national and international scope of the problem; tickets stolen in Atlanta may surface in Los Angeles on the day of the theft or, at most, a day later. Nevertheless, exceptional effort by a number of police departments has resulted in both prosecutions and a consequent decline in stolen tickets presented in those jurisdictions. Unfortunately, the effect is only local; the real thrust of prosecution must come at a federal level.

A federal grand jury was impaneled in Brooklyn, New York, in 1973 to hear evidence obtained by investigators from several federal agencies, including Postal Inspectors and the FBI. Over a period of nine months, the grand jury heard nearly 200 witnesses. Indictments were returned charging 19 persons with the theft of more than 7000 blank tickets, and two other persons were indicted for perjury. This grand jury investigation was significant because of the involvement of the FBI; the agency had previously been relatively uninvolved in stolen ticket investigations, chiefly because of problems of jurisdiction.

Under Title 18, Section 2314 [Interstate Transportation of Stolen Property (ITSP)] the FBI is authorized to investigate the interstate transportation of "stolen goods, securities, moneys, fraudulent State tax stamps, or articles used in counterfeiting." The definition of the word "securities" is given in Section 2311 as including "any note stock certificate, bond debenture, check, draft, warrant, traveler's check, letter of credit, warehouse receipt, negotiable bill of lading, evidence of indebtedness" and numerous other "securities," but airline tickets are not specifically named. Although legislation has been proposed from time to time to correct this, none has yet been passed. Under the ITSP statute, the aggregate amount of loss must be more than $5000.

In February,1975, the FBI acted voluntarily to announce that stolen air travel ticket numbers could and should be sent to the National Crime Information Center for inclusion in the NCIC computer's stolen article file. Although the first--and best--line of defense against stolen tickets is still the agent who checks the ticket against the blacklist, the storage of ticket numbers in the NCIC should be of material aid to law enforcement agencies in detecting possession and establishing the chain of evidence necessary for successful prosecution.

Some larger metropolitan police departments have detectives trained and knowledgeable in the intricacies of airline ticket fraud who are dedicated to fighting this type of crime. One of the best known is the Bunco-Forgery Department of the Los Angeles Police Department. All the detectives who specialize in this area are excellent. The New York Port Authority Police also has very fine detectives dedicated to airline ticket fraud.

On a federal level, the U.S. Postal Inspectors can become involved as can the FBI. The FBI has been of particular help in some highly involved and intricate ticket investigations extending around the world, and a number of Special Agents have become experts in this area. Prosecutions relating to airline tickets can also be supported under a number of federal statutes. Among them are:

Fraud by Wire, 18 U.S.C. 1343
Mail Fraud, 18 U.S.C. 1341
Consumer Protection, 15 U.S.C. 1644
Theft from Interstate Shipment, 18 U.S.C. 659
Interstate Transportation of Stolen Property, 18 U.S.C. 2314, 2315
Crime on a Government Reservation, 18 U.S.C. 7 and applicable enclave statutes
Electronic Funds Transfer Act, 18 U.S.C. 1951(a), (b)(1)

Among the successful federal prosecutions of crimes involving airline tickets are the following:

United States v. Jorge Gustavo Kraiselburd (S.D. Cal.). On May 2, 1983, Kraiselburd was sentenced to six years in prison. He had pleaded guilty to five counts of mail fraud arising from his participation in the fraudulent acquisition of travel agencies.

United States v. Beverly Caro (E.D. Va.). An indictment charging Caro with wire fraud was filed in April 1983 as a result of the defendant's having issued over $60,000 worth of tickets charged against invalid credit cards. The case was initially investigated as a possible crime on a government reservation -- Washington National Airport -- where Caro was employed.

United States v. Broffman, et al. (S D. Cal). Jerry Broffman, his wife Yadzia, and Ronald Park Doke were indicted in June 1983 for mail fraud. Broffman had Doke put mailboxes on the doors of empty stores to collect airline tickets after Broffman had called an airline and fraudulently charged tickets to a business at that address. Five airlines were defrauded in this scheme; one airline lost $24,000 in two months.

United States v. Andrew Walker (D. Conn.). Walker had acquired travelers' names and credit card numbers from discarded receipts at hotels and, fraudulently identifying himself as that person, placed telephone orders for tickets to Africa, the Far East, and elsewhere for several fictitious persons. Walker had the tickets mailed to him, or he sent a messenger over to pick them up. Walker accomplished this scheme in several jurisdictions. He cost the victimized travel agencies thousands of dollars and, in fact, bankrupted one agency. He was sentenced to four years in prison on each of two counts of wire fraud.

Chapter 18

TICKET STOCK SECURITY

The security of ticket stock begins at the printing plant. A set of guidelines developed for railroad ticket printers' plants and issued by the American Association of Passenger Traffic Officers in 1915 was furnished by Art Retzke, former Vice President--Ticket Division, Rand McNally. Clearly, good ticket printing security procedures have changed little in three-quarters of a century.

Inspection of Licensed Ticket Printers' Plants

In examination of ticket printers' plants by resident members, particular attention should be given to the following details to see that printers, including foremen of ticket departments, have thorough understanding of ALL the requirements and are observing them, and to insure this, two copies are to be left with the printer, one of which shall be framed and kept posted by him in the ticket printing department.

1. A separate or distinct department or enclosure must be used for printing all railroad tickets; it must at all times be under the exclusive control of a foreman; and must be kept securely locked when not being operated.
2. All employees not regularly employed in such department and other persons must be excluded therefrom.
3. All Association Safety Ticket Paper received must be stored in a substantially constructed room, vault or compartment, insuring its safekeeping.

4. All waste ticket paper, SPOILED TICKETS and SAMPLE TICKETS not needed, must be cut up or otherwise effectively destroyed with ticket paper destroying machine, under the supervision of the person responsible for safeguarding of all such paper.
5. Any SAMPLE TICKETS furnished railroads, or retained in printing establishments for future reference, must be printed or indelibly stamped in red across the face of each coupon, contract and stub, if any, "Sample" or "Specimen" and have the facsimile signature and the number space cancelled by punching holes through them.
6. Printing of sample, model, or patent tickets of any kind by licensed printers for individuals other than authorized officers of membership lines, not permitted without written consent of the Secretary of the A.A. of P.T.O.
7. Delivery of tickets from printers to railroad ticket stock departments must be safeguarded by proper methods for protection of all lines.
   (a) In delivering to railroad offices located in the same city, a strong box locked in transit should be used.
   (b) Delivery of tickets by foot messengers and in paper packages should not be done except in emergencies.
   (c) Printers' delivery wagons used for carrying tickets in packages should be covered and have locked screen compartments, and a trustworthy employee should remain on wagon while containing tickets.
   (d) Receipts in duplicate for delivery of tickets should always be required and delivery within office hours and before ticket stock rooms are closed.
   (e) Tickets forwarded to another city should be encased in a strong box or in a securely wrapped and sealed package and FORWARDED BY EXPRESS.

W.C. HOPE
Secretary

PRINTING PLANT SECURITY

When conducting a security survey of a printing plant, both the physical security and the internal procedures must be examined. In assessing the potential for security problems, the location of the plant is a consideration. If possible, the plant should not be widely known to be engaged in printing negotiables, and the fact that airline tickets are printed there should not be advertised in any way.

A long discussion on how to conduct a facility physical security survey is not needed here because the subject is basic to the knowledge of most security professionals. Briefly, however, the survey should include the security of the plant perimeter. Perimeter fencing and lighting, access to the building through ground-level windows, alarming of all perimeter entrances including skylights, and visitor control should all be examined. Egress from the building must also be considered. One recent plant survey disclosed that employees were free to leave the building through any one of several personnel doors, including one that opened on the company parking lot. None of the doors was locked or alarmed.

The easiest method of correcting such a deficiency--and it must be corrected--is to install exit alarm locks on each exit door. These permit the doors to be kept locked from both the outside and the inside, as they are equipped with a bar or level actuator that unlocks the door for egress in the event of a fire or other emergency. Use of the emergency exit sounds an alarm at the door. For nonemergency use, the lock is operated by key, and no alarm sounds. Exit alarms can also be wired to existing central alarm systems or connected to an alarm monitor at a central guard station. Another available option allows unalarmed access at certain doors during authorized hours by switching the designated door or doors out of the alarm system at the monitor.

A security officer who makes scheduled rounds is an asset, particularly as a "fire watch," but he or she cannot take the place of an alarm system. In the absence of an alarm system, once the security officer is overpowered, the burglars can take their time collecting and removing whatever ticket stock the plant is holding.

The survey must include the protection given to negatives and printing plates. Once the final copy of the desired ticket has been prepared in the composing room, photographic negatives are made from which the printing plate is etched. While printers use different plate-making processes, the printing plate is generally made of some type of photopolymer plastic on a stainless steel backing.

The degree of protection given negatives and plates should be in relation to the amount of risk that would be incurred if they were to fall into the wrong hands. After all, the only value of the negatives is for their use in preparing a plate. If unscrupulous individuals want to make a negative, they can easily produce a passable one by photographing a well-printed ticket.

Nonetheless, negatives and plates should be given at least a reasonable degree of protection. This means locking them in steel cabinets, which should be in a locked and alarmed area. When printing is in progress, those plates that are not on the press at the moment should be kept in a locked cabinet, and the shift supervisor should have the only key.

Access Control

A visitors' register should be maintained, and all nonemployees should wear a visitor's badge and be escorted at all times. Contractor personnel working in the plant temporarily should be provided with special identification badges. Employees should be encouraged and reminded to challenge strangers.

A ticket-printing plant is not the type of operation that should be open to tours by various civic groups; such tours are not compatible with the low profile essential to the plant's security. Most printing plants engaged in airline ticket printing recognize this and have done an excellent job in this regard.

Internal Procedures

In addition to physical security, the internal procedures of the printing operation should be reviewed. This pertains primarily to the handling and disposal of tickets printed at the plant but not delivered to the customer. Some of these tickets are produced during the "make ready" run immediately prior to printing. Because quality control is so important to the ticket-printing operation, particularly for tickets that will be read by optical scanners, as many as 300 to 400 tickets may be run off before the desired quality of printing has been achieved and the production run can begin. As a precaution against shortage due to spoilage, more tickets are printed than are required; this is standard printing practice, and this extra quantity is known as "overrun."

The method used for disposing of make-ready and overrun tickets should be of vital interest to the individual conducting the survey, because these tickets can most certainly be used or sold for use on airlines, and will be as readily acceptable as any other ticket. Destruction of make-ready and overrun tickets not delivered to the carrier can be accomplished in several different ways. Burning, the simplest solution, has been severely restricted by air pollution ordinances. The most practical means of disposing of scrap and overrun from ticket printing appears to be shredding. A common practice is to sell the shredded paper as scrap, not for the income to be derived but to avoid paying to have it hauled away.

Liability Agreement

In addition, the surveyor should verify the status of any written agreement between the carrier and the printer that establishes the dollar amount to which the printer could be held liable if the airline suffered a monetary loss due to negligence on the part of the printer, and the surveyor should know what this maximum figure is.

Employee Screening

Personnel selection and screening is another important facet of a security survey of printing plant operations. Because of the opportunity for theft and collusion that ticket printing affords, new employees should be thoroughly screened. At a minimum, the screening should include checking with previous employers, verifying periods of employment and schooling, and a credit check.

TICKET WAREHOUSES AND DISTRIBUTION CENTERS

The physical security of a ticket warehouse should be commensurate with the value of the items being protected; in the case of airline tickets, that value is substantial. In addition, the underworld is interested in stolen tickets; the market available to criminal syndicates can absorb large numbers of tickets, making a large-scale warehouse theft profitable.

A basic precaution is an after-hours alarm system protecting all perimeter openings, including skylights, and adequate exterior floodlighting. If items other than tickets are stored in the warehouse, ticket stock should be kept in its own secure area or strong room and protected around the clock by both contact and space alarms separate from the nighttime protection provided the building's perimeter. Access to the protected ticket-stock area should be restricted to as few people as is consistent with efficient operation.

Both accountability and frequent physical inventory are essential. In most warehouses, the inventory of ticket stock is not so large that frequent counts would be too time-consuming. Procedures and arrangements should ensure that if even one box of tickets is taken, its disappearance would be noted immediately.

CITY TICKET OFFICES

One of the underworld's prime sources for stolen tickets is the burglary of ticket offices. The simplest and most important precaution that can be taken against burglaries is to reduce the amount of ticket stock on the premises to a minimum. The loss of a one- or two-day supply of tickets has far less impact than, say, a one-week or two-week supply in terms of the number of tickets made available on the black market.

One method of protecting ticket offices against burglary is, of course, an alarm system. Because of the costs involved, not all ticket offices are alarmed, and the expense of installing and maintaining an alarm system must be weighed against all relevant factors. Has that ticket office had a history of burglaries? What is the crime pattern or trend in the area in which the ticket office is located? (The local police department should be contacted for this information as a normal part of any ticket office security survey.) In what type of neighborhood is the ticket office located? Is the location isolated at night?

When an alarm system is needed at a ticket office, the solution can be a simple matter for those airlines that have centralized reservations offices that cover a large geographical area. By making use of the existing leased telephone lines into the reservations office, the carrier can install a proprietary alarm system for less than a commercial installation would cost. In such a case, the alarm system is purchased by the airline, and it signals any alarm to the central reservations office over the leased telephone line. For example, an unauthorized entrance into a ticket office in Milwaukee would activate an alarm that registers on a panel in the carrier's airline reservations center in Chicago. The person on duty in Chicago would call the Milwaukee Police Department and advise law enforcement of the alarm.

During a security survey, the security of all perimeter openings should be reviewed, including skylights and vents as well as doors and windows. All aspects of key control must be reviewed, including the storage of unissued and master keys, who is responsible for them, whether those holding keys actually need them, how many keys have been lost, when locks were last changed, how burglary-resistant the present locks are, and so forth.

The individual examining the security of a ticket sales office should become informed of several other points. Who takes care of the janitorial and maintenance services? During what hours? Do maintenance personnel have keys to the premises, or are they admitted by an employee key holder? Is there a security patrol, or do police patrols pass the location regularly? With what frequency? What tenants are adjacent to or near the airline's office?

CONTROL OF INVENTORY EXPOSURE

The security of the ticket stock is of prime consideration. If master stock is located on the premises, the counter stock can be kept to less than a one-day supply, a substantial protection against loss due to armed robbery. Where it is necessary to keep master ticket stock in a sales office, no more than a one-month's supply should be kept on hand.

If a safe is not available, the minimum acceptable storage facility should be a heavy-duty steel cabinet bolted to a wall or floor and secured when closed by a steel bar, security hasp, and padlock. The record of office validator plates and the agent to whom each is assigned should be kept in a safe place, and the plates themselves should be locked up separately from the ticket stock. Where both tickets and validator plates are held in the ticket office safe, the plates should be kept in a locked inner compartment, separate from the tickets.

During working hours, accountable forms, validators, and plates should be so positioned at the counter as to be out of reach of customers. Blank or partially completed tickets, validators, and baggage checks should not be left on the ticket counter when it is unattended, however briefly. Some offices have locked, tamper-resistant ticket dispensers. These are not burglary-resistant; ticket stock should always be removed from these dispensers for overnight storage.

AUTOMATED TICKET AND BOARDING PASS

A relatively recent innovation in ticketing is the automated ticket and boarding pass (ATB). The ATB replaces the traditional ticket and boarding pass with a single piece of stiff paper that resembles a computer card. The large left-hand piece is the value coupon (ticket), and the right-hand stub is the boarding pass that is given back to the customer when he or she boards the aircraft.

A problem occurred when the ATBs first came into use because they looked almost identical to the boarding pass document. Some customers who checked in with an old paper ticket and received a regular boarding pass learned that they could sometimes give the flight attendant the boarding pass card which looked exactly like an ATB. If the flight attendant didn't notice the discrepancy and gave the passenger back the stub, he or she could retain the value coupon.

The newest version of the ATB has a magnetic stripe and is known as the ATB2. All of the information that's on the face of the ticket is encoded on the stripe. Magnetic-stripe technology will enable the airlines to detect stolen tickets. Future applications are to tie the system into existing baggage tag scanner devices to verify that the baggage is on the right aircraft. Boarding Control Devices (BCDs) that read the magnetic stripe on the back of the ticket are now being installed at many airports. The BCD is designed so the customer can insert the ATB into the BCD (are you following the acronyms?) which reads the ticket, separates it at the perforation, and returns the stub to the customer. The BCD also has a screen that displays information to an attendant if anything is irregular--wrong fare, wrong date, duplicate seat, etc. The BCD will tell the attendant how many people are on board, how many have paid F, Y, or Q fares, how many children are flying, where they came from, and where they are going. It will even display the passenger's name so the gate agent can greet them by name. Both Continental and United use magnetic stripes on their ATBs. American has agreed to put magnetic stripe technology in place by April 1996.

Another application of the ATB technology is the deployment of satellite ticket printers. A travel agency can electronically order a ticket printed out at a location that could be as close as the lobby of a traveler's hotel. Hoteleticket, National Ticketing Network, and Quick Delivery Airline Tickets (QDAT) already have printers in place in hotels. These firms are required to obtain airline accreditation as travel agencies in order to handle ATB stock. QDAT has printers in over 200 hotels now and ultimately expects to have about 1600 printers in the field.

AIRPORT TICKET COUNTERS

Although much of the security appropriate to airline ticket offices is applicable to airport ticket counters, other precautions are needed. Adequate controls must be maintained to account for tickets voided because of errors or changes. Unauthorized entry into administrative areas behind the counter must be prevented. This usually amounts to challenging unknown individuals, because there are usually no significant physical barriers to such access. An individual who was repeatedly discovered behind the ticket counters of different carriers claimed, whenever challenged, that he was an auditor for that particular airline, and he had forged identification for that particular carrier to back up his claim. No individuals not known to agents on duty should ever be assumed to have legitimate business behind the counter, no matter how confident--or brazen--their actions.

Where automated and computerized ticket printers are in use, special attention must be given to the security of ticket stock in the machine. Security departments should examine the machine for its resistance to various forms of attack, and to make certain the lock is sturdy enough to protect the contents against surreptitious or sudden attack. The security of ticket stock in the machines when the counter is closed or unattended presents more of a problem. One auditor arrived at an unattended ticket counter during the early morning hours to find tickets being busily produced by a computer-operated ticket printer. Tickets were draped over the counter and spilling onto the floor on the passenger side when the auditor arrived. There was no ticket agent in sight.

Although removal of the ticket stock from the machine at the end of business hours is certainly the best security method, other less time-consuming options are also employed. One is to remove the ticket stock from the print mechanism feed path so that the ticket stock will not feed out even if an attempt is made to operate the machine. In addition, all ticket printers are connected to a central computer by which they can be made inoperative for certain preprogrammed hours, such as from 11 p.m. to 7 a.m.

Neither method offers protection against the forcible attack on stock in printers. However, the precaution of removing and replacing ticket stock in a big station's many printers would involve so much time that this measure is not likely to be employed unless the location appears to offer opportunity for such attack. In any event, one precaution should be taken in every case, and that is to de-link the agent's CRT display set from the ticket printer whenever the ticket counter is closed or unattended.

TRAVEL AGENCIES

Travel agency ticket stock security was materially improved by new requirements incorporated in the Air Traffic Conference Sales Agency Agreement effective June 17, 1974. A further updating of these agency security requirements was approved in CAB Order 76-5-47, dated May 12, 1976. These new rules set minimum security standards that govern the handling, storage, and retention of ticket stock, accountable forms, and ticket validator plates by travel agents. The requirements were developed in the wake of evidence that safeguards previously in effect were not being followed in practice. Indeed, the Air Traffic Conference had reported that, of the 38 travel agency locations reporting burglaries during 1973, over 60 percent had neglected the minimum safeguards prescribed.

The real strength in the rules is that the security procedures are mandatory. If a travel agency fails to comply with these standards, the matter is subject to review by a compliance panel, and failure to comply can cost the agency operator the right to operate a ticket-writing agency. The operator can also be held liable to the carrier for tickets lost in a burglary, if it can be proven that the operator failed to exercise reasonable care in protecting the carrier's ticket stock at the time of the theft. And "reasonable care," it should be noted, is defined in the rules as compliance with the minimum safeguards set forth in the agreement.

Because of the ease of negotiability of frequent flyer award certificates, they require the same protection as ticket stock. They need to be inventoried frequently and kept locked up.

SATELLITE TICKET PRINTERS

A satellite ticket printer (STP) is a ticketing location approved by the Airlines Reporting Corporation (ARC) that produces tickets generated at a travel agency's home office or regular branch. Examples of potential STP locations are business travel departments at major corporations, bank branches, and retail stores. Some of the machines are attended and some are not. Because these printers contain ticket stock, serious concern has been expressed over the physical and electronic security of the STPs. Similarly, in October 1986, Southwest Airlines began testing the sale of tickets through automated teller machines at 7-Eleven stores in Texas. As these new methods of selling tickets emerge, they will require adequate protection.

DOCUMENT SECURITY RESOLUTION

A Document Security Task Force was established within the Air Traffic Conference, a division of the Air Transport Association, to formulate a Document Security Resolution. This resolution will standardize such matters as ticket layout and design, type, color and quality of ink, utilization of bar codes, carboning pattern on the back of tickets, watermarks, ticket printer design, blacklisting of documents, distribution of ticket stock from the airline central office to the final point of destination, indemnification agreements, and circulating information on lost or stolen documents.

Chapter 19

CREDIT CARD FRAUD

Credit cards are big business. The typical American charge customer carries nine cards, owes more than $2,000 on them, and pays an average interest rate close to 19 percent.[1] Visa has approximately 257 million cards in circulation, MasterCard, 163 million, and American Express, 37 million. Currently, 13.5 percent of U.S. consumer spending is paid by credit card.[2] The profitability of illicit use of credit cards has not escaped the criminal element in our society.

STOLEN AND ALTERED CREDIT CARDS

Stolen credit cards are obtained in a number of ways. Some are issued on the basis of false and fraudulent applications. Others are stolen before they reach the legitimate cardholder; in such cases, the thief may be a postal employee, a mail carrier, or a specialist who concentrates on rifling mail boxes. In areas where mailbox theft is particularly severe, postal carriers blow a whistle when they leave mail in the box to alert residents that mail has been delivered. If the recipient is usually at work when mail is delivered, the only secure alternative is to rent a post office box. Another form of attack on the mails has been the armed robbery of mail carriers.

A measure successfully used to protect against the theft of credit cards by post office employees has been to seal the mailing in packages by zip

---

1 "Charge It Your Way," *Time* (July 1, 1991):50.

2 Ibid. 51.

codes and register them at the post office of delivery, where the package is opened and the mailing sorted. The accountability that results has proven effective as a deterrent to the problem.

Actually, it is not difficult to obtain credit cards by fraud if one is determined to do so. This was made quite evident by one underground publisher. A circular advertising a book entitled *Credit* states that the book tells the reader (among other things):

- How to obtain AAA credit in just 30 days!

- How to receive all the credit cards you want!

- How to obtain almost unlimited consumer credit!

A companion volume from the same publisher, *The Paper Trip*, is promised to show its readers how to hide undesirable employment records; change one's age; adopt a new name; cover arrest, jail, or prison records; beat the draft; disappear for good; change citizenship to any country; leave bad credit records behind; avoid any and all creditors; and "much, much more."

Of course, it is no trick for the average citizen to get credit cards. Indeed, each day's mail seems to bring offers of new "preapproved" credit cards.

CREDIT CARD FRAUD ACT OF 1984

The Credit Card Fraud Act of 1984[3] is a part of a larger section of the U.S. Code entitled Fraud and Related Activity in Connection with Access Devices. The law defines an access device as:

> Any card, plate, code, account number or other means of account access that can be used, alone or in conjunction with another access device, to obtain money, goods, services, or any other things of value, or that can be used to initiate of transfer of funds (other than a transfer originated solely by paper instrument).

It further defines a counterfeit access device as "any access device that is counterfeit, fictitious, altered or forged or an identifiable component of an access device or a counterfeit access device." The law also defines an unauthorized access device as "any access device that is lost, stolen, expired, revoked, canceled or obtained with intent to defraud." As can be seen, the term "access device" includes a broader spectrum than "credit card" does. Section 1029 makes it a felony to produce, use, or traffic in one or more counterfeit access devices. The most obvious violation of this type is the *facsimile card*, which is a card that is manufactured to resemble a genuine credit card and is used to defraud unsuspecting businesses. These cards can be made from scratch which requires an experienced printer, or from stolen cards that are

---

[3] Title 18 U.S.C. §1029.

flattened and re-embossed with genuine account information. Account information may be obtained from many sources, including sales draft carbons retrieved from trash, dishonest bank employees, unwitting account holders through telephone solicitation, and unauthorized access to credit bureau computers.

Another type of counterfeit access device is the *white plastic card.* These are pieces of plain plastic stock embossed with genuine account information. Obviously, the use of this card requires a conspiring merchant who is part of the scheme. (The sales drafts imprinted with the white plastic cards are as good as cash to the merchant.) A variation of this is the "bust out," in which a business is opened for the primary purpose of processing a large number of sales drafts, produced with white plastic cards, in a short period of time and closing down before the bank or anyone else becomes aware of the scam.

It is a felony to traffic in or use one or more unauthorized access devices during any one-year period in order to obtain anything of value totaling $1000 or more. This violation covers a great number of offenses ranging from someone who finds a lost credit card and runs up $1000 in charges to a sophisticated, computerized telemarketing operation involving thousands of accounts. Telemarketing is a legitimate telephone sales technique in most cases. However, some fraudulent operators contact the holders of legitimate credit card accounts by telephone and, by selling shoddy merchandise or promising free prizes, trick the cardholders out of their account numbers. Once in possession of the number, the company uses the information to complete fraudulent sales drafts or sells the account information to another fraudulent company. It is a violation of this law to sell lists of account holders and numbers for use by card counterfeiters and others.

Also illegal under the act are such things as the use of stolen or fraudulently obtained personal identification numbers used in automatic teller machines (ATMs). These numbers used to initiate transactions in emerging systems such as home banking, direct deposit, automatic payment, and computerized home shopping systems are also considered to be access devices under this statute.

It is also a felony to possess fifteen or more counterfeit or unauthorized access devices or to produce, traffic in, or possess device-making equipment. This equipment includes all of the paraphernalia used to produce access devices, such as plastic sheets, presses, inks, silk-screening equipment, embossing machines, and electronic components such as the microcircuits in the new smart chip cards. In addition, it is an offense to attempt or conspire to commit any of the acts described. Of course, these offenses must be committed knowingly and with intent to defraud, and they must affect interstate or foreign commerce.

For offenses concerning unauthorized access devices, §1029 carries a maximum penalty for the first offense of 10 years in prison and a fine of $10,000 or twice the value obtained by the offense, whichever is greater. For counterfeit access devices or device-making equipment, the punishment can be 15 years in prison and a fine of $50,000 or twice the value obtained by the offense. If the offense should occur after a previous conviction under this statute, the offender can be imprisoned for 20 years and fined $100,000 or twice the value obtained by the offense.

PREVENTIVE MEASURES

Identifying the fraudulent credit card user is the difficult task of the airline employee at the ticket counter. Although most carriers offer a reward as an incentive for employees to pick up a blacklisted card, employees often lack the time to check each credit card presented against the hot list. The new automated ticket printers, however, automatically check the credit card number in the process of producing the ticket.

Experience shows that fraudulent credit card purchases most often involve tickets being purchased for someone other than the cardholder, open or group tickets, and tickets for which the flight date is some date other than the date of purchase. Ticket agents should be alerted to this and prompted to check credit card validity in these cases particularly.

To make the difficult virtually impossible, there have been instances of credit card alterations in which both the alphabetical and numerical characters are changed. For example, card AAX6479 issued to J. D. Nagle can become AAN8499 issued to J. Denagle. Until the alteration of the card or the nonexistence of the forged account has been discovered, there will be no entry on the hot list.

If, upon presentation, a credit card is found to be listed as stolen, the agent should make an attempt to detain the individual presenting the card long enough for a police officer to be summoned to interview the passer about where the card was obtained and to identify him or her for possible future contact. The purchaser is in possession of stolen property, which generally is just cause to detain and interview. Depending upon local ordinances and law and the result of the interview, the police officer may make an arrest.

If there is merely a pickup order on the card for other reasons, no attempt should be made to detain or restrain the person presenting the card; there is little possibility that a crime has been committed. Cards are often recalled because the cardholder has run up excessive charges or failed to make payments. Although such use may be fraud in the moral sense, it is not necessarily fraud in the legal sense if all charges were made in the cardholder's own name. The card should be picked up, however, and the customer told that the card has been voided by the issuer. It should be explained that the agent has instructions to keep the card and that, if an error has occurred, the card will be returned by mail. Each carrier should issue its own specific instructions for the proper handling of these incidents by its employees. A cardholder whose card is picked up may properly ask that the card be destroyed or cut in two in his or her presence (to ensure against possible misuse), and such a request should be complied with, taking care to avoid obliterating of the numbers in case the card is later needed for evidentiary purposes.

One available defense against credit card fraud is called the bank Identification number system (BINS). Each credit card uses the first three to ten digits of the account number to designate the bank or retailer that has issued the card. That string of digits is called the bank identification number. Because most fraud is carried out without the actual card, the criminal has no way of knowing what bank issued the card. If the user is asked, either on the phone or by a printed question on the mail-order form, for the name of the issuing bank, the person has to fake it or confess ignorance. In either case, the seller can cancel the sale. The retailer can also use the BINS number to make an authorization call to the bank to check the name and address of the cardholder. If that information differs from that provided on the sales slip, the sale should be voided.

In July 1990 the largest telemarketing fraud ever uncovered by Visa and MasterCard involved a scheme to recruit small businesses to launder more than $4 million in credit card charges. In this case, consumers received a post card telling them, "Congratulations! We are proud to announce that you are absolutely guaranteed" to win one of five prizes: a Chevy Blazer, $5000 in retail merchandise, a $5000 cashier's check, a watch valued at $500, and a $1000 Series EE Savings Bond. The operators asked callers to provide their credit card numbers, supposedly to check the winners' list, implying that Visa and MasterCard had sponsored the giveaways. Callers were then pressed into buying goods ranging from water purifiers, security systems, travel packages, and vitamins. In many cases, the goods were misrepresented or not delivered or multiple bills were sent to the consumer. Consumers who did not order the merchandise were often billed anyway. The losses from the alleged scam described in the suit ranged from $4 million to $50 million.

Another current scam is practiced by "credit doctors" who tap into credit-bureau databases to extract information using laptop computers and stolen access codes. They obtain names, loan histories and Social Security numbers of people with good credit records. Then, they sell the information to clients with poor records to use when applying for credit. Since late 1987 when police began investigating in Houston, purchases totaling nearly $20 million have been traced to phony credit data.[4] The scam works by accessing the good credit histories of people with names similar to those of their clients. Then, they turn over to the client the information needed to complete an acceptable credit application. This information may include Social Security number, credit card reference, employment, salary, and loan information.

I first became aware of this scam when a coworker asked me to check into a situation where a person who worked for another company in Silicon Valley, and who had a name similar to that of our employee, was using the Social Security number of our employee while piling up a terrible credit history. Obviously, it is most difficult to unsnarl a ruined credit history and to convince potential lenders to extend credit in such a case.

CARD EMBOSSING AND MAILING SECURITY

Some carriers contract their credit card embossing operation to outside vendors, while others do such work in-house. The location of an in-house em-

---

[4]*Wall Street Journal*, 14 August 1990.

bossing operation should be kept highly confidential. It should not be listed in company directories nor should there be signs indicating the operation's location in the building.

The area or room in which embossing is done should be constructed like a strong-room, without windows. When unoccupied, it should be alarmed and locked with a vault quality combination lock or a high-security lock that has a pick-resistant cylinder and restricted keying. The alarm system should include both contact alarms at the door or doors and a space alarm inside the room. An ultrasonic alarm would probably be the space alarm of choice. If the openings of air conditioning ducts entering the room exceed 96 square inches, they should be secured and alarmed.

During working hours, the entrance door should be secured with a push-button lock designed to allow the room to remain locked from the outside, while permitting egress without keys from the inside. Some means of signaling for entry and for identifying visitors from inside the room should be installed. The cheapest viewing method is to install a peephole device that provides an individual inside the room with a wide-angle view of the area outside the door. Newly designed peepholes don't require individuals on the inside to put their eye next to the door in order to see. A more sophisticated system could include a CCTV camera that is monitored inside the room.

Only those employees who have an absolute need to be in the embossing room should be admitted, and a list of such authorized personnel should be posted within the room. Visitors should be cleared with supervisory personnel and should sign a register. Janitorial and maintenance personnel should be accompanied by an escort on a one-to-one basis, rather than simply supervised by someone who is performing other work in the room.

Internal audit controls are most important to a secure embossing operation. Blank credit cards are potentially as valuable as imprinted ones, and they are "money in the bank" to an individual with access to an embossing machine. Electronic counters can count a long tray of cards in a matter of seconds. All new stock should be inventoried upon receipt before it is stored. Blank stock taken for daily production should be entered on control records, as should the cards spoiled in production, rejected and corrected. Supervisory personnel should make periodic inventories to ensure there are no discrepancies between actual and control counts. Cards returned for cancellation should be destroyed by shredding on a daily basis and not be allowed to accumulate.

During nonbusiness hours, both blank and finished stock should be stored in locked containers. The minimum security for stock storage should be a steel cabinet kept within the secure room and equipped with a steel bar, security hasp, and high security padlock. Given the added protection of the room's alarm system, it should not be necessary to store the cards and stock in a safe.

The stuffing and sealing of envelopes for credit card mailings should be given the same security and should take place in the same area as the embossing operation. Mailing envelopes should show only an innocuous return address and most certainly should not show the name of the company issuing the card. Usually, the return address is a post office box.

Most major issuers of credit cards follow the mailing of the credit card with a follow-up notice to the addressee, mailed several days after the card. The notice advises the cardholder that the credit card should have been received and provides instructions as to what notification should be given if the card has not been received. This prompt advisory notice gives the credit card recipient timely notice of a possibly stolen card, and--in turn--increases the chance the issuer will be advised of missing cards at a relatively early time.

Chapter 20

BAGGAGE HANDLING AND SECURITY

Customer complaints about airline service principally concern baggage problems. The annual cost chargeable to baggage mishandling, loss, theft, damage, and pilferage is increasing faster than any other security-related cost. This is regrettable, not just because of the expense, but because the problems these costs represent could be reduced with concentrated study and effort.

Besides the increase in the amount of baggage checked, the cost of each correction is increasing. When losses occur, passengers are claiming greater value for their lost or stolen items. Because planes are flying with more passengers and higher load factors, more baggage is being carried on each aircraft. The cost of delivering late luggage, investigation and claims handling are all increasing. It would profit the security manager to take a quick look at how much the airline is expending on replacing lost and damaged luggage, and pilfered articles, on passenger interim expenses, and on late arrivals. The company may be losing as much or more to these costs as is being lost to air cargo theft, for example.

According to baggage handling studies, the primary source of baggage mishandling is interline (between carriers) transfers and misconnects, with online misconnects a close second. Clearly, a direct correlation exists between aircraft on-time arrival performance and the incidence of mishandled bags. In addition to inadvertent mishandlings, there are the intentional thefts by baggage handlers, a category that includes the airline baggage handler who sorts bags, the contractor employee who handles interline connections, the ramp serviceman who works in a belly pit, and the taxi driver who delivers a late-arrival bag to the passenger's hotel.

CHECK-IN

Baggage check-in occurs when the passenger's baggage is first given into the custody of the airline. It is at this point that passenger tickets should be compared with baggage tags to make certain that both destinations are the same. Baggage should never be accepted for checking to cities beyond the passenger's destination or to a destination other than the passenger's. Ticket agents should take care to make certain that no baggage other than that which has been checked in through normal processing is placed on belts or allowed to be placed in the baggage handling system.

The security of unused baggage tags at check-in areas must be a high priority for air carriers; if tags can be acquired by unauthorized persons, the door is open to several security vulnerabilities.Where tags can be readily obtained, every opportunity exists for free shipment of freight, even shipment of contraband items, particularly when they are going to locations where there is no positive claim service--that is, where no claim stub is required in order to retrieve a bag or package. A vendor of mail-order merchandise was found to be forwarding merchandise to his customers by means of airline expedite baggage tags; the service was cheaper and faster than the mails. He obtained the baggage tags from unattended passenger check-in locations. Airline employees have also been found to be sending packages by this bargain method.

Tag theft also opens the door for placing explosives aboard an aircraft as passenger-unaccompanied baggage without even having to purchase a ticket to show. And, of course, tag theft provides every opportunity for setting up fraudulent claims for nonexistent lost baggage. Because of the potential threat and the risks inherent in failing to protect baggage tags, their security demands the careful attention of both security and operational employees.

One of the most vulnerable areas, of course, is the curbside baggage check-in cart used by skycaps. Generally, the security emphasis is on improving procedures, training, and reminding skycaps to protect the baggage tag stock. Most carts now have lock boxes designed so that they cannot be opened or closed except by key.

In 1974, a baggage mishandling survey by the Airport Security Council at Kennedy International Airport disclosed that as much as 90 percent of baggage not recovered by its owners is actually found by the carrier but cannot be identified with the claim.[1] Of mishandled bags with identification labels, 95 percent is returned. If the baggage tag is torn off and the bag does not reach its destination, the carrier tries to find the owner. However, each year literally thousands of bags never reached their owners because the owner could not be identified even through an examination of the contents. Further, the cost of misrouted baggage might well be called astronomical. In 1975, the last full year before the Civil Aeronautics Board required baggage identification labels, one major airline handled 43 million bags, of which only .8

---

[1] Airport Security Council, Baggage Handling Committee, Baggage Mishandling Survey, John F. Kennedy International Airport, 1974.

percent were delayed or lost. The cost to the carrier of that .8 percent was $4.5 million.

Effective September 21, 1975, the CAB issued an order that no carrier could accept a passenger's bag for checking unless the passenger's name and address were on the outside of the bag. Further, the carriers were required to make bag identification labels available at no cost to the passenger. Some passengers objected to having their name and address exposed; this complaint was resolved by requiring the identification but allowing it to be covered. The identification tags can be folded and the name and address written on the inside so that all that shows is the carrier's logo. If the bag is lost, the plastic I.D. tag is taken apart and the card with the name and address is recovered to identify the bag's owner. Early statistics on the result of mandatory baggage labeling indicated that one third less claims were being made. The number of bags that reached one carrier's central lost and found department was down by more than one quarter.

Passengers should be encouraged to lock their luggage. Although the standard brands of suitcases do have common keys and most locks can readily be picked open, the most elementary measures, such as locking the bag, will often deter a theft.

From time to time, airlines have experimented with issuing baggage tags to travel agents and with enclosing baggage tags with tickets-by-mail. The latter quickly proved unsuccessful, because passengers did not want to make the effort to fill out the tags with the flight number and attach it properly to their bags. Although some carriers still provide baggage tags to travel agents, the practice is now questionable because of the implications for security. If a blank-destination tag is used instead of a tag with a printed destination, the risk of the bag being lost is much increased. The printed tags were established for the primary purpose of avoiding bag mishandling due to handwritten tags. This would help to keep a bag being sent to Fresno (FAT) from arriving in Fairbanks (FAI) and a bag being sent to Phoenix (PHX) from arriving in Portland (PDX). The security risk focuses on the protection against misappropriation that the travel agent provides for the baggage tags. Another problem is that agency personnel might provide the tags of one carrier to a passenger traveling by another carrier. If baggage tags are to be supplied to travel agencies as a service, the tags should be issued in an accountable number series on the written assurance of the travel agency that the tags will be given adequate security.

BAGGAGE HANDLING

Baggage handling remains one of the unresolved problems of the industry. Passengers sometimes wait 30 minutes to recover their luggage after a one-hour flight. Passengers counter this wait by carrying their baggage on the plane with them. Carriers are, and have been, experimenting with a variety of methods for speeding up the baggage-handling process. Some of the solutions are working, and some are not. Ambitious programs of computer utilization have sometimes led to "overcomputerization" and the creation of more problems than were solved. At one airport, passengers complained that the computerized baggage-handling system, "chews up their bags as voraciously as if they were hamburger," according to a *Time* magazine article on airline industry baggage problems. Understandably, most of the problems were said to be attributed to the newness of the system and the inevitable need to debug the system. With both air carriers and airports working to efficiently expedite baggage handling, solutions will probably be forthcoming.

Computerization can also and perhaps more readily be applied to baggage accountability. United Airlines has a computer system for recording, storing, and reporting all records dealing with misrouted, pilfered, and damaged baggage. In addition, Total Apollo Baggage System (TABS) introduced a unique factor in tracing misrouted baggage. When a piece of baggage is found that has not been put on its scheduled flight, the station that has found the bag enters the new routing of the bag into the computer. When the agent at the destination enters a missing baggage claim for a passenger into the computer, an automatic match process takes place; while the passenger is still there, the CRT advises the agent that the missing bag has been found and is on its way.

DAMAGED BAGS

When bags are discovered to be damaged, whether from being torn on a conveyor or by falling from a transfer vehicle, the suitcase or bag should be immediately recoopered[2] with tape or cord to keep the contents from becoming a temptation to baggage handlers. When a torn bag is discovered, conveyors should be promptly inspected to prevent further damage and to ensure that the cause of the damage will be eliminated.

THEFT DETERRENCE

Baggage staging areas and sorting rooms should be well lighted and arranged so that a supervisor can survey the entire area. Transfer vehicles will be obvious obstructions to vision, but the room should not be laid out with hidden corners, extensions, or alcoves. Adequate lighting is also essential for ensuring that handlers can read baggage tags easily and accurately.

If a carrier is experiencing pilferage from baggage (and most are), the baggage staging area is a prime location for the application of closed-circuit television to provide more constant observation. The use of CCTV in the baggage staging area will probably increase as a result of the Air Carrier Standard Security Program because a section concerning baggage acceptance requires that procedures be in place that "prevent or deter unauthorized access to checked baggage from the time of acceptance until delivery at destination to preclude tampering with or adding to contents." Carriers are subject to a civil penalty if this is not done.

---

2 Transportation industry term dating back to the coopering or making of barrels. To recooper is to repackage.

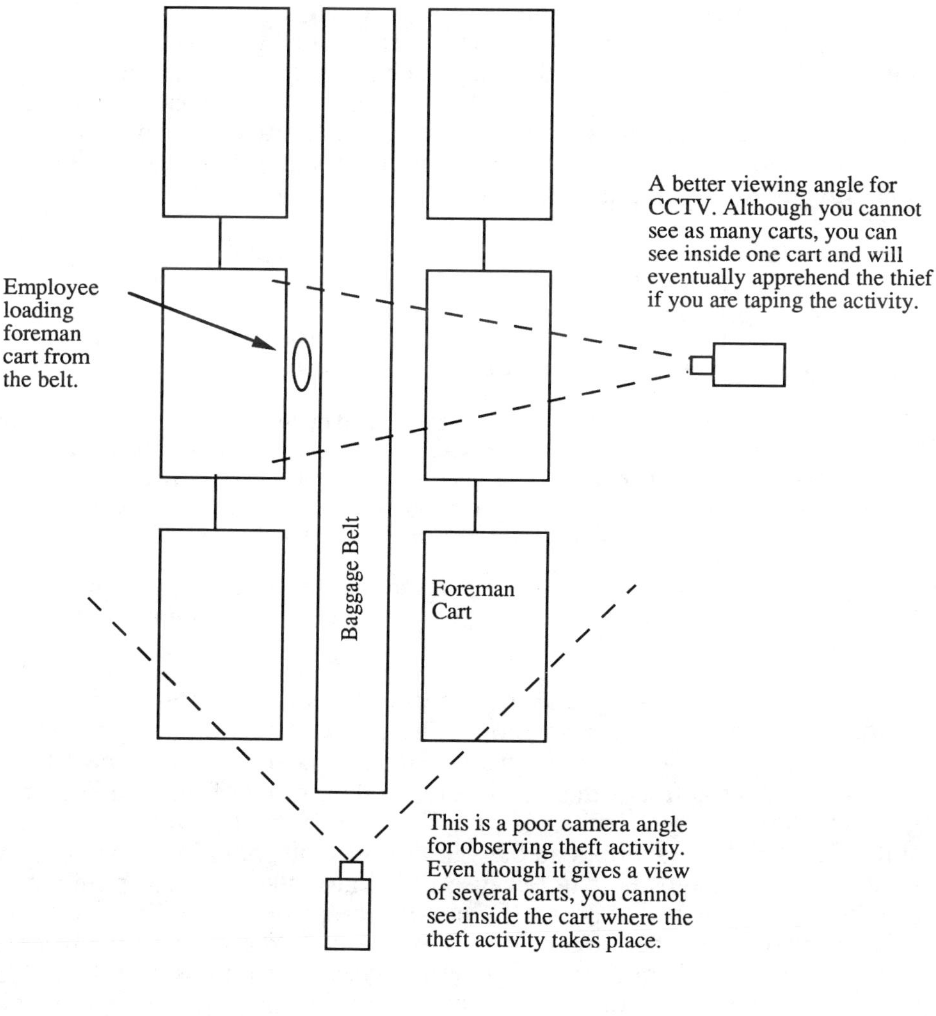

Figure 20-1. CCTV camera placement in baggage staging area

Other techniques will deter baggage theft. Staging areas should be isolated from public access areas. If a passerby can walk through a swinging door or step over a low fence and gain access to the luggage, remedial physical changes should be made. As in the case of cargo security, the routes taken by transfer vehicles carrying baggage should be established so as to avoid passing public parking areas and traveling on public roads, and vehicles should be enclosed to avoid weather damage to baggage.

Obviously, there is no legitimate reason for a baggage handler to carry luggage keys, but instances of this have repeatedly come to light in the course of baggage theft investigations. The best solution, and one that is much needed, is a company rule stating that, except when traveling, employees carrying luggage keys may be subject to disciplinary action, including termination. Suggested wording is as follows:

> In order to avoid any inference of involvement in pilferage from baggage, employees who handle baggage for (Blank) Airlines are not permitted to have luggage keys in their possession or control while on the premises of (Blank) Airlines or upon airport property, except while traveling as a passenger. Violation of this rule can result in disciplinary action up to and including termination for a first offense.
>
> This rule includes luggage keys carried on the employee's person or found in his or her locker, vehicle or work station. It also includes keys to the employee's own luggage. Because many such keys are universal in nature, no luggage key of any type may be brought to the airport without the express authorization of your supervisor unless you are traveling.
>
> Employees excluded from this regulation are those who have a legitimate need for having luggage keys, such as those who work in customer service, lost and found, ...

The amount and quality of supervision is as important for baggage handling as for any other operation. There must be enough supervisors to ensure that incoming interline transfer bags are passed on to the next flight, that bags delivered from other carriers reach their proper flight, and that baggage handlers see their supervisors with some frequency throughout their shift. The employees who load bags in the belly pit should be hesitant to attempt pilferage from the bag because they know the supervisor frequently checks on activity at that point. (Of course, the baggage handler outside the belly pit may act as the lookout.) Supervisors should also be in the baggage areas with enough frequency to deter employees from putting an inbound tag on an outbound bag, in order to be able to pick the bag up later in the claim area.

Theft and pilferage can also occur when bags are in the custody of a tour operator. An investigation into numerous pilferage claims coming from charter groups on excursion to Las Vegas resulted in police surveillance of the drivers who picked up the luggage of returning charter groups for delivery at the airport. The surveillance disclosed that at least two of the drivers were pulling in behind a building and going through the bags for easily pilferable items.

IDENTIFYING BAGGAGE THIEVES

When placing CCTV cameras for the purpose of apprehending an employee who is stealing from bags, the temptation is to cover as many foreman[3] carts or containers as possible with one camera view. The problem with this technique is that the theft takes place inside the foreman cart. As bags come down the baggage belt, they are put into the cart designated for a specific flight. If the employee leans into the cart and the camera is not looking directly into the cart, the only view provided of the employee is from the waist down because the upper body will be hidden in the cart. A company may suspect that an employee who spends too much time leaning into a cart is up to no good, but it will have no evidence. From an evidential standpoint, it is best to focus the lens on one cart,usually from a concealed overhead position. (See Figure 20-1.) Eventually, these efforts will pay off if indeed a thief is on the payroll.

At stations where baggage theft is known to be occurring, investigators often use decoy baggage to identify the thief or thieves. Desirable items and, possibly, cash are placed in the dummy bag. (If currency is used, the serial numbers are recorded.) A powder blue dye visible only under ultraviolet light is used to tag items and bills, and care is taken to ensure that the bag must be opened before the handler can come in contact with the powder. The salted baggage is checked and loaded normally, but removed before takeoff. Should a check reveal that cash or bait items are missing, the ultraviolet lamp is employed. As the search for missing items can extend to out-of-the-way areas and cars, battery-powered ultraviolet lamp units have proven to be the most useful choice.

BAGGAGE CLAIM

Positive baggage claim, the practice of requiring passengers to present baggage claim checks to a security attendant before being allowed to take luggage from the baggage claim area, appears to have proven itself a worthwhile and profitable measure for large stations that have numerous daily flight arrivals. The point at which the law of diminishing returns sets in for this practice depends upon the station experience with lost baggage. From an economic standpoint, a station that has no significant history of stolen or misclaimed bags offers little justification for the introduction of positive baggage claim. If, for example, it costs $250,000 per year at a large station to provide security coverage for positive claim, it might be preferable to absorb a few thousand dollars in bag claims and avoid the cost of additional labor. Indeed, that is the approach now taken at most airports. Where a history of thefts exists, however, the practice of positive claim is certain to prove most effective in drastically reducing such losses, even if instituted for a short time. If a station experiences a dramatic increase in bag thefts, it would appear prudent to

---

[3] A foreman cart is the standard 4 wheeled cart utilized by air carriers for transportation of cargo and baggage to and from the aircraft. It normally is open on both sides with canvas and chain coverings.

institute positive claim for a period of 30, 60, or 90 days until the problem resolves itself.

Whatever the system in use, the baggage carousels should be frequently observed by supervisors, particularly where positive claim is not used. Bags unclaimed after other passengers have departed with their luggage should be promptly removed and placed in storage, in either a locked room or by securing the bags with a padlocked cable run through the handles of the bags. Persons making late pickups of bags should be checked to ascertain whether they have the proper claim checks. Baggage thieves often watch the claim area until only a couple of bags remain. When it becomes obvious the owner is not coming to pick them up immediately, the thieves step up to take the bags. The baggage claim area should always be staffed by airline personnel whenever baggage is present, both to assist passengers in recovering their luggage and to deter others from taking luggage that does not belong to them,, whether intentionally or by mistake.

BAGGAGE LIABILITY

International baggage liability is governed by the Warsaw Convention. This convention was held in 1929 for the purpose of prescribing liability rules for carriers and passengers involved international air transportation. A protocol to this convention which increased the liability limits was signed at the Hague, Netherlands, on September 28, 1955, and became effective August 1, 1963.

The carrier is responsible for loss, destruction, or damage of checked baggage under the terms of this convention. The carrier is also responsible Warsaw Convention for consequential damage, such as damages occasioned by delay in the transportation of baggage by air. Consequential damage liability, which does not apply to domestic baggage, is a nebulous and difficult area, and damages are sometimes difficult to prove. Consider the hypothetical salesman who claims that the carrier lost his presentation kit for 48 hours and thus he was unable to make a satisfactory sales presentation to his client and lost a million dollar sale. The damages under the Warsaw Convention are relatively small and limited to approximately $9.07 per pound ($20.00 per kilo) for checked baggage and $400 per passenger for unchecked baggage. The tickets of one major international carrier stated that unless otherwise noted on the baggage check, the weight of each piece of baggage shall be deemed to be 70 pounds (32 kilos) and that "this presumption limits our liability to no more that $640 per piece of checked baggage."

Domestically, the maximum limit is $1250 per passenger. The term "per passenger" is important to note. If a man and wife are traveling together, then there is a possibility of a $2,500 claim if more than one bag is missing. Several credit card companies offer additional automatic insurance if the airline ticket was purchased with their credit card. Homeowners insurance may also provide coverage.

When a passenger uses two or more airlines to arrive at a destination, the carrier who had the last leg of the trip is responsible for the loss. This can occasionally cause some apparent injustices when, for example, a passenger flies coast to coast or internationally on carrier A and then takes a twenty-minute commuter flight to the final destination on carrier B. Carrier A received the lion's share of the revenue, but carrier B must pay the loss.

BAGGAGE CLAIMS INVESTIGATION

Much can be done to locate pilfered articles if their loss is reported promptly. Sometimes, pilferage is discovered as soon as the bag arrives at the destination perhaps because the bag is not securely closed or shows signs of having been disturbed. When an immediate check of the contents of a bag discloses the theft of one or more articles, telephone tracing should be started at once. It is possible and even probable that individuals who handled the bag at the origin station may still be at work. A physical search of the origin station baggage area and its surroundings can be quickly conducted with a locker check in an effort to locate the missing items before they are removed from the premises.

Most carriers will compensate passengers for reasonable expenses until the bags arrive. Carriers will generally provide passengers with a remain overnight (RON) kit that contains necessary toiletry articles, and they may reimburse for reasonable articles of clothing, such as a shirt or underwear. However, when possible, passengers should check with the carrier to see if substantial expenses will be allowed. There have been numerous cases of businessmen who bought a new suit and then had it disallowed.

The problem with this suggestion is that in some cases, it is almost impossible to get through by phone to the baggage claim desk at some airlines. Occasionally, passengers either get a constant busy signal or are placed on hold for unreasonably long periods of time. Some carriers require that calls be made to their reservations line, and can update the passenger on any progress made in locating the bag. Many carriers have recognized the marketing significance of providing exemplary service to passengers in the area of lost or delayed baggage and have provided adequate personnel and telephones to handle these claims.

As a part of claim settlement, the passenger should be asked to furnish as much supporting and substantiating evidence of the loss as possible. This can include bag tags, copies of airline tickets, receipts for the purchase of lost items, and similar documentation. The carrier may arrive at settlement values by calculating depreciation on the items being replaced, particularly for articles whose replacement cost may bear little relation to their value, such as garments. Particularly in the case of substantial claims, the passenger should be asked to mail the claim form; if fraud is later proven, there may be a case for using the mails to defraud.

Another measure that has been of assistance is that, since 1969, Eastern Airlines has operated a Baggage Claims Central File for the U.S. airline industry. Baggage claims filed with participating carriers are entered into the computer, where they are stored for 24 months. If the passenger has filed other claims in that period, the carrier with the latest claim is notified. Certainly, the existence of other claims does not mean the claim is fraudulent;

it may only be that the passenger is a frequent traveler and an unlucky one. Nevertheless, the information should ensure that the claim is reviewed with more than the usual scrutiny.

Most airline passengers are honest and would not involve themselves in a fraudulent claim, but some means is needed of detecting those who try to take advantage of airlines. I worked one case where the claimant made duplicate copies of his claim and proof of loss forms and sent them to five different carriers on which he had traveled, claiming to have lost the same luggage on each carrier. Similar claims files in other industries, particularly automobile insurance, have proven their value in alerting investigators to individuals who are repeatedly involved in accident and personal injury claims. Insurance fraud rings have been uncovered that have made establishing false claims into a fine art.

Any number of unusual circumstances may arouse the curiosity of the claims employee. Personnel involved in claims settlement should be cautious of those who press for early payment, particularly before there has been enough time in which to locate the bag. The total number and size of items claimed to have been stolen may arouse suspicion when such a volume of articles could not fit into the suitcase involved. The number and value of the articles claimed may be inconsistent with the length and duration of the claimant's trip. The impressions of the employee who takes the report can also be valuable; often, something said in conversation will impress the employee with the authenticity of the claim or with a doubt of its authenticity.

The carrier can productively endeavor to trace the missing suitcase or luggage to its last known location. A striking example of this was an investigation in which I became involved. The investigation of the $3500 claim had dragged on for several weeks. It was alleged that the bag had been delivered by the airline to a nearby hotel for delivery to a person claiming to be the wife of the passenger making the claim. The woman had the proper baggage claim check number, but not the claim check; however, the bag was delivered to the hotel in care of her. The passenger claimed no knowledge of this person. In an effort to identify the fictitious wife, the first stop was at the hotel. There, it was discovered that the woman--whoever she may have been--had not claimed the missing bag; it was still safe in the custody of the hotel.

An informal survey of hotel baggage storage rooms in a resort area revealed that many unclaimed bags were being held in hotel storage rooms rather than returned to the air carrier liable for baggage claims on the luggage. As a result of this survey, 35 lost bags were returned to airline custody. The explanation offered by the manager of one hotel was that the unclaimed bags were being held for sale to hotel employees. Certainly, a physical check of the location to which the bag was last delivered is an appropriate precaution to take before paying a claim for a lost bag.

Not all items lost or pilfered from luggage are reimbursable. While tariffs vary by carrier, airlines generally are not liable for loss, damage, or delay in the delivery of fragile or perishable articles, money, jewelry, silverware, negotiable papers, securities or other valuables, business documents, or samples which are included in the passenger's checked baggage, whether with or without the knowledge of the carrier.

DISPOSITION OF UNCLAIMED BAGGAGE

Although most missing baggage is located within 72 hours, much of it is never found. Airlines Computer Tracing System Resolution 30.24 states, "On-hand baggage shall be retained by the Central Lost and Found Office for a minimum period of 90 days after receipt in the Central Lost and Found Office." Implicit in such a rule is the understanding that the baggage may be sold after that time.

The total amount and value of unclaimed items is staggering; although may items are of little value, there are large numbers of cameras, sporting goods, furs, quality clothing, computers, and even television sets. Holding this baggage for the 90-day period requires controls to prevent theft by knowledgeable carrier personnel. If no inventory controls or records are kept on what comes in and goes out, the theft-bent employee will soon deduce that the items can be taken with impunity. From one or two items the dishonest employee will progress to trunk-loads and will furnish goods to friends and associates.

The controls established should be designed to prevent the premature salvage of attractive items from baggage prior to the expiration of the 90-day holding period. Where accountability controls are in effect, Lost and Found personnel must falsify or destroy records in order to steal, a situation less inviting to the light-fingered than an absence of records. To entrust employees with an operation involving so many attractive and valuable items without instituting adequate controls is to choose to create far too much temptation. In such cases, management has failed in its obligation to its employees.

Unclaimed luggage must be disposed of eventually. Some carriers have experimented with selling it to employees on special sale days. The logistics of setting out the contents of the suitcases, pricing and categorizing of merchandise, handling valuable furs, jewelry, etc., has caused more problems than most carriers want. Several have opted instead to sell unclaimed baggage in bulk to firms that specialize in disposing of unclaimed items or selling items for insurance companies.

The airline may consider keeping those items found in unclaimed baggage that are usable in the day-to-day operation of the airline. Appropriate items include calculators, computers, cameras and tools. Control procedures must be established to ensure that these company-use items are properly entered into company property records to prevent their conversion to personal use.

Chapter 21

MISAPPROPRIATION

Although air piracy, terrorism, and the antihijacking program have captured the attention of the public in recent years, airline security people were kept quite busy long before hijacking became a serious problem. From the beginning, air carriers have been concerned with the same internal problems faced by any business where employees handle cash and negotiables, deal with the public, and contract for the purchase of equipment and supplies. In addition, carriers have fought problems peculiar to airline security, including cargo and baggage theft, stolen and counterfeit tickets, and external threats to the safety of passengers and crews.

Although security has been a force in being with the airlines for many years, only recently has security's significant contribution to the profits of a carrier been recognized. The Security Committee of the Air Transport Association was not organized until 1969. Except for the Airport Security Council, which was formed by the airlines a year earlier and limited to the New York metropolitan area, the ATA Security Committee was the first significant attempt by the carriers to band together in an organized effort to meet mutual security problems. Thus, although security departments had long existed in the organizational structure of most air carriers, the organization of the ATA Security Committee represented an important milestone for air transport security. Within two years, the ATA saw the need to create the post of Director of Security for the Association, with the responsibility to direct industry-wide efforts in all phases of air transport security. Harry J. Murphy was appointed to the position, and he played an important role in strengthening local ATA Security Committees at each airport and making them more effective. Following Murphy's death, Richard F. Lally became Director and then Vice President of Security for ATA.

Although the antihijacking programs placed an added burden upon carrier security departments, they also allowed them to obtain, in most cases, the necessary funds and manpower to accomplish the job and to strengthen their departments. Now that the gearing-up phase of the antihijacking program has ended, airline security people are beginning to turn their new resources and new recognition toward older operational problems that have stubbornly resisted earlier efforts. Among these are the problems of theft and embezzlement--larceny by trust--which can be categorized as misappropriation.

FLIGHT KITCHENS AND COMMISSARIES

The cost of food taken from flight kitchens can amount to a sizeable operational loss if control measures are not taken. When beef prices rise even higher than usual, a box of frozen prime steaks is a highly desirable prize to those who would rather steal their luxuries than pay for them. Surprisingly, the heaviest losses in this area can be due to food eaten by catering employees while engaged in meal preparation. For this reason, most U.S. air carriers provide food-handling employees with a free meal during each shift to reduce hunger and make it easier for the carrier to enforce a rule that workers are to eat nothing from the food preparation area. Although the cost of providing free meals is substantial, most carriers believe it costs much less than absorbing the losses that would occur from workers eating from the ingredients of the flight meals. Although some security people have seen this practice as a modified form of bribing people not to steal, one free meal per shift is a standard benefit in the U.S. restaurant industry, and food handlers have come to expect it.

Appropriate controls can reduce food loss to a minimum. For example, the buyer ordering food from a vendor should take care to include the quantities ordered on each purchase order, a logical measure not always taken. When the delivery is made, a responsible employee other than the buyer should be present to verify that the items and quantities are as ordered. The buyer should also be present on delivery to check for quality, as may the chef, but the employee responsible for ordering should not be the one assigned to receive and inventory the merchandise into the facility.

The technique of "portioning" helps. Accountability is easier if steaks are ordered preportioned, rather than ordered in bulk and cut, because preportioned steaks can be counted at any step. As a matter of fact, every item that comes in or goes out of the flight kitchen needs to be accounted for. If a steak is thrown out because it was burned, for example, this should be noted. Of course, there are a variety of reasons why a count of meals prepared and meals dispatched must be kept. Food inventories must be current not just for security purposes, but because the chef must know what foods must be ordered and in what quantity.

As in any operation where a high risk of pilferage exists, flight kitchen employees should be required to use a single door to enter and exit the operational area. A supervisor should be present at the door so that obvious packages will be noticed and should periodically ask to check lunch boxes and other containers on a regular or random basis. A special pass form should be provided to supervisors for authorizing the removal of approved salvage.

A substantial source of loss in this as in all restaurant operations, is silverware inadvertently thrown into waste receptacles with food scraps. The question of recovery must be resolved by determining whether the loss exceeds the costs of attempting to recover the silverware. Although magnetic drums through which garbage is passed so that silverware may (with less than 100 percent effectiveness) be separated out, are available for this purpose, opinions vary as to whether the savings are always worth the time and expense of using these devices.

The liquor storage area is another highly vulnerable and attractive feature in most flight kitchen operations. The primary control needed here seems to be simply to get back to the basic principle of accountability. If the liquor supply is properly stored, it can be inventoried twice each shift; once when the designated worker begins the shift and again before he or she leaves. If the count does not tally with the current balance, calculated from what has come in and gone out during the shift, the inventory worker should remain until the records can be made to balance with the count. In this way, accountability can be tied to a specific shift, and the responsibility for it is tied to a specific person. If a loss occurs, it will be discovered before the end of that shift. The knowledge of employees that a close liquor stock inventory is maintained on the basis of two physical counts per shift acts as a strong deterrent to liquor thefts.

IN-FLIGHT SALES

Few outside the industry are aware of the amount of cash flight attendants and supervisors receive and handle during a flight. Airline tickets are sold, headsets are rented for watching movies, and liquor is sold, as are some meals in economy class. On international flights, anything from cigarettes to perfumes and jewelry may be available for sale during the flight.

Any time cash is handled by employees, whatever the business enterprise, adequate controls are necessary or embezzlement and theft will occur. The key to cash-handling control is to establish a basis for accountability at each step. If carrier management is unwilling to pay the cost of cash accountability, then they must accept the price of the losses inherent and inevitable when cash is handled without accountability.

Liquor

This is particularly true of in-flight liquor sales. The liquor must be counted when it leaves the kitchen, and that count must be verified on the aircraft by the responsible flight attendant before signing for the liquor. The seal number on the liquor cart should also be checked against the accompanying records to make certain no one has tampered with the contents. The basic inbound controls revolve around reconciling the number of units returned and the amount of cash turned in. Each carrier accomplishes this reconciliation in its own way. Some lock the money from liquor sales in the cart with the liquor, so that the two are counted together when the container is opened. Auditors recommend that two persons be assigned to any such count, a standard banking practice. Although this will deter thievery, it does not prevent it. I investigated one case where two cashiers who counted cash from ticket sales together were both stealing without the knowledge of the other.

Other carriers have the flight attendant seal the money in a deposit envelope. The envelope may be placed in a deposit safe at the inbound terminal by the flight attendant, particularly if he or she is terminating at that station. The attendant who is continuing may hand the deposit to a passenger agent at the terminal to make the deposit for her. When this is done, the flight attendant should receive a receipt for the sealed envelope. Of course, the passenger agent cannot be asked to sign for the actual amount of cash in the envelope unless the agent is able to count and verify the amount in the presence of the flight attendant. In actual operation, there is rarely, if ever, time for such a count.

No matter who deposits the envelope in the safe or deposit receptacle, that person should sign the safe's deposit register to show that he or she deposited the cash receipts envelope from that specific flight at the time logged. When the safe is opened and the envelopes are being gathered for delivery to the bank, each sealed envelope should be tallied with the appropriate entry in the register of envelopes deposited. Should any deposit envelope be found to be missing, the disappearance should be investigated immediately.

Removing a deposited envelope from the safe can be ruled out if management has taken the precaution of having a deposit "drop" within a safe that requires two keys to open, with one key held by the armored car service. This requires a safe with an "anti-fish" deposit slot through which envelopes may be dropped into the double-locked compartment. The remainder of the space within the safe remains accessible by combination, but the deposit drop cannot be opened until the armored car service arrives. Where numerous people have access to the safe and the deposit envelopes are not secured separately, accountability is nonexistent.

When cash count discrepancies are found in liquor deposits, a notation should be made of the discrepancy in the file of the flight attendant who signed for the count. Occasional discrepancies are bound to occur but, if discrepancies are faithfully noted, a periodic review of attendants' files will reveal patterns and trends. Excessive discrepancies should receive special attention, and investigation should be initiated where it appears to be indicated.

Tickets

The controls maintained over airline tickets held in stock and sold during the flight are much the same as those used for tickets sold on the ground. If an attendant pockets the cash from a ticket sale and destroys the auditor's coupon, the embezzlement will be discovered as soon as the used ticket is found to have no matching auditor's coupon. From that point, investigation is a relatively simple matter of determining on which flight the ticket was sold and interviewing the person who was authorized to sell tickets on that flight. In most cases, only one attendant or supervisor has that responsibility.

Headset Rentals

Money collected for the rental of audio headsets presents a more difficult problem, one that carriers have not as yet completely solved. Industry practice and tariffs require that a fee be charged on certain flights for the use of the headset needed for listening to movies and, in some instances, for audio entertainment. Complications occur when headsets are free to passengers who want only to listen to audio entertainment but are rented to those who want to watch the movie.

Some carriers attempt to solve this problem by providing headsets of one color for passengers who want to listen to audio and headsets of another color for those who are paying to watch the movie. The audio headsets are picked up from the passengers before the movie is shown and passed out again after it is over. When the two-color system is used, the movie headsets should be brought on board in sealed bags with a recorded count of the headsets. The rental fees turned in for each headset used should reconcile with the number returned unused. Most commonly, headsets remain in the seat pocket and the flight attendant sells the "pigtail" (the device that attaches the headset to the receiver.) Again, accountability is the key.

Where there is only one color of headset, the method some carriers use for accountability is to have the attendant supervisor make and record a count of headsets in use during the movie; the fees collected are later roughly reconciled with this count.

The primary method for accountability for headsets at the present time is the use of receipts. When headsets are sold, passengers are issued a receipt. They are told that they should keep the receipt so that, in the event there is a problem with the move, the airline will know to whom to refund the money. Actually, the money received is balanced against the receipts issued. For example, if 150 receipts were issued at $4 each, then the flight attendant will be expected to turn in $600. If the flight attendants do not give out a receipt, the money is theirs to keep. Flights that appear to be experiencing problems with low receipts are occasionally audited by security and auditing personnel. They listen to see if an announcement is made explaining the receipt system so that passengers will expect to receive a receipt. They also observe the sales to see if individual flight attendants are handing out receipts.

Flight attendants have, upon occasion, used headset rental money to balance shortages in liquor deposits and, in some cases, for their personal use. When one carrier placed audit controls on its headset rental procedures, this revenue immediately increased by 40 percent. When the controls were temporarily removed, the receipts fell by the same amount, again showing what occurs when the opportunity to steal is joined with a small risk of detection.

Another problem relating to headset rental is the passenger who attempts to avoid the headset rental charge by purchasing an airline-type headset or an adapter from an airport gift shop or through mail order or by stealing a headset for use on a future flight. This robs the airline of the revenue it is due for providing the movie. On November 9, 1990, an attorney who was a passenger on a United Airlines flight used his own headphones to listen to a movie. He was asked by the flight attendant to either stop using the headphones or to pay the rental fee. The flight attendant also suggested that if he kept watching the movie, the airline would call security when the flight landed. The passenger told her to tell the captain that he was a lawyer and that if anything happened on the ground, he would file suit. When the plane landed at Los Angeles, a citizen's arrest on the charge of petty theft was performed by the plane's captain, according to newspaper reports, and the passenger was turned over to the airport police.

The various schemes that can be used to defraud carriers of income from liquor sales or to steal the liquor itself are fairly well known among flight attendants. If liquor is served from full-size bottles rather than from miniatures, the flight attendant can take liquor by simply pouring it into a flask carried in a purse or suitcase. Attendants have also collected empty miniatures and refilled them at home from a larger bottle. The attendant then "opens" the refilled miniature in front of the passenger and pours the drink. The passenger is unaware that the bottle was not sealed and he receives the drink that was ordered. The attendant pockets the money received for the drink.

THEFT OF AIRLINE PROPERTY

Another constant drain due to theft involves the loss of airline property. Most of this property is probably taken from aircraft, and it is difficult to assign responsibility for these thefts. One major carrier bought 16,490 blankets at a cost of $120,128. Taking into consideration the airline's monthly "retirement rate" for blankets, it became apparent that the carrier's annual pilferage rate was approximately 15,000 blankets, a value of $110,000. As a test, two flights were checked before and after a round trip from the coast to coast. One airplane returned with 21 blankets missing, and the second was short 18 blankets. I once followed a passenger, who was wearing an airline blanket over her shoulders, off of a flight. She begrudgingly gave up the blanket but not without some words.

Another carrier published in its house newsletter the following figures for equipment replacement for a single year. In an industry as financially pinched as the airline industry, such losses--and these are for a single carrier and not one of the largest--are a significant factor. Although breakage might account for some portion, it obviously can not account for all of the following:

5,000	pillows ($3,500)
12,311	teaspoons ($27,000)
4,349	coffee cups ($66,000)
3,000	bread-and-butter plates ($51,000)
	glassware ($44,000)

	wine and champagne glasses ($33,000)
	china ($330,000)
	silverware ($140,000)
	knives ($42,000)
4,570	salad bowls
46,095	salt and pepper shakers

THE DISHONEST FEW

Unfortunately, the recitation of even less than all the schemes THAT can be and have been used to defraud airlines may imply that all airline employees are out to defraud their employer at every opportunity. Such is not the case; the great majority of airline employees are hard working and totally honest. The fact that most are honest provides all the more reason for the carrier to exercise diligence in eliminating attractive opportunities to steal and in supporting the effort required to identify and remove from the payroll those few employees who lack the integrity to remain honest in the face of temptation.

The offense committed by an employee through an act of theft is often dismissed by thoughtless observers as trivial because the air carrier is "only a corporation." In reality, the offense is against the loyalty and integrity of coworkers toward whom the employee ought to feel some kindred spirit. Some people, it seems, subscribe to the philosophy that stealing from a corporation, an insurance company, or the government is a mitigation in itself, as if such thefts should be seen as different from a theft from a roommate, for example. Such a philosophy is rationalization, an attempt to excuse or justify acts the individual knows are wrong.

If such a philosophy is allowed to spread until it becomes the prevalent attitude of an entire group of employees, then dishonest employees will no longer attempt to conceal their actions from their peers. Their dishonesty will become common knowledge, and stealing from the company will become an acceptable practice within the group. Where such a situation has developed, severe and concentrated action by management is required to reverse the trend.

Section V

AIR FREIGHT SECURITY

Chapter 22

AIR CARGO SECURITY PROBLEMS

Man has transported goods by ship and by beasts of burden since the dawn of civilization. From before recorded time to the nineteenth century, the only innovation made in the carriage of goods was the introduction of wheeled carts, which are thought to date from 35 to 40 centuries ago. Little more than a century encompasses a worldwide revolution in the carriage of freight. The result, the capability of moving larger cargoes more rapidly between more points, has already had its effect on history in making possible the development of landlocked resources and the support of great urban populations. In the long view of history, today's principal methods of transporting goods appeared only yesterday. Railroads have been carrying freight for little more than 140 years; trucks, only in this century; and aircraft, only in the last 65 years. The carriage of cargo by air is the most recent of the three innovations, but its peacetime importance alone, particularly in commerce, has meant great and continuing change.

GROWTH OF CARGO CARRIAGE

The first commercial business of U.S. airlines was the carriage of mail, which began in 1918, eight years before the introduction of commercial passenger traffic. At first, cargo carriage increased much more slowly than the carriage of air mail. For example, it was not until 1956 that air freight passed air mail as a source of revenue for United Airlines. Today, air cargo carriage increases at a rate that can only be called dramatic. In testimony before a congressional committee in 1969, one witness documented the ton/mile increase for air cargo from 1958 to 1967 at 459.52 percent. In 1973, U.S. carriers had their first billion dollar freight year. Current figures show that cargo ton miles increased from 11,469,193 in 1988 to 12,180,139 in 1989, an increase of 6.1 percent,[1] in one year, indicating the industry is still maintaining a healthy growth.

WHAT CARGO GOES BY AIR?

The 250,000-pound payload of a 747 freighter permits carriage of extremely large shipments of freight, such as a 52,000-pound printing press or several limousines. At the opposite end of the scale, one of the fastest-growing cargo services has been the carriage of small packages weighing less than 50 pounds, although the introduction of the facsimile machine has had an impact on that service.

The speed of air freight makes it ideal for the shipment of perishable freight such as flowers, strawberries, and tropical fish. Race horses have long been flown throughout the world. The U.S. Department of Agriculture is designing improved livestock containers for wide-body aircraft. For example, pens for transporting sheep are planned; each pen will hold about 40. Air freight enables the manufacturer to get merchandise to stores and users in the shortest possible time, improving the service and accelerating receipts.

INHERENT SECURITY PROBLEMS

A number of security problems are peculiar to air cargo operations. For one thing, air cargo has a greater proportion of high-value shipments than any other method of freight carriage; as a result, the potential for loss is much greater.

Identifying the point at which a loss occurred, always a problem, is even more difficult for air cargo, due to its greater exposure. In all other forms of freight carriage, shipments are picked up, transshipped, and delivered at loading docks. In addition to this, air cargo must be moved (usually in "trains" of carts) from the carrier freight terminal to the flight line, where it is loaded onto the aircraft. This results in increased exposure and additional handling. When the freight is to go in the belly of a passenger airplane, the exposure is further increased because the freight must then go to the passenger terminal ramp, where it is exposed to other persons. At many airports, the tractors (tugs) towing the carts travel to and from the flight line along a route that passes unlighted (and sometimes unfenced) parking lots or dark adjacent streets. This creates opportunities for goods to be quickly passed to a waiting accomplice or stashed for later pickup by the driver.

Another risk-creating factor special to air freight is that air cargo shipping is basically a nighttime operation, despite efforts of carrier marketing

---

[1] *The Annual Report of the U.S. Scheduled Airline Industry, Air Transport, 1990.* Air Transport Association of America (June 1990).

people to sell space in the lower deck of wide-body passenger aircraft. In the U.S., 90 percent of all air mail and 90 percent of all air cargo--domestic and international--arrives or departs between 10:00 p.m. and 7:00 a.m. Darkness has always been attractive as a cover for criminal acts, and air cargo theft is no exception to this general rule. All of the problems that are peculiar to air cargo operations are, of course, only added to problems common to freight carriage generally: pilferage, fraudulent pickups, theft by drivers, and so forth.

EMERGENCE OF CARGO THEFT PROBLEMS

The public first became aware of air cargo theft problems in the late sixties. One of the first major disclosures came during hearings of the New York State Investigation Commission in December 1967, which revealed that labor racketeers had infiltrated the local pickup and delivery segment of the air freight industry in New York City and that organized crime was exploiting the drivers who served the New York area airports.

According to the testimony, employees of air freight truckers serving New York metropolitan airports were required to join Local 295 of the International Brotherhood of Teamsters. Owners of trucking firms wishing to avoid harassment and labor problems were told to join an association called the Metropolitan Import Truckers Association (MITA). The hearings confirmed that racketeers did in fact hold office in Teamsters Local 295 and in MITA and that several persons of similar background were retained as consultants to air cargo trucking companies. MITA, later renamed the National Association for Air Freight, Inc., was said to be under the influence of an Anthony Di Lorenzo, identified by law enforcement as a member of a national crime syndicate. Di Lorenzo, another officer of the association, and 12 member air freight trucking companies were indicted on charges of antitrust violation. In addition, the U.S. Justice Department successfully brought suit to have MITA dissolved and to prevent the creation of any similar organization.

These disclosures and others (including a few spectacular holdups of valuable shipments) led carriers to establish the Airport Security Council in May 1968 to deal with theft and pilferage at the three major airports serving the New York metropolitan area: Kennedy, LaGuardia, and Newark.

The following year, additional findings were published by the Senate Select Committee on Small Business, after hearings held during the spring and summer of 1969 on the effect of cargo loss and theft in air commerce. One finding met with general agreement: The growth of air freight had been so rapid that many air cargo terminal facilities were inadequate for security purposes, and carrier handling methods had also failed to keep pace. In his closing report on the hearings, Committee Chairman Senator Alan Bible accurately cited some of the problems impeding the accountability and disclosure essential to an effective attack on the problem of air cargo theft:

> Losses are not reported systematically. Standards of reporting have not been established. There is no good way of telling at what point in the sequence of many events from the point of origin to ultimate destination

> the loss occurred. The relation between extent of protection and value of shipment is haphazard. No data are available on a systematic basis as to the circumstances surrounding events of loss. The techniques used by thieves can only be guessed at; all we can say at this time is that they are successfully practiced, at a rapidly increasing rate.

Because most of the major U.S. airlines were at this time in an economic squeeze of major proportions, the climate for increased expenditures for theft prevention was not good. In addition, it was obvious to the industry that effective action would require systematic preliminary study; each carrier's specific problems and the range of appropriate countermeasures available would have to be identified before effective cargo theft-prevention policy and programs could be developed.

Henry De Geneste, Police Superintendent for the Port Authority of New York and New Jersey, revealed that in 1988, airlines and forwarders reported $3.9 million worth of stolen freight at Kennedy Airport, compared with $1.1 million in 1987. De Geneste suggested that part of the dramatic increase might be attributable to employees who are stealing to support drug habits, but he also indicated that complacency on the part of airlines and cargo agents was a significant cause of the increase.

The Los Angeles Police Departmenthas expressed growing concern about the rapidly increasing number of cargo thefts at ports and warehouses in the Los Angeles area. For containers alone, they reported $2.4 million worth of merchandise stolen in 23 thefts in 1985. From 1986 to 1988, between 50 and 58 container thefts occured each year amounting to annual merchandise losses of $2.6 million to nearly $7.1 million. For the first half of 1989, this figure had risen to 60 container thefts involving merchandise valued at $3.5 million. The police commented that this type of crime is popular among criminals because it is a relatively low-risk, high-profit crime.

GOVERNMENT RESPONSE

Prompted in part by the national press coverage given to the most dramatic statements of committee witnesses regarding cargo theft, federal and state agencies became interested in attacking the problem. Although the only evidence of serious crime that had been brought out had clearly concerned the pickup and delivery trucking segment of the industry alone, it became apparent that these agencies generally saw a need for the regulation of the entire air cargo industry, with particular attention to air operations. The FAA and the CAB identified regulatory avenues by which they could make their own contributions, and the Department of Transportation became deeply involved in studies concerning all aspects of air freight. Many other agencies were equally interested in developing roles to play, posing a substantial danger that overlapping jurisdiction and conflicting regulation could develop. The activities of some agencies concerned only their own responsibilities. In cooperation with the carriers, the Treasury Department instituted a program to upgrade the security of cargo subject to Customs supervision, and the U.S. Postal Service moved to improve the security of mail shipments.

Table 22-1 Air Cargo Claims Report for U.S. Scheduled Airlines, 1979-1989

Year	*Freight Revenue*	*Amount of Claims Paid*	*Number of Claims Paid*	*Average Claim Cost*	*Theft Claims Loss Ratio**
1979	$2,188,330,000	$19,034,364	47,707	$398.98	0.95
1980	$2,326,509,814	$16,124,678	41,638	$387.26	0.69
1981	$2,695,206,454	$15,893,978	39,489	$402.49	0.59
1982	$3,621,113,003	$15,252,141	36,637	$416.30	0.42
1983	$3,987,740,636	$14,619,260	35,622	$410.40	0.37
1984	$4,171,372,043	$15,949,783	39,165	$407.25	0.38
1985	$4,463,955,049	$17,640,827	47,643	$370.27	0.39
1986	$4,649,970,143	$16,034,561	51,109	$313.73	0.34
1987	$6,459,970,687	$18,710,000	71,557	$261.47	0.29
1988	$6,973,109,685	$20,286,009	77,506	$261.73	0.29
1989	$7,849,052,003	$23,680,729	93,751	$252.59	0.30

*The theft claims loss ratio is the percentage of the total amount paid for claims that are theft related losses, and is expressed in cents per $100.00 of freight revenue.

Source: Air Transport Association of America, Air Cargo Claims Report. 1990.

Publication of DOT Standards

In March 1973, the Department of Transportation published the first of its Cargo Security Advisory Standards. The standards, which DOT held to be equally applicable to every form of freight carriage, address such subjects as internal accountability procedures, seal accountability, and high-value storage facilities.

In 1975, a White House Executive Order was issued, raising questions as to whether more federal regulations were needed to increase the effectiveness of the National Cargo Security Program. This order set a date in 1976 when the Secretary of Transportation would report to the President on the effectiveness of the federal cargo security program. In addition, the order required that the FAA train many of its field personnel in air cargo security.

Cargo Security Working Groups

One of the contributions of the Department of Transportation was the establishment of Cargo Security Working Groups in 15 large cities that had continuing cargo theft problems. The program was developed to strengthen government-industry cooperation in recognition of the numerous problems that have arisen in this area. Problems regarding law enforcement cooperation and questions of jurisdiction have been numerous, in some cases because of apathy on the part of the carriers and in others because diligent efforts on the part of many agencies have met with little success. In some instances, law enforcement has offered assistance that the carrier approached has declined on the grounds that their particular problem was internal and could be resolved by their own resources. At times, a carrier's theft problem has failed to qualify for assistance; some agencies have refused to act unless the value of the thefts met or exceeded prosecutor-established minimums, sometimes $10,000 or higher.

In most cases, the Cargo Security Working Groups were organized and led by the U.S. Attorney in that city, who called on the cooperation of the area's law enforcement agencies and the carriers so that a concerted attack could be made. In September 1975, however, the Department of Transportation appointed City Campaign Coordinators (usually DOT officials) to replace the U.S. Attorneys, who had become discouraged about the productivity of their involvement in the program. The program ultimately disappeared.

It is essential that carrier security personnel discuss their security and theft problems with each other and with the law enforcement people who have responsibility for law violations at the airport. If one carrier knows it is having problems with short deliveries from a particular freight forwarder or a certain driver, this should be communicated because other carriers at that airport are probably having the same problem, though they may not be aware of it yet. Caution must be exercised, of course, in identifying specific firms and individuals to other carriers. If the evidence available does not constitute legal proof, evidence that a carrier had communicated such suspicions to persons other than law enforcement could be held to constitute libel or slander. The advice of corporate legal counsel should be sought in regard to the manner in which such exchanges of information might be made.

Air Cargo Security Legislation

Many legislative and regulatory approaches have been proposed, and some have been enacted into law. One of the more interesting pieces of legislation proposed as a result of the cargo security hearings is the Civil Remedies for Victims of Racketeering Activity and Theft Act. This proposed legislation was never enacted into law, but would have allowed the carrier to file a civil suit for treble damages against anyone who steals goods or purchases or resells goods stolen from interstate commerce. The act was primarily designed to make it possible to reach fences, without whose services and resources few large or high-value thefts are practical; indeed, major thefts of merchandise are usually undertaken only after arrangements have been made for fencing the goods.

The act would have established that the sale of goods at less than fair market value would be *prima facie* evidence that the seller was trafficking in stolen goods. In addition, the act would have provided for the issuance of federal injunctions requiring persons engaged in the sale of stolen goods to divest themselves of the business fronts used in illicit marketing activities.

Such civil action is no bar to criminal prosecution, and it would offer significant advantages. For one thing, the burden of proof in civil litigation is far less than that required in criminal prosecution; the standard is not that guilt must be proven beyond all reasonable doubt, but simply that guilt must be established by the preponderance of the evidence.

Although the act would only have covered goods in interstate shipment, several states have passed or are considering similar laws to cover goods shipped intrastate. California, North Carolina, Minnesota, and Colorado have similar laws in effect. Civil remedy laws, where enacted, offer a valuable tool to any carrier willing to make use of them. A suit for triple the fair market value of the merchandise stolen, in addition to drying up potential outlets for stolen goods, would give employees second thoughts about the attractions of pilferage. It is no exaggeration to say that a judgment for treble damages represents a substantially greater punishment for the dishonest employee than does most criminal prosecution, which quite usually results in no more than probation and a suspended sentence.

PRIMARY PREVENTION

No amount of regulation, legislation, or communication can solve cargo theft problems unless supported by well-founded and consistently maintained programs to deter and detect theft. Indeed, it was the lack of such programs that instigated legislative and regulatory intervention. Basic security is, as always, the primary need. The next seven chapters discuss aspects of air cargo security and security procedures essential to effective theft prevention.

Chapter 23

PHYSICAL SECURITY OF THE FREIGHT TERMINAL

Cargo theft is, as stated earlier, an internal problem, and its solution lies primarily in improved procedural controls rather than in building higher fences to keep people out. Yet, fences also serve to control movement of employees, and they can be of considerable assistance in controlling the ingress and egress of both employees and outsiders.

Fencing choice and installation are discussed in Chapter 11, but the point cannot be made too often that air freight terminals, too, should be fenced on as many sides as possible. Quite obviously, they cannot be fenced on a side facing an adjacent airport ramp, but fencing should separate air freight terminals from public access roads and all parking lots.

Fencing high enough to discourage passing or throwing items over the top usually means an eight-foot fence with a top guard, but this is only sufficient if the fence is kept clear. Stacking boxes next to the fence or parking cars or lift trucks by the fence defeats the purpose. Although space is at a premium at most airports, if a fence is to be an effective barrier, a well-lighted clear space of 20 to 30 feet must be maintained on each side. Where freighters are loaded next to public access roads, adequate fencing and lighting are particularly essential.

In any case, the opportunities for an off-duty air freight employee or other accomplice to drive alongside the fence and receive goods passed over the fence must be reduced. If this means of theft is a frequent and continuing problem, the only solution may be to cover the suspect area by low-light-level closed-circuit television, which is capable of clearly distinguishing activities that might otherwise go unseen.

Gates into the freight facility should be limited in number to no more than are absolutely essential to the efficient operation of the terminal. These gates should be staffed by guard personnel, and employees leaving or entering the premises should have to pass a guard post on their way in or out of the facility. Patrols by local and airport police departments and by company guards can provide added perimeter security.

Los Angeles International has installed a new perimeter barrier of crash-proof fencing to replace 5900 feet of existing fencing. The total cost of the first phase of construction will be $1,134,600, and the FAA will share 75 percent of the project cost. The new perimeter fence will include a four-foot-high, poured-in-place, K-rail concrete barrier, of which two feet will be underground. A six-foot-high chain-link fence with double-arm barbed wire holders will make up the upper portion of the fence. The total height of the fence will be nine feet, three inches, and it will have light posts installed at 70-foot intervals adjacent to the fence. The airport is also developing plans for an electronic, computer-enhanced, intrusion-sensitive fence.

EMPLOYEE PARKING

The parking lot should be separated from the air freight facility by a fence, and employees should not be allowed to park on the terminal side of this fence. Employees should have no occasion to go to their cars during a shift; indeed, some carriers enclose the employee parking lot and secure it during the shift. Economics may dictate that a guard be stationed at a personnel gate only during periods of shift change; in such cases, employees leaving during a shift should be required to use the terminal's truck entrance.

The fact that many employees have access to the entire airport as part of their work assignment presents special problems requiring a variety of solutions. Employees may park their cars in a lot some distance from the terminal but convenient to the route they will be taking later in their shift, perhaps with a piece of attractive freight that can be off-loaded near the car. Indeed at some terminals, like-minded employees have been observed parking their cars in distant and inconvenient lots and then riding back to their air freight terminal four or five in one car, as if they had come to work in a car pool.

From any standpoint, parking is a difficult problem to solve. A primary step toward control seems to be a company rule that requires employees to park in specified company lots and threatens them with dismissal if they park elsewhere. This gives security a tool to work with, provided the rule is enforced. During an investigation, license numbers and descriptions of cars belonging to known and suspected employee thieves should be distributed to the security department, airport police, airport law enforcement, and the FBI. Parking lot patrols should be instructed to look for these cars.

Requiring company decals on employee cars provides another method of recognizing "misparked" cars. Where this parking problem is of concern, it might be best to require such a decal on the employee's windshield in order for the employee's car to be admitted to the company parking lot. Of course, the dedicated thief might come to work in a borrowed car in order to be able to park in another lot without detection. This method can only be detected by surveillance, beginning with those instances where several workers leave the terminal in one car; that car pool may only be going as far as other parking lots at the airport.

Employee parking, at some airports, may be at a satellite location from which employees travel to their place of work by shuttle bus. These satellite lots should be under the observation--direct or by CCTV--of a guard to ensure that any unusual activity will be noticed and to protect the employee, the car, and its contents.

Carrier regulations should require that employees agree, as a condition of employment, that any car they drive to the airport may be searched, whether the car is on company property or not. Where company regulations only provide for the search of an employee's car when on company property and a suspect car is parked at the airport but not on company property, security's hands may be tied. The usual provision is that an employee refusing to agree to a search by company officials will be subject to dismissal for insubordination. Rules like this must, of course, be announced to employees when they are hired. In the case of employees hired before the rule goes into effect, notice of the new rule must be given to these employees prior to the institution of searches.

DOCK SECURITY

Areas where cargo is stored should be well lighted, whether they are inside the terminal building or outside. Properly placed dock lights maintain good visibility inside trucks parked at the dock and make it difficult for persons in the truck to tell if they are being observed.

Loading dock doors should be kept closed except when a truck is backed up to the dock for loading or unloading. Where hot weather is a problem and it is necessary to keep the doors open for ventilation, gates of chain link or wire mesh can be used. These doors or gates provide adequate security while allowing ventilation. Standard alarm contacts linked to a light display can be employed to show which dock doors are open.

For some time, the Airport Security Council has recommended an exceptionally good dock practice. The council suggests that a barrier or at least a four-inch-wide yellow painted line be located about ten feet back from the edge of the dock and that signs be suspended over the line advising that no one other than employees is allowed beyond that point. Cargo should not be left within that area of the dock; where efficient operation requires that the rule be breached, cargo left in this accessible area should never be unattended. The lines may require frequent repainting, but they serve a useful purpose and should be maintained. The longevity of the stripe may be increased by etching the concrete prior to painting, although this is relatively expensive. Taped lines have also been used, but these appear to be little more durable than plain painted lines.

TERMINAL ACCESS

Keeping unauthorized individuals on the dock side of the line can be a problem. When terminal construction or remodeling is planned, rest room facilities should be provided where truckers can reach them without walking through the freight terminal. Where no such facilities are available within the ten-foot limit, truckers should be escorted while within the terminal. Coffee machines can be provided in an anteroom located so that truckers need not walk through the terminal to reach them. It is equally important that such facilities not be accessible from within the terminal.

The question of who is authorized to have what access must be resolved. At some terminals, almost anyone connected with the air freight industry is allowed freedom of access within the terminal, though there is no logical reason why employees from other airlines or freight carriers should be allowed beyond the designated transfer area. Unless they are specifically assigned to air freight duties, not even employees of the carrier operating the terminal should have any more access to the terminal than an outsider. Air carrier employees working for supply or stores should not be allowed to meet freight shipments at planeside or to walk through the freight terminal looking for incoming shipments; they should be required to come to the same pickup areas as other consignees.

The wearing of badges assists in the identification of employees authorized to be in the air freight facility, particularly if the badges are color-coded and carry a photograph. Identification based upon uniform alone is inadequate; cargo employees' uniforms have been stolen in the past. Yet, at some stations, the wearing of a pair of blue or white coveralls may grant unlimited access.

Employees must be encouraged to challenge those not wearing badges. Some airports employ a security patrolman in civilian clothing to attempt to penetrate air cargo facilities on the terminal and report the result to the airport's Security Committee. Although the idea of such tests is not new, they are not commonly used. The practice clearly tends to keep employees alert to the possibility of unauthorized intrusion.

GATE HOUSES

Gate houses at all vehicle entrances and exits should be staffed during business hours by operators of facilities handling a substantial volume of high-value cargo. Gate houses should be situated in from the gate so that exiting vehicles may be halted and examined on terminal property. All gate houses should be equipped with telephones or other communication devices. The area surrounding gate houses must be cleared of any items that restrict the guard's line of vision. Signs that advise drivers and visitors to stop and list the conditions of entry onto company property should be posted prominently on the exterior of all gate houses. Included in these conditions should be a notice that all vehicles and personnel entering are subject to search.

Gate house personnel should check that the proper name is displayed on the side of every truck in either stencil, paint, or decal. Crayon, marker, masking tape, and other nonpermanent placards are not acceptable means of identifying trucks. Temporary placards used by lease operators or owner operators may be accepted if the operator of the truck can produce an acceptable Trip Lease Agreement showing affiliation with the trucker of record. All lease equipment drivers should produce lease papers and a driver's license prior to being allowed to pass through the gate. Straight trucks, where the body and cab are one unit, may be admitted when the name of the company appears on the body of the truck, provided it is the company of record. All others should have at least a lease agreement or letterhead authorization to be approved. Truck registration should be presented at the gate, and the guard should check that the registration matches the vehicle plates.

SECURITY REVIEW OF NEW CONSTRUCTION

Procedures should be established within the carrier's organizational structure to ensure that plans for new construction or significant remodeling of air freight facilities are routed to the security department for sign-off approval. Much can be accomplished--and at less cost--if security is planned into construction rather than added afterward. The Security Manager should review the plans for alarm locations, access control, fencing, lighting, restricted areas, and a variety of other security considerations.

Openings, including doors, roof, and windows, should be kept to an absolute minimum. All access points, particularly roof openings, should be alarmed. Windows accessible from the ground should be covered with 11-gauge wire mesh. Employees should enter and exit through specific doors where their activity can be monitored. Other doors should be for emergency use only and alarmed with both silent and local alarms.

Particular consideration should be given to situations where leased space is shared with other tenants in a cargo terminal. Often, the interior space is divided by means of a six- or eight-foot cyclone fence. That is much less than adequate when there are fork lifts around that can easily give someone a boost over the fence. Such fences should extend from true floor to true ceiling, as differentiated from a false or suspended ceiling. Consideration should also be given to draping the fence with tarps to block the view. The temptation of highly desirable air freight may be too much for some people to overcome. Alarms may also be appropriate. I've seen one terminal shared with a bay occupied by mechanics who repaired city vehicles. There was a floor to ceiling fence, but it was in such a state of disrepair that it was easy to pass from one side to the other.

CLOSED-CIRCUIT TELEVISION

Closed-circuit television (CCTV) is used in a number of air freight terminals, sometimes effectively and sometimes ineffectively. Although CCTV has often been accused of being an ineffective deterrent, the fact is that it is often used ineffectively, without proper planning. For example, if thefts are primarily caused by collusion between an air freight employee and a driver, CCTV at the loading platform can be of little assistance; a shipment being stolen through collusion will look no different to the TV camera than a shipment being legitimately loaded. On the other hand, if the problem is pilferage at the dock, a camera at that location can be of some assistance.

The breakdown area where carts and containers are unloaded is an excellent point for camera coverage. It is at this location that particular items are spotted for theft and perhaps set aside or placed back in the cart to be towed to the cart storage area, where the item can be picked up at the employee's leisure. If limitations of space require that carts be held outside the terminal awaiting break-down of shipments, a television camera should cover this area.

The placement of cameras is unique to each terminal and must be tailored to fit the terminal's particular needs. It may be that a low-light camera with a remote-control zoom lens should be placed on top of a building or at some similar vantage point to command a view of the cart road and the activities of those driving carts from the air freight terminal to the ramp. At some airports, there are locations where employees are known to cache stolen items or to transfer them to confederates; these should receive CCTV coverage. The U.S. Customs appraisal room is frequently a good place for a camera if the station is experiencing a pilferage problem.[1] If security is allowed to participate in the early planning of a terminal, its construction can be planned so as not to impede a clear line of sight to key areas that might be the object of CCTV surveillance.

Anyone considering the purchase of CCTV equipment should consult with several vendors and study the variety of equipment available. Cameras with remote panning controls can be enclosed within a reflective dome that conceals the direction in which the camera is pointing. Cameras can be mounted on tracks suspended from the ceiling, allowing them to move horizontally; this installation could be particularly useful for looking into trucks at the dock. Portable CCTV cameras, available with or without battery-pack power, can be used by security to meet temporary needs without having to make permanent installations.

One of the often overlooked aspects of CCTV is the monitoring of the cameras. At some locations, the monitors are installed in the supervisor's office and office personnel are instructed to watch the monitors when they are not occupied with other tasks. The problem is, of course, that these personnel are not only extremely busy but they would probably be accused of not working if they were to be seen watching the monitors for any length of time. Closed-circuit television is a security tool, and the monitors should be watched by people hired for that specific purpose--in most cases, security guards.

Problems do exist in the watching of TV monitors. Extended periods of watching tend to mesmerize the individual or at least to render him or her

---

[1] J.E. Thorsen, "Air Cargo Security: A Concept that works...An Interview with Arthur A. Ives," *Security World*, 10:31 (June 1973).

inattentive. Persons assigned these duties should be rotated frequently, from hourly to as often as every 15 minutes, depending upon such factors as the number of cameras being monitored and the subtlety of the action under observation. Several cameras can be monitored more effectively if only one or two cameras have their own monitors and the other cameras appear in sequence on a single monitor. This automatic switching operation can, of course, be stopped by the operator for a longer look. Unusual activity on one camera can be switched to a stand-by monitor while the remaining cameras continue to sequence. A guard can watch TV monitors in connection with other duties only if those other duties are not too demanding and allow sufficient time to observe the monitors.

For areas in which little or no activity is scheduled, various kinds of "video alarms" can be employed. These alarms can activate a TV tape recorder, a light, or an audible signal. The scene is scanned constantly, and any movement in the area covered by the video alarm sensor will trigger an alarm condition. One such device stores the path of intrusion visually on the monitor for a period of time.

A valuable accessory for most CCTV systems is videotape recording. The tape may record constantly (in real time) or in time lapse, recording the scene every few seconds. Not only does the videotape document infractions in an indisputable way, but a lapsed-time recording of an entire shift can be speeded up and reviewed by a guard or supervisor in a relatively short time. Depending upon the recorder selected and the mode of recording used, arrangements can be made for unattended videotape recording for such lengths of time as an entire shift, a full 24-hour period or a weekend without changing the tape. In addition, the tapes can be erased and used again. They should be kept for two to four weeks, however, in the event a loss is discovered. It's extremely frustrating to discover a loss and find the pertinent activity has already been taped over .

One method of protecting high-value items or freight that is particularly theft prone, such as a pallet load of VCRs or computers, is to place the material in a designated area of the terminal or the ramp where it can be monitored by CCTV connected to a lapse-time video recorder. Make certain that all sides of the area can be seen and that the lighting is adequate, then run tests to make certain that specific individuals can be identified on the monitor. If freight is missing, it is simply a matter of playing the tape and identifying the culprit.

GUARD FORCE OPERATIONS

A uniformed guard force, properly used, can assist in reducing cargo theft at an air freight terminal. It is essential that guards be permanently assigned to air freight duties and trained in air freight methods and procedures. If their suspicions are aroused by a piece of freight in an unusual location, they must be able to look at the airbill or other documentation and know whether it is being properly handled. They should be able to look at receiving and shipping documents and know whether the shipment is in its proper location and whether the count is accurate. Because of the training and judgment required, a contract guard force may not suffice unless the agency can guarantee some degree of permanency and a low turnover in guard personnel.

Fixed guard posts are necessary at some locations such as gates. Attempts have been made to get guards to leave their posts through a ruse; guards must be trained to verify phone calls or other irregular instructions with their supervisor before acting upon them. A guard not assigned to a fixed post should be free to roam throughout the freight terminal and into the aircraft belly pits, if need be. Because much theft and pilferage occur within these pits, particularly when the open loading door to the aircraft is away from the terminal, guards should make random checks of these areas while on their walking tours.

Guards should be cautioned against fraternization with air freight employees. If possible, locker space for guards should be entirely separate from that of the other employees at the location. The guards are working in a crime-prevention capacity and must accept the fact that personal friendships with employees could hinder their authority and theft-deterring function.

The use of armed company guards to follow high-value shipments as escorts to and from aircraft should not be overlooked. If the volume of freight precludes their use on every high-value shipment, escort guards should follow all shipments over an established dollar amount appropriate to that location. At some stations, airport police will provide an escort for shipments over $25,000 or a similar set dollar amount. In February 1976, a new provision of the Trade Practice Manual, Standard Freight Practices, mandated that a police escort shall be requested for shipments containing articles of extraordinary value worth more than $25,000.

One carrier recently reported a dramatic drop in cargo theft losses at Kennedy Airport following increased security measures; claims paid for lost and stolen merchandise were cut by 75 percent. A major reason cited was the hiring of independent security guards, mostly off-duty police officers, to patrol the freight terminal. These security efforts cut the claims that the carrier paid from 2 percent to 0.5 percent of cargo revenues since the new program was implemented several years ago -- providing bottom-line savings of $3 million in 1988 alone.

CONCLUSION

Even after implementing a majority of the security measures described here, security managers are still not likely to rest easy because they know that physical security is not the total answer to theft prevention. Obviously, physical means cannot make a station theft-proof, particularly if it is to operate efficiently and profitably. An essential balance between physical security and these operational requirements is necessary. Achieving operating efficiency within reasonable security restraints provides the foundation on which the company must build for profitable operation. Any imbalance on either side will be reflected in profitability.

Chapter 24

SECURITY PROCEDURES FOR AIR CARGO HANDLING

ACCOUNTABILITY

What loss-prevention measure is simpler and what procedure is more basic to good freight management than the elementary practice of counting and weighing freight when it is received and when it is delivered? Unfortunately, counting and weighing require labor and time, and this basic tenet is often violated in the interests of time saving and efficiency.

The following quotations from actual security inspection reports illustrate what regularly occurs in actual operations:

> In ten instances, it was noted that drivers delivering freight unloaded their own freight and obtained a signed delivery receipt without the freight being counted by our own employees. In some cases, one of our employees was in the area while the truck driver unloaded freight, but the employee did not count the freight. In other instances, no employee was in attendance during the unloading and, in these cases, the drivers were observed proceeding to a receiving desk, presenting an airbill, and obtaining a signed receipt without our employee actually seeing the freight.

> The tendering of freight for delivery was observed between 2:00 a.m. and 7:30 a.m. During this period, only one supervisor was on duty and he was observed on the floor only twice. During most of this period, no freight agent was on duty at the delivery dock to take care of forwarders or truckmen there to pick up freight. When freight for delivery to the customer was brought up by ramp servicemen, it was tendered without any count. On one occasion, a truck driver went unaccompanied to the area where his freight was stored, sorted it, said to a ramp serviceman, "This is

> mine," and the freight was delivered to him without any count. On another occasion, a similar air freight driver went unaccompanied to an area where his freight was stored, pointed it out to a ramp serviceman, and said, "All of this is mine." After he put the indicated freight on the truck, he returned one piece to the ramp serviceman with the comment that it was not his.

The result is that when a shipment is short at delivery to the consignee, there is no certainty whether the missing items actually were received by the carrier, were received but disappeared in transit, or were loaded for delivery and removed after that point. Almost all claims arising from the mishandling of freight during receiving and delivery operations relate in one way or another to this problem of accountability. It is in this area that security needs the support of top management because local management is probably already understaffed and the imposition of stricter requirements for counting all freight may require the hiring of additional personnel. Also, the order to weigh and to make counts without exception is more likely to be obeyed if the local manager and the dock personnel know that the order comes from the company's policy makers.

DETERRENTS TO PILFERAGE BY COUNT

Counts are also essential in order to stop pilferage of loads by truck drivers. At its simplest, pilferage attempts may take several forms. Drivers may accept delivery of an entire load but claim the shipment is short one or more boxes by their count. If the supervisor signs off one short on the shipment to avoid the labor and time of unloading the truck and making a recount, the box belongs to the driver. If the shipment is recounted, the driver cannot be accused on theft--he or she merely miscounted. The only way to counter this scam is to ensure that it will not be attempted; supervisors must insist, as a matter of policy, that whenever the count of the driver and the count of the air freight employee do not agree, the shipment must be recounted, without exception.

On occasion, it will be noted that this and similar schemes are unusually frequent with a particular driver or, perhaps, with a particular forwarder. In such cases, supervisory or security personnel should be present on every occasion that driver or a driver of that forwarder is at the terminal. Truckers who frequently "misplace" items or cartons should be brought to the attention of the security department of the driver's firm. Other air carriers operating at the airport might also be notified, perhaps through a network of those interested in the security of air cargo, such as a Cargo Security Working Group.

Other "short count" variations include the driver who delivers a shipment one short to the airline and claims a complete shipment. If the clerk doesn't count the shipment, the driver has netted one article. If the clerk does count the shipment and the count comes up one short, the shortage is a mistake; indeed, the driver may even promptly "find" the missing carton.

In another variation, the driver may deliver a complete shipment, unload it onto the dock, and insist that the clerk count and verify the shipment. Should the clerk then leave the area--after making the count and before the driver's departure--the driver can simply put onc or more of the cartons back onto the truck before leaving. This is one of several reasons why drivers should never be left on the dock unescorted.

Caution must be exercised in counting palletized freight; where there are supposed to be 12 cases on each layer, the trucker may have stacked the bottom layers so that they contain as few as nine cases to the layer. Unless a verifying count is made while the boxes are being stacked, rather than waiting until the stacking is completed, the deception may not be caught.

TRUCK SEALS

If seals are used on departing trucks, the guard should verify that the seal on the truck is the one indicated by number on the airbill. Otherwise, the trucker can use a seal he has obtained himself, replacing it with the correct seal after rifling the cargo. (Obtaining the necessary extra seal, as well as the designated seal before it is used to seal the cargo, usually requires the cooperation of an air freight agent, and collusion is to be suspected when such seal errors are found.)

COLLUSION

Theft becomes easier and its detection much more difficult when collusion is involved, whether it is collusion between two employees or between an air freight employee and a trucker. As with any theft problem, collusion gets progressively worse until the responsible individuals are apprehended. The size of the risk multiplies as more individuals become involved. At one large warehouse, even contract guards were a part of the theft. Successful minor pilferage--if not checked--leads to collusion, the source of the largest thefts and the greatest theft losses.

When an employee has a collaborating truck driver, the ease of theft is remarkable; the freight is simply loaded aboard the cooperating driver's truck and driven off the premises. Onlookers who see large boxes of wearing apparel or console television sets being stolen assume they are watching a normal pickup or delivery.

Where no procedural controls against this kind of theft are in effect, this method can account for a major portion of a carrier's thefts. However, effective security measures will decrease the incidence of this kind of theft and may even stop it entirely.

A rotation system should be established as a routine to preclude the same trucker and air freight agent working together on a daily basis. At many terminals, however, the practice is just the opposite; a particular agent is assigned to work with a particular trucker so that the carrier may provide better service through the agent's familiarity with the trucker's particular needs and problems. Any advantage of such an agreement is, of course, lost if it results in the two conspiring against the carrier to their mutual profit. In any event, it creates the opportunity for theft.

Spot checks conducted by supervisors are another measure useful in countering collusion. During freight loading, the supervisor approaches the freight agent during any stage of the loading, but preferably near completion, and compares the freight on the truck with the paperwork. During receiving operations, similar spot checks determine that proper exceptions have been taken and that the truck contains no undocumented freight from the dock. Each supervisor should be responsible for conducting several of these checks during each shift so that employees know to expect them on a regular basis. If thievery through collusion is taking place, the spot check will locate it before losses become astronomical.

Another procedure useful in combating collusion is to divide responsibility; that is, to have one agent pull (and count) the freight and another release the freight to the trucker. In addition to providing a double count of the freight, this division involves two of the carrier's employees in the delivery, rather than just one, and it is more likely to bring to the truck only the freight that should be loaded.

Other procedures, such as special storage areas for high-value freight and attractive merchandise, are also valuable in preventing collusion. These are discussed in Chapter 25.

FRAUDULENT PICKUPS

A fraudulent pickup occurs when freight is delivered to an individual not authorized to receive it. It is often the basis for false claims of loss as well as for theft. In its simplest form, it may involve a college student who sends personal effects home by air freight, and has a friend pick them up, then calls at the air freight office a day or so later to request the shipment. Shipments released to an unauthorized individual can force the carrier to pay the claim.

In its more flagrant application, fraudulent pickup may involve an entire truckload of cargo. If the carrier releases the delivery to someone who appears to be a driver from XYZ freight without requiring appropriate documentation and identification, the company may have to pay a sizeable claim. A driver can use any of a number of ruses to obtain the freight, such as claiming to be filling in for the regular driver, who is ill. The ruse is not material to the carrier's liability. A major fraudulent pickup is a preventable occurrence. It is one of those things that should never happen, but occasionally does.

At those stations where fraudulent pickups have become a problem, dual-lens cameras (which simultaneously photograph the driver and the papers--driver's license, pickup order, and airbill) have been of great assistance, principally in deterring such attempts. Both the identity of the individual and the individual's authority to receive the shipment must be established in every case of delivering a shipment. If both cannot be established, the shipment should not be released. Demanding identification of the driver and comparing the identification with the pickup order will often bring to light a forged pickup order. Particular attention should always be given to drivers who are not regulars known to terminal workers and to individuals who arrive in unmarked trucks or cars. Do not hesitate to call the consignee and verify the identify of an unknown driver.

PROTECTION OF DOCUMENTATION

Some fraudulent pickups involve valid but stolen documentation, which demonstrates the need to protect shipping and receiving documents. Air freight pickup counters should never be left unattended, particularly where documentation is kept behind the counter. Blank airbills, too, must be protected; an airbill in the hands of an individual who knows how they should be filled out can be costly. In one case, an agent setting up his thefts by taking the airbills off outbound shipments and substituting new airbills marked "Hold for Pickup" at his station.

CONTAINERIZATION AND SECURITY

Today, about 25 percent of air freight's volume moves in containers packed by shippers at their own facilities. Containerization offers a number of advantages to the shipper. By "unitizing" shipments and delivering them to the carrier in containers, the shipper receives rate reductions because of the reduced carrier handling costs. The shipper can also eliminate packaging costs that might otherwise be incurred, because the cartons are now protected by enclosure in the container. The risk of theft is materially reduced when freight is containerized because the handling of individual items is eliminated. The container can be delivered to the terminal, loaded directly from the shipper's truck onto handling equipment, and taken out to the aircraft without being exposed to freight handling at the terminal. A number of different containers are available. The choice depends on the type of aircraft to be used and the volume of freight to be shipped.

While containerization is not a cure-all, it certainly can contribute heavily toward the reduction of theft and pilferage, if properly handled. One major carrier showed an annual loss ratio for all wearing apparel of 4.76. During the same period, however, the carrier's loss ratio of wearing apparel on hangers was .16. The difference was almost wholly attributable to the fact that wearing apparel on hangers is transported in sealed containers.

Containers received from forwarders, for example, should be sealed by the carrier upon receipt, unless the containers already have a shipper's seal on them. ("Sealed," in this context, means affixing a numbered tag to the container and securing it to prevent undetectable opening. The loop created with the seal must be severed before the secured door, gate, or chain curtain can be opened.)

The mere fact that a container is sealed and even locked with padlocks, however, does not negate the need for security precautions. Although seals deter pilferage from containers and padlocks virtually eliminate it, neither prevents the theft of the entire container, and any container filled with high-value items demands extra security precautions.

During peak periods, the carrier can eliminate excess handling by loading directly from arriving trucks into containers at the dock. The full containers are then sealed and moved to staging areas, avoiding bulk storage in the terminal and the rehandling of freight.

Inbound containers that arrive at a destination station over the weekend to await a Monday pickup should be sealed if there is not a seal already in place.

PACKAGING

Packaging may seem an unusual subject to be included in this chapter yet packaging should be inspected at the time of receipt . If the packaging is not adequate, the shipment should be refused. Packaging problems rank alongside the intentional opening of cartons and containers as prime pilferage factors. Not only should the exterior container be strong enough to hold the shipment, but the item inside the container should be properly braced, blocked, or cushioned from shock. The weight of a piece of freight should not exceed the burst strength of the carton in which it is packed.

Carriers have occasionally been successful in convincing some manufacturers that shipping their products in boxes that do not advertise their contents on the outside will cut down on the theft and pilferage of their merchandise. However, "blind" packaging and other disguises are not a total cure because air freight workers are quite adept at associating specific merchandise with certain shippers, and most carriers learned long ago that listing watches on airbills as "horary goods" doesn't prevent them from being pilfered. Nevertheless, it is a worthwhile precaution to eliminate as much unnecessary identifying advertising as possible from shipping cartons. Small electronic items packed in a plain box have a greater chance of arriving at their destination than those packed in cartons that announce their contents. However blank the cartons, though, regulatory tariffs require that airbills accurately describe the contents of shipments so that the carrier can provide security protection appropriate to the shipment.

Some manufacturers have a legitimate and important reason for wanting their products identified on the outer carton by brand and model number. Products of these manufacturers are held on stockroom shelves. If the customer selects a particular brand of stereo set from several models on display, the clerk goes to the stockroom to get a stereo set of that brand and model. One can image the chaos, the handling time, and the costs that unmarked cartons would present under such circumstances.

DAMAGED CARTONS

All cargo received, whether from a shipper or from another carrier, should be promptly inspected for damage. The damage is not always obvious. Most air freight supervisors are familiar with the diagonal razor slit that allows of a pilfering hand to be inserted into the carton but that is almost undetectable unless the box is flexed in a certain manner. Less sophisticated employee opportunists are rarely so discreet; they merely put their fist through the side of the box, leaving a sizeable round hole.

When damage is found or when it appears entry may have been gained to a piece of freight being received, specific exception to the shipment must be made on the proper forms. In the United States, the required form is Form AC-12, Register of Air Freight Shipments, more commonly known as the "Pickup and Delivery Manifest." If the exception entered is nonspecific, though, such as "box damaged," the information is virtually useless for claim processing purposes. Exceptions must be specific as to the extent, type, and location of the evident damage, and an inventory of the contents must be made at that time. A camera should be used to make a photographic record of damaged or improperly packaged merchandise. A Polaroid™ camera is best for this, as it instantly provides pictures that can be affixed to the documentation, and ensures that the photographs do indeed show the damage clearly. Such photographs are of great value in claims handling. In addition to the Pickup and Delivery Manifest, an Over, Short, and Damage (OS&D) report should be made on every damaged shipment passing through the terminal.

Cartons found to be damaged must be recoopered as soon as possible. If the broken cartons cannot be recoopered immediately, they should be placed in a secure storage area until they can be resecured.

CARRIER TRANSFER VEHICLE SUPERVISION

The ease with which carrier employees can move about the airport contributes to air cargo security problems. The necessity for moving freight back and forth between one part of an airport and another is an exposure unique to the air cargo industry. At one U.S. airport located at one of the 15 cities identified by the DOT as having particularly severe cargo theft problems, freight in transit between the air freight terminal and the passenger terminal travels a distance of five miles, part of it on public roads.

Losses from transfer vehicles can be heavy. Not all losses are due to theft; losses result from cargo falling off a cart because it was overloaded and from old or damaged equipment that cannot be covered or sealed. The design of transfer vehicles may vary from the open carts that are lifted intact into the belly of a DC-8 to a foreman cart with canvas and chain link sides. Lower deck (LD) containers are also vulnerable to pilferage if they are not sealed.

Control of transfer vehicles is important to a secure cargo operation. Specific routes must be prescribed for transfer vehicles, and any tractor or tug seen off this route should be subject to inquiry. Locks or seals should be considered for use on these carts, particularly when they are carrying attractive merchandise or articles of extraordinary value. There may be resistance to this practice from those who insist on moving freight in the shortest possible time with the least amount of handling, but placing a padlock on a cart--whether at the freight terminal or at the ramp--takes little extra effort. As long as the key or the combination to the lock is restricted to supervisors, locking carts will reduce pilferage enroute.

Records (such as a log) should be kept that indicate such things as when a vehicle train leaves the freight terminal and when it arrives at the ramp or staging area. Any trips that take an unusual length of time should be investigated. Control of these trips by two-way radio is productive; a call to the ramp as the vehicle leaves the terminal and a call from the ramp as the vehicle arrives, lets drivers know their trips are accurately monitored.

CONTAINER SEALS

Seals have been used by the trucking and rail industry for the protection of cargo for many years, but only with the advent of containerization and the sealable container did they come into limited use in the air cargo industry. Now, a number of freight forwarders and shippers seal their containers before delivery to the air carrier. This practice considerably benefits the air carrier because an unbroken seal at the destination eliminates any possibility of carrier liability for shortages discovered after arrival.

When a sealable container is delivered to the air carrier without a seal, air carrier freight personnel should, as previously mentioned, affix a seal. The seal number should be entered on the airbill, and the shipper's representative should be asked to sign or initial the airbill to signify that he or she was present when the seal was affixed. Should the representative refuse to sign the seal number entry, that should be noted on the airbill. If the driver has pilfered the container enroute to the terminal or if the shipper's count is inaccurate, the carrier is protected against claims as long as the designated seal is attached and secure upon delivery to the consignee.

High-value items (see also Chapter 25) should be loaded into sealable containers whenever possible, sealed with one or more seals as necessary (the LD-3 container, for example, requires four seals), and the seal numbers transmitted as part of the advance advice message sent to the destination station. If the practice is to be effective, the seal numbers must also be checked and verified at the destination.

If theft and pilferage are believed to be occurring between the air freight terminal and the passenger ramp, then seals must be checked as the freight leaves the terminal and as it is loaded on the aircraft. At the destination, the seals must be checked as the container comes off the aircraft and again as it arrives at the terminal.

One steamship company has tested new fiber-optic seals on containers aboard its container ship. The seal, manufactured by E. J. Brooks Co. of Newark, New Jersey, contains fiber-optic strands that are photographed by a special camera at the start and end of a voyage to check whether the container has been tampered with. Presumably, if these seals can be used on ocean containers they could also have applications for aircraft shipping containers.

Seal Accountability

Unless proper accountability of seals is maintained, this precaution can be rendered ineffective. Much helpful information on seal accountability is contained in the Department of Transportation's Cargo Security Advisory Standard titled "Seal Accountability and Procedures."

All seals should be ordered by a single central source within the company, such as the purchasing department. The type of seal that appears to be particularly appropriate for air cargo containers is an unbreakable nylon strap that can be cinched tight. This assists in holding the container hasps together and thus prevents opening the hatch wide enough to get a hand inside.

By means of imprinting and color, the seals should identify not only the airline, but the station where the seal was affixed. Identifying each station's seals precludes the substitution of a bogus seal at the destination station; for example, if a container has been shipped from Dallas to New York, the substitution of a New York seal for the Dallas seal will be readily seen.

Seals should be secured within a locked room, cabinet, or drawer. A record of to how many seals are sent to each location should be kept by the central procurement department, and the seals should be signed out to the stations by blocks of numbers.

The air freight manager at the terminal should assign the responsibility for the safekeeping, issuance, and record keeping of the station's seals to no more than one person per shift. Applying and removing seals should be done only by those employees the supervisor has specifically authorized to do so. Adequate seal accountability and records are essential to pinpointing where a loss occurred and to fixing responsibility.

Seals should be checked for integrity as the containers are placed aboard the aircraft. If such a program can be sold to management, it is an excellent idea to check seal numbers also at this point. This is a time-consuming precaution, however, and may need to be held in reserve for those times when pilferage enroute to the aircraft is suspected.

Once again, the practicalities of selling a seal program to management enter the picture. If management opposes such a program because of the potential costs of controls and record keeping, a less secure, but adequate, procedure would be to establish the secure storage of seals at the terminal and to have the seals issued to each authorized freight agent in recorded blocks of numbers.

The primary control is to list the seal numbers on the airbill and in the advice message. If there is collusion between the individual agent who issues the seals and the runner, there is a risk of the agent giving the legitimate seals to the runner and fastening bogus seals on the container. Enroute to the terminal, the runner might remove the bogus seals attached by the accomplice, steal the freight or high-value item, and then fasten the legitimate seals on the container. Whether or not such an act is suspected, it is an excellent practice to spot-check containers as they leave the freight terminal to make certain the correct seals are indeed on the container.

Greater accountability will also solve this problem: If an agent seals five containers with four seals per container, an inventory of his or her seals should show the number issued less only 20 seals, and any seals unaccounted for would require an explanation.

ROUGH HANDLING AND DELAY

Although claims resulting from rough handling and avoidable delay are not always thought of in connection with the security of shipments and the losses due to theft and pilferage, they deserve comment because they contribute heavily to claim losses. Although the security department's scope of responsibility varies with the company involved, if security is construed to involve all aspects of loss prevention and the protection of corporate assets, rough handling and avoidable delays deserve the attention of security.

Inspecting freight upon receipt to make certain it is properly packaged is one method of reducing losses due to rough handling. If the contents of a box are not properly padded or braced, or if the weight of the contents exceeds the burst strength of the carton, or if the crating is not sufficiently strong, then the carrier may have a claim in the making before it even moves the freight. Poorly packed shipments should be either properly recoopered or refused. Restricted articles must also be inspected to confirm they are packaged in strict accordance with the Restricted Articles Tariff.

Some very basic measures can contribute substantially to reduced damage claims--if they are followed. It seems basic that a box with an up arrow should not be stacked with the arrow pointing down, yet it is not unusual to see this in a terminal. Heavy cargo should be stacked on the bottom with lighter cargo toward the top. Fragile cargo should be on top of the stack and not on the bottom. Repeated instances of failing to follow these simple rules may be an indication of poor supervision.

I was present as courier for a classified military shipment when the forklift operator dropped a crate containing a delicate instrument worth $250,000--a loss preventable by a small amount of elementary caution and planning. Pure carelessness, horseplay, and ignoring labels all contribute to losses due to rough handling. Quite obviously, damages do not occur because a plane touched down too hard; they are the result of improper ground handling.

Losses due to delay primarily involve perishable commodities such as flowers, fruits, tropical fish, or other cargo for which time is a critical factor. Exposing perishables to temperature extremes or shipping them to a destination without regard for the weather there also creates losses in perishable cargo.

In many cases, a delay loss can be avoided by a knowledge of the carrier's capacity to handle the freight and a management policy that allows freight managers to turn down time-critical freight they can't handle. The acceptance of flower shipments is probably the best example. Capacity limits should be established for peak periods such as Easter and Mother's Day; once a station has accepted as many flowers as it can handle in time to get the flowers to the destination without deterioration or damage, then no more should be accepted. Any additional flowers accepted are simply "bought" by the carrier; they cannot arrive at destination in good condition and will certainly result in a claim payment. If such shipments are accepted, the carrier should not sell shipper's insurance to the shipper, and the shipper should be advised as to why the carrier cannot insure the cargo. Similarly, if tropical fish have a 48-

hour oxygen supply and it will take a total of three days to deliver them to their destination, the shipment should be refused.

Delay also creates opportunities for theft. If boxes of stereo equipment or cameras sit around a terminal for several days, some of them are almost certain to be stolen. Richard Lally, then Director of Transportation Security for the Department of Transportation, summed it up on one occasion by saying, "When a terminal becomes a warehouse, thefts will increase."

Chapter 25

HIGH-VALUE SHIPMENTS

The Department of Transportation broadly defines "high-value cargo" as "cargo [that], because of its monetary value, utility, desirability, or history of frequent theft, requires greater protection than other commodities normally handled at that facility."[1] This definition correctly covers a wide range of cargo, from diamonds to garments, and from gold to guns.

ARTICLES OF EXTRAORDINARY VALUE

A narrower definition of "high-value cargo" was used by the now-defunct Civil Aeronautics Board in requiring specific procedures for the handling of certain high-value items generally referred to as "articles of extraordinary value" (AEV). These items were specified in CAB Tariff 96, Rule 2, and carriers still refer to this general list in defining what should be treated as high-value items. The list includes artworks; bills of exchange; bonds; bullion or precious metals; clocks, watches, and parts; currency; deeds; doré bullion; evidences of debt; furs, fur clothing, and fur-trimmed clothing; gems, cut or uncut (including diamonds for industrial use); gold bullion, coined and uncoined, cyanides, dust or sulfides; jewelry (other than costume jewelry); money; pearls; platinum; promissory notes; negotiable securities; silver bullion, coined and uncoined, concentrates, cyanides, dust, or sulfides; postage or revenue stamps; and stock certificates.

[1] 49 CFR 85-3.5, Cargo Security Advisory Standard, "Internal Accountability Procedures."

Special procedures applicable to high-value cargo generally include advance arrangements of not less than 24 hours before shipment, delivery or shipment to the carrier not more than three hours prior to the departure of the flight and written assurance that the consignee will accept delivery within three hours and/or not later than 24 hours after the scheduled arrival time at the destination. If no reservation is available, perhaps because the flight is full, the shipment should be refused with instructions to the shipper as to the procedure to be followed relative to reservations. Under CAB tariffs, if the consignee did not accept the shipment within three hours or could not be located, the carrier had the authority to hire--at the expense of the shipper and the consignee--either an armored vehicle to deliver the shipment or an armed guard to protect the shipment until the consignee accepted it. The three-hour limit is sometimes referred to as "free time." The authority of those tariffs no longer exists, but those same provisions may be written into a carrier's operating procedures.

The U.S. Customs' Tariff Act, Section 490 (b), states:

> At the request of the consignee of any merchandise, or of the owner or master of the vessel or the person in charge of the vehicle in which the same is imported, any merchandise may be taken possession of by the appropriate customs officer after the expiration of one day after the entry of the vessel or report of the vehicle and may be unladen and held at the risk and expense of the consignee until entry thereof is made.

All of these options are designed for the security benefit of the customer and the carrier. It is in the interest of all that valuables remain in the custody of the airline for the shortest possible time. When this is not possible, the customer is asked to assume the expense of the added protection. If so, the shipper needs to be made aware of this eventuality. The vehicle for such notice is the tariff. Once having been advised of the rules, the choice is the customer's.

Upon receipt, high-value shipments should be carefully and thoroughly inspected for damage, piece count, and proper packaging. Unless the flight is leaving almost immediately, the item should be locked in the high-value security cage and entered in a log maintained for the security cage.

Small shipments of high-value cargo may, at the discretion of the airline, be held for a sufficient period of time to consolidate such shipments in a suitable container, provided the period for consolidation does not represent unreasonable delay. In the application of this recommendation, the reservation procedure for high-value cargo should be coordinated with this consolidation feature and shippers advised of possible shipping delays. Where size and weight permit, individual pieces are usually consolidated into a single carrying container. This may be a pouch, locker, or box that is secure and simple for a runner to handle. The container should be sealed and a record made of the seal number on documents remaining at the origin station and on documents accompanying the shipment.

When it is time to take the item to the aircraft, it should be signed out of the cage, and the employee to whom the package was released should be entered in the log, preferably by his or her own signature. Signature accountability should be maintained all the way to the aircraft; if accountability cannot be traced, investigating a theft after the fact is of little value. Even if the runner who signed for the package claims to have laid it on the belt loader or to have given it to the crew loading the aircraft, it will be a rare occasion if one of the ramp crew remembers a specific package. If, however, the runner is required to obtain the signature of the individual at planeside to whom the package was delivered, the possibility of theft immediately lessens; individuals are less likely to steal--or be careless with--a package if they know that their signature is the last one on a receipt. Where such a hand receipt system is used, the time should be part of the receipt. On shipments of particularly high value, generally over $25,000, special escorts should be arranged, either through airport police, carrier company guards, or armed contract service.

Stations where radio communications are available have proven that calling the ramp supervisor as the high-value shipment leaves the terminal to advise that a runner is enroute with a high-value shipment is an effective practice. The runner, aware that the time of departure is known, is less inclined to make side trips or stops on the way. When the shipment arrives at planeside or at the ramp, the supervisor calls the freight terminal to advise that the shipment has arrived. A record should be made of the time the runner leaves the terminal, and the time should be entered when the runner arrives at the aircraft.

Delivery of the AEV item to the aircraft should be as late as possible; lengthening the time the shipment remains in the security cage reduces the time the shipment is vulnerable to theft. Once the item is placed aboard the aircraft, the aircraft should be kept under surveillance until it taxis away from the ramp. Occasionally, a supervisor has placed an AEV item in the belly pits, closed the doors, and left the aircraft. Closely following the supervisor's departure, a ramp worker or a worker loading another pit on the same aircraft has opened the hatch, removed the AEV item, and disappeared with it. So far as the written records are concerned, the item was safely loaded aboard that flight, and the shipment joins the long list of items known to have been loaded aboard an aircraft but missing upon arrival, leaving one with the conclusion that shipments disappear in midair. Under no circumstances should an AEV shipment be left without constant surveillance. I've even heard of one case where ramp servicemen followed the aircraft in a pickup truck while the plane was taxiing out. A man standing in the pickup bed under the moving plane opened the belly pit and stole the high-value shipment.

The procedures followed at the station of origin must be followed at the destination station. High-value shipments must be met, and there must be signature accountability for the shipment from the aircraft to the high-value security cage. The destination station should have been notified that an AEV shipment was enroute by means of an advice message, usually through the company's internal telex system. Some carriers prefer to use the telephone for the advice message, believing that the telex message may inform the wrong people of the shipment. The disadvantage here is that, should a loss occur, a telephone message leaves no "audit trail" for the investigator.

Sometimes, a shipment is not met at destination because the destination terminal did not receive an advice message. When investigating thefts arising from such circumstances, the investigator should attempt to ascertain who was responsible for sending and for relaying the message; the advice message may have intentionally not been sent or may have been destroyed or waylaid after receipt. Sometimes, the message is merely laid on someone's desk at the destination station and becomes lost in a pile of papers. Such negligence is serious and requires prompt corrective action. A station that fails to send an advice message should be immediately advised of its failure to do so. Information regarding the shipment of high-value cargo should receive special handling to ensure the greatest possible protection and the most limited dissemination of the arrival information.

Air carriers, particularly international carriers, may wish to adopt a signature service, wherein a form such as a Special Handling Notice is used to indicate a high-value shipment. This form, which serves as a backup to the advance notice to the destination system, is carried in the pouch containing the shipping documents that accompany the flight. The Special Handling Notice is to be signed by an authorized person at each point of interface from origin to destination. When the compartment of the aircraft is opened during the stop-off, the notice is to be signed to indicate that the high-value items have not been tampered with at the stop-off point. The form may also include information regarding the special handling of perishables and other cargo requiring prompt or particular service.

When cargoliner or freighter service is available, AEV items should be boarded onto freighters at the freight terminal rather than be transported to the passenger ramp. This procedure reduces the exposure to theft on the route between the freight terminal and the ramp. Moving sensitive cargo over extended routes unnecessarily exposes it to additional risks of theft. If freighter carriage is not possible or practical and this type of freight is being boarded onto wide-body aircraft, it should be loaded into sealable containers and sealed.

Circuitous routings should be avoided; high-value freight, particularly, should be sent on direct, nonstop, daylight flights whenever possible. If a shipment is going from Chicago to Seattle, it should not be sent by way of San Francisco simply because a plane for San Francisco is being loaded and it would get the cargo out of the Chicago terminal faster. Indirect routing exposes the cargo to additional risk of theft and to additional handling. Specific flights can be designated as high-value flights, and high-value cargo sent only on those.

If an anticipated AEV shipment fails to arrive as schedule, immediate telephone tracing should be started, followed by whatever written message traffic is necessary. If the shipment appears to have disappeared, the appropriate law enforcement agency should be notified immediately. In most cases, this would be the FBI, although U.S. Customs will want to be advised if the theft concerns in-bond shipments. The air carrier's own investigations or security department people should also be notified so that they can participate in the investigation and maintain the company's liaison with the law enforcement agency.

Split shipments of AEV items should be prohibited. A split shipment occurs when part of a shipment goes out on one flight and part on another. If the destination station is expecting one shipment of five boxes and receives only three of the boxes, the station cannot know without tracing whether they have a split shipment and the other two boxes will be along on a later flight or whether the two boxes have been stolen. While split shipments are difficult to eliminate entirely when dealing with regular air freight, they should not be permitted with AEV items.

Shuttling air cargo by truck is a common practice in the airline industry, generally occurring between a large metropolitan city and smaller communities nearby or between several large airports in one general area, such as Kennedy, LaGuardia, and Newark in the New York metropolitan area. Any shuttling of AEV items creates an additional and unnecessary risk of theft and should be avoided; such items should be shipped into the station specified on the airbill, rather than into a general metropolitan area.

Upon arrival at the destination, AEV items should be immediately located and placed in the high-value security cage. They should never be allowed to sit in carts or containers until a breakdown crew is ready to open the container.

Security people disagree on the subject of marking high-value items. Some believe that high-value items should be clearly marked so they will be obvious and visible to all workers and thus will be sure to receive special attention. The opposite view is that high-value goods should not be conspicuously marked. Those supporting the latter view draw an analogy with the request that manufacturers not advertise their goods on their shipping cartons. The resolution of the marking dilemma is in too many cases, a poor compromise: marking the goods so that everyone knows they are high-value goods and failing to provide special security precautions. In such a case, the cargo simply becomes a well-marked target for theft. I lean toward the belief that such cargo should be marked so as to be readily identifiable. In most cases, the easiest way to mark such items is with highly visible tape.

One effective preventive measure in use by many carriers is to refuse to accept any item for shipment packaged so that it is smaller than one cubic foot in size. This eliminates the small, easily-pocketed, high-value theft, and makes the article much more difficult to conceal.

ATTRACTIVE MERCHANDISE

Considerable attention is given to the protection of AEV items, and elaborate procedures are often developed for the protection of this cargo. On the other hand, shipments regularly include other types of merchandise just as likely to be stolen. The Department of Transportation's definition of high-value cargo rightfully includes classes of items that, because of their desirability or history of frequent theft, need greater protection; yet, the procedures of most carriers call for extra precautions only for tariff-defined AEV items.

Wearing apparel, the largest volume item moved by air freight, is perhaps the most glaring example of this. This single commodity accounts for some of the heaviest claim losses sustained by air carriers and forwarders--in some cases, as much as over half of all the company's theft-related losses. Yet, just as often, no special precautions are taken to protect those shipments of wearing apparel that can be identified as being particularly theft-attractive, on the grounds that wearing apparel is shipped in such quantity that controls would slow the shipments. Air cargo management occasionally protests that these goods are not suitable for safe storage because the merchandise is packed in large cartons requiring too much space. These very practical operational problems are discussed in Chapter 27 in connection with the problems of selling a security program to management. For the present discussion, however, it is enough to say that the problem of "attractive merchandise" and the necessity for placing special security controls on it must be recognized .

Attractive merchandise can be any commodity in which the carrier is experiencing heavy losses; generally, these items are desirable, readily resold products such as stereo equipment, video cameras, electric power tools, and appliances. Risk-attracting commodities must be specified in the company procedures, and it must be agreed that these commodities will receive special handling. If possible, the controls placed upon attractive merchandise should be generally the same as those used for AEV items. This includes special advice messages to downline stations, storage in secure areas, and signature accountability out to the aircraft.

Obviously, if a major portion of the freight shipped from a station qualifies as attractive merchandise, these procedures are not feasible; tight controls cannot profitably be placed on all cargo shipped. In most cases, however, attractive merchandise is no more than about 5 percent of a station's total cargo volume.

The economics of the terminal's loss experience may require the carrier to select for protection only the top 2 or 3 theft-prone items until the problem they represent is under control. If the insurance claims department or company computer can provide the necessary information, it is even possible to concentrate on particular shippers. Using wearing apparel as an example again, not all wearing apparel is attractive merchandise; air freight thieves are usually most discriminating. Indeed, air freight managers can identify from experience which brands of wearing apparel will be targets and which will pass through the system untouched. Clothing consigned to high-fashion stores or shipped by manufacturers of high-quality garments is particularly vulnerable, and this is the merchandise upon which controls must be placed.

HIGH-VALUE SECURITY CAGE

A specific room or cage designed for the secure storage of high-value cargo should be available at each air freight terminal. The secure area should be large enough to handle all the terminal's high-value freight at any one time. If all attractive merchandise is included in this category, the room may need to be quite large. The physical security and the construction of the cage or room should be commensurate with the value of the goods that might be stored there. Some excellent guidelines are given in the Department of Transportation's Cargo Security Advisory entitled "High Value Commodity Storage."

The cage or crib should be so located as to be constantly visible to supervision; it should never occupy an isolated location. When closed-circuit television is available, cameras can be used to observe the cage. Nothing other than high-value items should be stored in high-value rooms or cages. Their use for storing such items as cleaning supplies, cases of soft drinks, and company material requires entry by people who have no legitimate need for access to high-value cargo. Damaged cartons awaiting recoopering should not be placed in this cage unless the merchandise they contain is truly valuable; another lockable room should be provided for the storage of all over, short, and damaged material.

A control log should be maintained for every high-value cage. When the shipment is placed in the cage, an entry should be made in the log noting the date, airbill number, a description of the item, the number of pieces, their weight, the identity of person placing the shipment in the cage, time placed in the cage, destination, commodity, and value. When the item is taken from the cage, the log entry should include the date, time, person removing the freight, individual to whom the freight is released, and the outbound flight the shipment is moving on. The contents of the high-value cage should be periodically inventoried by a supervisor, ideally at least once a shift.

The cage should have adequate locks, and this, of necessity, brings up the matter of key control. Only specific supervisory personnel should be authorized access to the key. Equipment on the market can record when the lock was opened and by whose key. Such an arrangement places the responsibility for the key on the key-holder, who will be less likely to loan the key to someone else or to leave it where it could be taken. Whenever a key to the high-value cage is lost or misplaced long enough for a duplicate to have been made, the cage lock should be changed immediately. Some carriers use a double-lock system that requires keys from two individuals in order to open the high-value cage. Requiring the presence of a second person is an added control because no one is in the high-value cage alone.

It is possible to gain entry to some high-value cages without leaving any evidence, by merely pressing down on the guide rail at the bottom of the sliding door to allow the door to be pulled away from the cage. In cages so constructed, it is necessary to lock both the forward and the rear edges of the door, usually with a padlock and chain on the rear edge. The chain should be pulled snug so that it holds the door closely to the frame.

Where the size of the security cage does not permit the storage of attractive merchandise, another solution must be found. One measure that has proven beneficial is to store such items in three-sided bins, which are then usually placed on the top tier of a three-tier storage rack. This eliminates ready access to the shipment and--by requiring the use of a forklift to lower the bin to the floor--reduces the opportunities for unobserved access.

Chapter 26

THE "PEOPLE FACTOR" IN CARGO THEFT

According to a Department of Transportation study, 10 percent of cargo theft of all kinds can be attributed to hijacking, 5 percent to burglary, and 85 percent to "opportunity," that is, to goods leaving the premises during working hours in the hands of persons or on vehicles authorized to be there. In the air freight industry, however, premises are generally occupied around the clock and burglary is rare to nonexistent. Hijacking is almost as infrequent and only occurs as the theft at gunpoint of a high-value shipment during transfer. Thus DOT's estimate of 85 percent for theft-by-trust is a demonstrably low figure for air cargo losses; indeed, 95 percent may be a conservative figure for the amount of thievery that is accomplished by the carrier's own employees. Quoting Pogo, "We have met the enemy and he is us." This proportion is highly significant for the security manager; it is obviously of little value to take elaborate precautions against outside intrusion if almost all of the losses are caused by dishonest employees.

PERSONNEL SCREENING

As a general rule, the higher the position within the company for which a candidate is being considered, the greater the effort expended in verifying his or her background. No applicant is hired as a vice president of a company merely on the basis of unverified entries on the company's application form, and this is as it should be. Previous employers are contacted, education is verified, and considerable expense may be incurred in determining the qualifications--and possible disqualifications--of the applicant.

Obviously, it cannot be suggested that the same in-depth screening be required for all air freight workers; far more freight agents are hired than vice presidents. Some basic verification of background and contact with prior employers should be regarded as essential, however, and these checks should be performed before employment, if possible, and certainly before the employee's probationary period expires.

An inherent part of a good personnel background investigation is a criminal records check with police departments in the communities in which the applicant has lived, if this is possible. Indeed, several proposed cargo security legislation bills have included provisions that persons with prior records of theft should not be eligible for employment to handle air cargo. On the other hand, criminal records have never been legally available to the security personnel responsible for background checks. In the past, such information has often been obtained by employing a private investigative agency for a fixed fee per name check. Investigative agency personnel, in turn, obtained the necessary information through friendly police contacts. However, indictments returned in New York some years ago against several private agencies and their police contacts deterred this practice to an extent, though there were reports that such checks--in lesser numbers--were still being obtained.

It cannot be in the public interest for freight carriers to unknowingly employ (and expose to severe temptation) persons who have been convicted of larceny, burglary, robbery, or receiving stolen good. Precedents for not employing individuals with records of conviction in this type of crime exist, but in other industries. Nationally, precedents include applicants for employment by banks. Some state and local codes require that record checks be made by the high-risk employer of, for example, armed guards and alarm technicians.

Nevertheless, in May 1975, the Department of Justice Order No. 601.75, Criminal Justice Information Systems, was issued. It prohibited the dissemination of criminal records information to unqualified recipients. Excepted were instances where records were made available in accordance with some statute, as in the precedents discussed above. As the Department of Justice has no legal control over state, county, or local criminal justice agencies or over any federal investigative agency other than the FBI and the DEA, the power to deny to criminal justice agencies those services that fall under Justice Department direction was used to provide the enforcement power the order needed. Hence, the order was said to apply to interstate, state, or local criminal justice agencies who were receiving Law Enforcement Assistance Administration (LEAA) funds (since expired) or who exchanged records with the FBI. The latter provision was the most inclusive; the FBI's National Crime Information Center (NCIC) is the nation's central file of criminal history records; it stores federal crime records and information contributed by police departments throughout the U.S. Without NCIC computer access, a department would be unable to run a national records check on a suspect. In addition, the order provided that the individual making the unauthorized disclosure could be fined up to $10,000.

One significant provision for those in security was the order's specific prohibition against disclosing of the existence or nonexistence of criminal history information "for employment purposes," thereby further ensuring that the honest worker's clean record would remain useless and that it would be as secure from exposure as if it were a long criminal history.

Happily, the Justice Department appears to have had second thoughts regarding the order. In March 1976, the order was modified to permit the disclosure of "conviction data, criminal history information...in court records of public judicial proceedings," and arrests for which prosecution is pending. As can be seen, the ground rules for backgrounding and screening are subject to change, and it is necessary that the security department stay constantly abreast of current developments. For example, decisions arising from charges of discrimination have limited the questions that can be asked on applications. An employer cannot require an applicant to list arrests (as opposed to convictions), on the grounds that arrests are more frequent among minorities than among other groups, and because an arrest is not an indication of guilt.

Most arbitrations have held that the employee cannot be discharged unless the activity can be shown to be incompatible with the position sought or held. Thus, I have known of a situation wherein an airline was not allowed to discharge a pilot who was a "flasher" and had convictions for indecent exposure, of a convicted child molester who was allowed to remain as an employee and of an air freight agent who had three convictions for burglary. None of the above circumstances were deemed sufficient grounds to terminate the employee. In the case of the pilot, the airline refused to let him fly but paid him wages as if he were an active employee. Drunk driving arrests have generally been held to disqualify an applicant whose employment would require driving a truck or company car or flying a plane. Alarm system installers in jurisdictions where a police bond (which involves a records check) is required are not disqualified by prospective employers for charges of nonsupport, through they would be for a single count of burglary or robbery. Access by security departments to applicants' records of criminal conviction does not and should not mean that any conviction will disqualify an applicant.

An important applicant check still available to the employer is the local credit bureau, and a credit history check should be performed for all individuals hired to work with air freight. Security practitioners are well aware of the risk represented by employing an individual with habitual credit problems. Although a finance company will not loan someone $200 without running a credit check, some carriers omit a credit check even though the individual they hire will be in a position to steal far more. Experience has shown that an employee with a poor credit record, sometimes with a history of garnishments, is less likely to be able to resist opportunities for theft than is an employee with a record of credit responsibility.

A number of companies specialize in conducting personnel background checks for employers, and this is generally a good approach to follow unless the carrier wants to hire the staff and train them to accomplish these checks. The cost of hiring an outside service is reasonable, and it is a good investment of company dollars. I recall one incident where an employee was taken away in handcuffs after having been discovered in a sizable embezzlement. A background check, which should have been accomplished before she was hired, was then done after the fact. It was found that during the time the applicant stated she was a student in dental school, she had actually been an inmate in the state women's prison.

PROCEDURES FOR NEW EMPLOYEES

All company rules and security procedures should be reviewed with new employees, particularly those relating to company policy and discipline in cases of theft and pilferage. The employee should then sign a prepared statement affirming knowledge and understanding of these policies and acceptance of them as a condition of employment.

PROMOTION ON SENIORITY

The present system of union seniority guarantees that the employee who stays with the company will inevitably be promoted to a position of at least minor supervisory responsibility, solely because of longevity, rather than for his or her capability for the job. Because of this, the personnel department should include in their consideration of an applicant whether this person will have the capability--given proper training--of filling a supervisory position. Unfortunately, capacity for promotion is not commonly considered, and some carriers are paying for hiring mistakes made in years past that were discovered too late for correction. The employee's potential for promotion also plays a role in theft prevention. The employee who feels there is no hope of advancement tends to have less loyalty to the company and hence may be more inclined to steal.

SUPERVISION

Although the support of top management is the first essential of an effective cargo security program, the support and the emphasis placed on security by the first-line supervisor certainly ranks a close second. It is the supervisor who must translate a cargo security program from regulations and recommendations into actual practice. The supervisor's attention to and interest in security are critical to the success of the program. Or, to put it another way, if the supervisor demonstrates indifference to security regulations and rules, there can be little hope that the terminal will be free of theft.

A problem encountered in some terminals is lack of supervision; there simply aren't enough supervisors to go around. Many terminals have no supervisors at all on weekends and during other slack periods, despite the fact that interested supervision is a strong deterrent to theft. If the supervisor is present on the floor, theft attempts will probably be delayed until he or she is out of the area, perhaps at the plane. Money saved by not paying supervisory salaries is probably lost in theft occurring because a supervisor is not present. Certainly, the supervisor's prime function is not that of a guard, but the deterrent effect of the supervisor's presence alone may more than compensate for his or her salary.

The selection of supervisors is most important; caution should be taken in the promotion of lead air freight agents to supervisory positions. Leads who have participated in theft in the past find themselves unable to enforce regulations against their former coworkers whom they now supervise, because these friends know that the new supervisor participated in thievery when an agent or lead. Indeed, the new supervisor may not have stopped these activities and may still be heavily involved in theft. Management should be aware of these possibilities and, when there is evidence of such a problem, should consider promoting a lead from another station to the vacant position.

A major problem of supervision is not knowing where employees are. This problem is particularly acute because of their characteristic freedom of movement in the airport environment. Where theft problems are particularly severe, such precautionary measures as restricting employees to the cargo terminal area and forbidding them to go to their cars during their shift may be appropriate.

Supervisor's Responsibilities

The duties of a supervisor should be outlined and detailed in a job description. Before and after writing the description, the tasks required of the supervisor should be analyzed to make certain that these employees are not spread so thin that they cannot adequately perform what is expected of them. There is a tendency to require the participation of the supervisor in any out-of-the-ordinary activity. If caution is not used, the supervisor's presence may simultaneously be required at three different aircraft and the high-value cage. When a realistic appraisal of duties demonstrates that the supervisor cannot accomplish all that is required, additional supervisory help must be obtained or some duties delegated.

It is important that supervisors make frequent inspections of the air freight terminal. These can ordinarily be accomplished as part of a normal tour of duty because they generally involve simple alertness to unusual situations. On-hand areas should be inspected to make certain no outbound freight is in the inbound storage area, whether in a bulk hold area or in the on-hand delivery area. Similarly, the outbound rollers should be checked for terminating freight to make certain no package has been purposely misdirected or temporarily cached.

Most terminals have an on-hand area where freight is held awaiting pickup. A system commonly used is to store the freight on ten rollers numbered zero through nine. Freight with an airbill having a last number of one is stored on the number one roller; if the last number of the airbill is two, the shipment is stored on the number two roller, and so on. Freight on the on-hand rollers should be inventoried at least once a day, and once each shift, if at all possible. A notation should be made of the time the inventory was taken to assist in narrowing down the possible time of a disappearance discovered later.

The terminal's cart storage area is another location that supervisors should check periodically during their shift. At the breakdown area, where freight is sorted, air freight employees get their first chance to survey incoming freight for attractive opportunities for larceny. If they find a piece of cargo they want to take, the easiest action is to put the wanted piece back into the cart, which will be taken away to the cart storage area until such time as it is needed. Unless the supposedly empty cart is checked, a stashed package can remain there awaiting pickup at a more convenient and less obvious moment.

Whenever a considerable number of flights arrive in rapid succession (usually in the afternoon), carts loaded with inbound freight may accumulate outside the terminal, waiting for the next shift's breakdown crew. At larger terminals, as many as a hundred carts may be waiting outside the terminal. The contents of these waiting carts are vulnerable to theft, particularly after dark, and they may even be waiting in an area accessible to employees of other airlines. At a minimum, the cart waiting area must be extremely well-lighted to ensure excellent visibility. An available guard should patrol this area. Closed-circuit television can be used to advantage here also; one camera with automatic daylight to low-light-level capabilities can be invaluable, particularly if it is equipped with pan and zoom for focusing on suspect activity.

The shift supervisor can be the determining factor in a successful cargo security program. Well-trained, alert supervision alone can provide the continued emphasis on loss prevention that is necessary to profitable operation. Of all the factors mentioned in this chapter that have a significant influence on reducing losses, supervision is most important.

Chapter 27

A MANAGEMENT APPROACH TO CARGO SECURITY

Historically, air carriers have derived approximately 90 percent of their revenue from passenger traffic and 10 percent from the carriage of cargo, including mail. Logically enough, attention has been divided between passenger security and cargo security in much the same proportion. However, the rapid growth of air cargo and its greater potential for avoidable loss call for considerably more than 10 percent of a carrier's security effort to hold its losses within acceptable limits.

The one ingredient absolutely essential to the success of any loss reduction program is the demonstrated interest and support of top management. This is the *sine qua non* of cargo security. Where the manager controls budget and policy, he or she may recognize the problem and that is all that is needed, but the security manager and lower-echelon management look to policy-level management for leadership. How is this support to be obtained?

SELLING CARGO SECURITY TO MANAGEMENT

Most cargo or freight managers will agree to the implementation of as many security controls on as much merchandise as may be suggested--as long as the proposed controls can be provided within existing staffing allocations. Unfortunately, each new control, each new requirement for accountability, and each demand for escort service or for needed supervision requires extra manpower. Only if top management is willing to authorize the additional personnel can these measures be effected. The security manager will meet considerable resistance in persuading an already short-handed air freight manager to place additional controls on freight if his or her management will not authorize the hiring of additional personnel.

Upon the security department rests the responsibility for selling its loss-prevention program to policy-level management by providing factual data. Security should be prepared to tell management what controls are needed, their cost in equipment and labor, and the probable savings to the company. Other factors should also be included, such as the certain consequences of failing to act. It is not surprising that management may be less than enthusiastic about a security program costing $60,000 to correct a $50,000 theft problem, unless they are made aware of the fact that tolerating an annual theft loss of $50,000 will inevitably result in steady and substantial loss increases; theft never stays at the same level.

To sell a program to management, some compromise may be needed. If management listens--as indeed they should--to their people in air freight marketing and terminal operations, they may hear strong objections to the proposed program. However, if security can retain the major thrust of its program (which should be in the area of adequate supervision and cargo accountability, then a major breakthrough has been made. Once the people in the field have had experience with the program and have found it workable, they will be more ready to accept the additional measures and refinements which were not approved as part of the initial program.

SELLING BY THE NUMBERS

Management is usually most receptive to those figures that reflect profit or loss. Comparing claims paid against net profit, rather than the traditional (and far less persuasive) comparison of claims versus total revenue, can be of material assistance in selling a cargo security program to management.

If a cargo security program is expected to save the company $100,000 a year in claims and claim-related costs, its value can best be assessed in terms of how much revenue the company would have to generate to earn $100,000. At a 5 percent profit rate, an additional $2 million in cargo revenue would be needed to achieve the profit that could be realized by a claims cost reduction of $100,000. When these facts are applied to larger carriers, who annually pay out as much as $500,000 in theft, pilferage, and shortage claims, the potential for carrier profit and revenue are even more striking.

Loss Ratio

The primary means of measuring the effectiveness of cargo security within an airline is through the loss ratio, which equals the total legal liability claims paid divided by the cargo revenue. A loss ratio of 1.5 means that, for every $100 in cargo revenue, the carrier has paid out $1.50 in cargo claims. In the seventies it was fairly standard that the carrier's loss ratio would be somewhere between 1.0 and 1.5. This figure dropped below 1.0 in both 1975 and 1976. Since then it has progressively declined until it reached .30 in 1989. Anything over 1.0 indicates that immediate added attention should be given to cargo losses. The recent industry figures appear to confirm that carriers with genuinely effective cargo security programs should experience ratios of considerably less than .50.

The loss ratio includes factors other than theft, pilferage, and nondelivery; claims paid for delay, rough handling, damage to to water and extreme temperatures, and other factors are involved. In evaluating the overall loss ratio, it is helpful to break these figures down into their individual categories in order to determine the percentage of loss due to theft and pilferage, for example.

Loss ratios can be computed on the basis of point-of-sale revenue or upon the poundage of freight handled by the station. The latter method more accurately reflects the loss experience of a station where a high percentage of traffic involves transfer. The poundage, or pounds-on/pounds-off method, utilizes the system's average revenue per pound times the total pounds handled on and off at the given station. Claims paid are divided by the resulting figure. This method should have no effect on the system loss ratio, but it will have an effect on local station ratios.

Loss Allocation

Allocating the responsibility for a loss is often difficult, and it is frequently necessary to divide the loss arbitrarily between the shipping and the receiving stations. Occasionally, however, a highly relevant breakdown in procedures may be evident at one of the stations, and in such cases, that station should be allocated the loss.

AVOIDING THE LOSS OPPORTUNITY

Carriers receive the same rate per pound for hauling an automobile transmission as they do for carrying diamonds, gold, silver, or platinum, even though the cost of the additional security necessary for the safe transport of the latter can make its handling unprofitable for the carrier. Because carriers are not free to refuse freight or to impose a rate differential to meet security costs, carrier marketing departments should be advised not to actively solicit these low-weight, high-value, high-loss commodities; under present freight rates, such "sales" may constitute losing propositions of major proportions. In time, federal regulations may be changed to allow carriers to charge more realistic rates for handling items that have a high incidence of theft, such as watches and wearing apparel, but this change does not appear to be imminent.

RETAINING MANAGEMENT SUPPORT

Having analyzed the problem and having recognized that the problem does exist, the next order of business is deciding what to do about it. Although no simple solution exists, it does appear that there is a logical way to proceed and that certain factors can be identified as essential to any loss prevention program. For example, it is undeniable that air freight theft cannot be solved by piecemeal efforts. Nevertheless, management all too often takes no action on the information developed by a completed theft investigation. The investigation should accomplish more than identifying the perpetrator; it should determine how the loss occurred and recommend ways to prevent recurrences. If anything of lasting practical value is to be gained from the investigation investment, management must respond to these recommendations. Every investigator has heard air freight managers say, in one form or another, "Don't bother me with your security problems. Just pay the claims and move the freight." Fortunately, such managers are no longer a majority.

Security professionals learned long ago that interest on the part of top management can never be taken for granted. Security's programs must be repeatedly sold to top management, and the security professional must be a top-notch salesperson who is well informed on company problems and policies. The security manager must gain the attention of top management, convince them that there is indeed a problem requiring a solution, and sell a solution to that problem.

As compared with other company programs, security proposals historically have been hard to sell because of the difficulty of relating the proposed program to a specific profit. Usually, the security manager can only show that the plans and procedures suggested are well designed to reduce known losses in specific areas and that their implementation is compatible with the company's cost and operational considerations. Once approval has been gained for a program, factual arguments should be gathered by maintaining statistics to document the effectiveness of the measures instituted under each program.

Chapter 28

CARGO THEFT INVESTIGATION

SIX SERVING MEN

Occasionally security practitioners face problems that seem almost insurmountable because they don't know how or where to begin the attack. They sometimes fall victim to the tendency to see something as more difficult than it really is. I will always be grateful that I learned, early in my investigative career, the Rudyard Kipling verse[1] that so neatly summarizes the questions that must be answered in an investigation:

> I kept six honest serving men
> (They taught me all I know):
> Their names are What and Why and When
> And How and Where and Who.

The "insurmountable" problem of air cargo security becomes more vulnerable to security management if some of these questions can be answered.

WHO IS STEALING AIR CARGO?

First, the question of*who* is committing the thefts under investigation must be considered. This question has been answered in general terms by the Department of Transportation study cited earlier, findings that are confirmed

---

[1] Rudyard Kipling, *The Elephant's Child*, Stanza 1.

by experience: 85 percent or more of our theft losses are accomplished during working hours by employees, by authorized drivers, or by both. Therefore, investigators are not looking for burglars, and they shouldn't need to be looking for unauthorized visitors.

WHERE ARE THE THEFTS TAKING PLACE?

The next question--"*Where* are they stealing?"--is not always separable from "*What* is being stolen?" A most important factor in this determination is the proper utilization of the resources available. For this question--or these related questions--the most effective tool is an analysis of claims reports (which, for the most carriers, can be done from computer runs).

At one airline, cargo shipments originate at any one of 100 or more stations. An analysis of claims loss statistics revealed that more than 80 percent of the losses were occurring at 13 stations, while the remaining 90 or so stations accounted for less than 20 percent of the losses. Similarly, other statistics showed that 10 percent of baggage pilferage losses on the system were occurring at one station. Clearly, sound management required that primary attention be directed toward those stations where the major portion of their losses were occurring. This priority does not neglect other stations; once the causes and means of loss at these key stations have been identified and countered, the information gained is easily adapted and applied to other stations.

To attack the problem of cargo theft without such an analysis results in a "shotgun approach" that might, with luck, hit the target. Even when such an approach does isolate and clear up some thefts, later claims loss analyses may reveal that the time and labor invested was thoroughly out of proportion with the amount of loss.

Included in the category of "where" must be all the points at which theft can occur, a question requiring personal knowledge of those locations at the airport that could be used as transfer points, drop points, and so forth. This introduces the question of *how* the items are being taken. For this, as the song from *The Music Man* puts it, "Ya gotta know the territory!" Unless one is willing to get out and become familiar with that terminal's operations and employees, it's difficult--if not impossible--to come up with the answer to "how."

HOW IS IT BEING STOLEN? (INFORMATION AND INTELLIGENCE)

Much depends upon the security practitioners being in the terminal often enough and being regarded by the terminal workers as trustworthy enough to encourage confidences. In any investigative activity, the value of inside information is almost incalculable. Compared to the time and cost of mounting, staffing, and sustaining an investigation until suspect areas and activities can be identified, "information received" is a major economy.

Developing inside information needn't require bringing in an undercover operative. Where the individual responsible for security has taken the time to get to know the freight handlers and to be at the terminal on a daily basis, such information may be volunteered in confidence. The identity of individuals committing thefts is often common knowledge among employees, not all of whom regard theft as acceptable conduct. Employees may feel a sincere desire to see theft stopped, and--if they believe that their information and identity will be kept confidential--they will often make an opportunity to pass the information on. Once an informant's identity becomes known, however, there will be no more volunteers. Hence, any disciplinary action or criminal complaint for theft must be based on instances observed in the course of a subsequent investigation.

Not all employee sources are moved by a simple desire to see theft checked. Occasionally, employees will indicate that they are open to exchanging information for cash. Extreme caution should be used in such cases. Repercussions from other employees can be particularly severe if such an arrangement becomes known and can even endanger the informant.

When cash for information offers are accepted, experienced investigators make it clear to the source that payment terms are strictly "cash on delivery." An alternative--and probably more satisfactory--arrangement for handling the profit-motivated informant is to put the individual in touch with the appropriate law enforcement agency, usually the FBI or U.S. Customs. Not only do these agencies often have funds available for good information, but this releases the company from responsibility for an operation that, if done by company security personnel, carriers a potential for creating serious employee relations problems.

Somewhere between the altruistic informant and the information-for-cash source falls the company reward program for information leading to the arrest and conviction of cargo thieves. Little factual information appears to be available on the effectiveness of this alternative. Although I have talked with officials of companies that have reward programs, I have found none who was enthusiastic about them. A complete study of the problems with reward programs, including their effect on morale, the union response (if any), and their effectiveness, would be a significant contribution to the security field.

WHAT IS BEING STOLEN?

How important is it to know what is being stolen? As we noted in Chapter 25, although carriers have traditionally placed security emphasis on AEV and high-value shipments, analyses of claims statistics reveal that carriers experience heavy losses in commodities not included in those categories. One airline found that the information obtained by claims analysis was so significant that it developed a new category called "attractive merchandise" and revised its cargo security regulations to provide added protection for these goods. It is important to concentrate on those stations suffering the highest losses, but it is equally essential to place added controls upon those commodities most frequently lost to theft. Knowing what is being stolen can also assist a specific investigation by ruling out one or more of the common methods of theft.

WHY ARE EMPLOYEES STEALING?

The answer to why cargo is stolen is harder to determine. Some cargo is taken because it is tempting to the employee; he or she sees the electric power tool, the video camera or the stereo equipment and takes it for private use. Next, he or she may begin taking things for family and friends. Later, the employee may come to realize the possibilities of selling merchandise to acquaintances, and the theft increases accordingly. Eventually, he or she will begin stealing "on order" for a friend or acquaintance who wants a bargain in a specific item such as a camera or watch. From this point, it is a short step to being put in contact with a "fence" who can offer an outlet for the stolen merchandise. The outlet may be a bar or even a meat market. (The public seems oddly willing to purchase, say, a color television set from such an unlikely purveyor as a butcher.) Soon, the employee is stealing in large quantities and most probably acting in collusion with at least one driver.

WHEN DO THEFTS OCCUR?

The last of Kipling's six serving men is *when*. Again, procedural controls are essential if there is to be any determination of when freight was stolen. For a productive investigation, the security manager needs to know when the carrier had possession of the freight and when it was found to be missing. Once the span of time in which the theft occurred can be narrowed to a particular shift, patterns and trends should be examined to see if the same employee's name appears with significant frequency. An analysis of theft losses will enable an investigator to determine if the goods are disappearing on weekends, on any particular day, or on a particular shift, or if the thefts follow any other significant pattern. Once the pattern is established, surveillance can be employed to identify the employees involved.

SECURITY SURVEYS

Considerable insight as to how thefts are occurring can also be gained from security surveys of freight terminals. If the terminal is typical and most of the operational activity occurs during the nighttime hours, a good portion of the survey should be conducted at night. One such survey quickly spotted the highly relevant fact that many of the cartons that were disappearing from the terminal were too large to fit into a car's trunk; hence the cartons had to be going out across the dock and onto a truck.

Surveys reveal factors that relate to how thefts occur. Most surveys confirm that thefts are a product of breakdowns in established procedures. That procedures exist is almost immaterial; if they aren't followed, whether by intention or through neglect, follow-up action is essential. The reason why procedures are breaking down should be determined, and the individuals responsible should be identified. To accomplish this, major losses must be investigated by the carrier's security organization; no law enforcement investigation produces this information.

The resulting internal investigative report should identify the individuals who failed to follow procedures, and management should make certain that these employees receive a letter in their file that ties their negligence to the loss. After an individual has received several such letters, management may want to reassess the importance of this person to the company's operations, or--at the least--to consider transferring the employee to a different job without access to air freight. All too often, employees escape responsibility for causing a loss merely by saying "I forgot" or "I was too busy." Accepting such excuses only invites further theft. Whether the forgetting was intentional does not matter; the habitual violator of procedures is virtually as much of a risk as a thief is. At the same time, such reports may be an indication that a supervisor has responsibility for too many activities, and that another assessment of employee practices should be made after this problem has been met.

ASSIGNING RESPONSIBILITY

What is management's view of all this? Usually, it is that the dishonest employee is a worthless thief who is taking advantage of conscientious management. But is management blameless? Cargo security is successful only when management provides as temptation-free an environment as possible. Management does indeed have an obligation to make it difficult for employees to steal. When management not only leaves attractive merchandise in view of the employee but also makes it easy for them to take merchandise without detection because of a lack of supervision or procedural controls, then management has its own substantial share of responsibility for the loss. It is too much to expect the average employee to remain honest in the face of temptation coupled with an evident lack of concern that advertises a low probability of being caught. The employee won't be caught because the negligent carrier won't know the item is missing until it is too late to pinpoint where the loss occurred--perhaps not until an attempt is made to deliver the shipment to the consignee. Without a clue as to the point of loss, there can be no investigation; the insurance department will merely pay another claim. As long as this apathy on the part of management continues, thefts will increase drastically--and it will be management, as much as the employee, who will be responsible.

Chapter 29

CONTROL BY AUDIT AND SURVEY

Although management textbooks may differ in terminology, all agree that any program or delegation of responsibility must include an audit or control function, an element that is most simply described as "follow-up." The pride of planning and organizing a quality cargo security program fades rapidly if a lack of follow-up allows the program to dissolve due to inattention.

Every program must provide for effective follow-up action, and this audit operation should be built into the initial program. In its most detailed form, the follow-up function can be a list of questions that an auditor is to ask in the course of an inspection and a list of specific areas the inspection should cover. Preferably, program audits should be conducted by persons familiar with air cargo operations because then the inspector will be capable of offering positive suggestions for improving operations and checking for compliance with specific procedures.

At present, air freight terminals are inspected by representatives from the Federal Aviation Administration, the U.S. Customs Service (where in-bond custody is involved), and the carrier's insurance underwriter, as well as by employees of the carrier who are responsible for loss reduction. Managers of air freight terminals have occasionally been less than enthusiastic about advice received from individuals who they felt were not experienced in air cargo operations or did not understand special local problems.

The program audit function is best performed by the carrier's own security department. Even then, those performing the audit should be acquainted with air cargo operations, if at all possible. If this is not the case, however, it has been my experience that most terminal managers will be more than cooperative in providing the investigator with the opportunity to "learn the business." This entails a good deal more than an afternoon's tour of the cargo terminal. To become familiar with every facet of that station's operation may require days or weeks spent at the terminal. An additional opportunity to learn operational procedures is provided by each loss investigation. Once an individual with a background in security and loss prevention becomes knowledgeable in air cargo operations, he or she will be in a position to make significant contributions toward the reduction of air cargo losses.

The following guidelines for the security inspection of air freight terminals may be used for audit and inspection purposes. The list is not all-inclusive; rather, it offers general guidelines that direct the investigator into known problem areas. In addition to using the inspection guidelines, flow charts should be developed for use in isolating weaknesses in the system. An analysis of these charts may lead the investigator to recommend eliminating certain functions, combining others, or even changing the flow of cargo or documents to achieve a more efficient and secure operation.

GUIDELINES FOR SECURITY INSPECTION OF AIR FREIGHT TERMINALS

General

1. Where is the facility located (with respect to security advantage or disadvantage, remoteness, etc.)?
2. What are the hours of operation? What are the peak hours?
3. If a facility is not open 24 hours a day, what security is provided during unattended hours?
4. What is the status of the general housekeeping? Is the terminal well laid out and organized or is there congestion?
5. How many flights per day does the facility service? What is the size of the station? How much cargo is moved?
6. Is there a history of theft problems? What has the claim ratio been?
7. Talk to the air freight manager or knowledgeable supervisors about the station. They may have information regarding specific problem areas (and recommendations for solving the problem), who the thieves are and how goods are being taken out, specific merchandise vulnerable at that station, etc. Use this information as input, but not necessarily as the final word on the matter; managers have an understandable tendency to want their station to look good on an inspection.
8. Is the facility shared with other tenants?

Physical Security

1. Inspect the fencing. Is the entire perimeter area fenced? What is the height of the fence? Is there barbed wire on top? (The fence should be at least 9-gauge, two-inch mesh, eight- to ten-feet high,

with a two-foot barbed-wire top guard at a 45° angle.) Ensure that objects cannot be passed beneath the fencing.

2. Is the area well lighted? Is the total perimeter lighted? Are the cargo areas outside the terminal lighted, including the area for holding incoming freight carts and the storage area for empty carts? (Outside lights should be a minimum of 400 watts per fixture, located 30 feet above ground level.) Is lighting protected from vandalism by wire screens, covers, or extra-strength glass?
3. Is employee parking effectively isolated from the terminal? Do employees pass a guard post to get to their cars?
4. Are cars parked adjacent to the perimeter fence? (There should be at least a 20- to 30-foot clear zone on both sides of the fence.)
5. If the building is not open 24 hours a day, are alarms needed?
6. Can cargo doors be opened from the outside? How are they secured?
7. Is closed-circuit television used, and if so, is it being used effectively? Where are the cameras? Where are the monitors? Who is watching the monitors? Has anyone ever been caught through the use of the CCTV equipment? If it is not being employed as yet at the facility, would its use be helpful? Where and why?
8. Are guards present? Are they carrier employees or contract guards? How many work each shift? Where are they stationed? What is their function? Have they been given any training in air freight documentation and procedures? Do they make any patrols? What information do the guards record on commercial vehicles entering the premises? Do guards ever accompany high-value shipments?
9. Is a badge system in effect? Are badges being worn?

Shipping and Receiving Operations

1. Are all pieces of all shipments counted when receiving or delivering freight at the dock? What action is taken if the company count disagrees with that of the trucker? Consider spot-checking a load or two and conducting your own count. Do supervisors spot-check loads (a protection against collusion between employee and trucker)? Are any shipments weighed?
2. Are loading dock doors shut when a bay is not in use? (This is not always practical in the daytime but is a necessity at night.) Are wire mesh screens available for securing bays in warm weather?
3. Are dock lights being used to illuminate the interior of trucks being loaded or unloaded?
4. Is access to the terminal by truckers and customers restricted? Are outsiders allowed to use rest rooms or canteen facilities within the restricted areas? If so, are they escorted?
5. It is recommended that a barrier or a four-inch-wide yellow line painted on the floor be located about ten feet from the edge of the

dock. Has this been done? Outsiders should not pass this point without the permission of the terminal manager or other designated person. Is this restriction being observed?

6. Is any cargo left in the accessible area of the dock? If so, is it left unattended?
7. Are cargo-handling areas and loading zones prominently posted as restricted areas? Are FBI posters ("Theft from interstate shipment is a federal Crime") displayed?
8. Is each trucker or customer making a pickup properly identified to ensure it is a legitimate pickup? What identification and documents are required? Has there been a problem with fraudulent pickups? Are document-recording cameras needed?
9. Is identification required from unknown shippers?

Terminal Operations

1. If the terminal has an area where arriving freight is held until it can be moved into the breakdown area, what security is provided for the waiting carts? Who has access to the holding area? Is the lighting adequate? Is there an application for CCTV here?
2. Inspect the empty-cart storage area and the carts in it. Is there cargo in any carts? Is lighting adequate? Do supervisors regularly check these carts to make certain they are empty?
3. Are special measures taken to protect attractive merchandise, such as wearing apparel, not specifically designated as high-value items? Is a special area set aside, or fenced off for it, or is a separate room provided?
4. Follow the route taken by the cart trains between the freight terminal and the passenger terminal. Does it include travel on public streets? Along parking lots? Is it consistently well-lighted?
5. Is cargo in the carts properly secured and covered so it will not fall off or bounce out?
6. What handling is given to shipments of company material (comat)? Compared to the handling given regular air freight are there any differences? Is there a history of problems with comat?
7. How is company material delivered to consignees? Do company employees other than those assigned to work there have access to the freight terminal? Do other employees, such as Stores, attempt to go through carts at the ramp before freight is brought to the freight terminal, or at the freight terminal after it arrives, to pick out company material or freight consigned to them? They should not have this type of access to air freight.
8. Are employees from other airlines restricted in any manner on the premises? Do employees compare manifests with the actual freight being transferred? Check the preparation of inspection reports for incomplete or damaged shipments being transferred to your airline from another carrier.

9. What handling, storage, receipt, and dispatch is given to U.S. mail?
10. What security is afforded cargo at the ramp while awaiting disposition?
11. If a shuttle truck is used between area airports, is freight manifested?
12. Are open carts used? (They should be replaced, for protection against weather, theft, and pilferage, with totally enclosed carts that have chained gates and canvas curtains on one side.)
13. Is there a program in effect for checking employee lockers?
14. Inspect AC-12 Pick Up and Delivery Manifests. Are exceptions being taken? Are they specific, noting damage, type of contents, and count (necessary to reduce exposure to theft and pilferage)?
15. How are incomplete shipments handled? Is a part lot file maintained? Is there an established program for tracing incomplete or split shipments, or is tracing done only when brought to the company's attention? This is particularly important with items highly vulnerable to theft; are these items given priority follow-up action?
16. Review procedures for handling in-bond shipments. Are in-bond storage facilities adequate from space and security standpoints?
17. Do supervisory personnel perform random inspections of the freight terminal to identify mishandled, misplaced, or stashed shipments? Are the inspections performed by a company guard? By an air freight agent?
18. Review cash handling procedures, particularly at the customer desk. Is the front desk left unattended when agent goes to look for freight? Where is the cash kept? What are cash limits? What is the frequency of deposit? Is there an armored car pickup? How are deposits to safe made? Are deposits witnessed? Who has the safe combination? Is the combination changed periodically as well as when an employee who knows the combination is transferred, reassigned, or terminated? Is the safe vulnerable to theft by rolling it away? Is the safe bolted to the floor?
19. Review trash-handling procedures to ensure that cargo is not being removed by this means.

High Value Cargo: Articles of Extraordinary Value (AEV)

1. Tariffs provide that carriers will handle as AEV all shipments of specific commodities listed in the tariffs that are valued at over $5000. Some carriers set a company floor of $1000 for items requiring special security. What is the qualifying amount at this facility?
2. Does the station have a high-value room? How secure is it? What type of key control is there?
3. Is a log maintained for signing freight in and out of the high value room? Does each entry logged include the airbill number, shipper, consignee, time, and description of the item?

4. Are inventories taken of high-value items on hand? How often? By whom?
5. What control is exercised over an item in transit between the high-value room and the aircraft? Are additional signatures required? Who must be present when a high-value item is placed on an aircraft? Runner? Supervisor?
6. What is the procedure for meeting a high-value shipment? Where is the shipment actually retrieved: at the aircraft or at the terminal? Who takes the item to the high-value room? If the item comes in a container, is it handled differently? Does the item ever sit in the container until the breakdown crew gets to that container?
7. How are furs handled? Are they sent as a separate shipment? Do they have their own airbill? Is the piece count verified? Are they kept in a locked security room? Is a pelt-handling log maintained?
8. Is special identifying tape used for AEV items?
9. Are advance arrangements made in accordance with CAB 96 Tariff? (For items valued at more than $5000 advance arrangements are to be made not less than 24 hours before tender of shipment. The shipper must tender the shipment not less than three hours before the flight.)
10. How is the shipment delivered to the consignee on arrival? What efforts are made to ensure these shipments are picked up immediately? (Tariffs provide that under most circumstances, if the consignee does not pick up a shipment within three hours after notification of arrival or the actual arrival of shipment, whichever is later, the carrier has authority to hire an armed guard at the expense of the shipper and the consignee or to hire an armed vehicle to deliver the shipment to the consignee.)
11. Compare the physical security of the station's high-value cage against the DOT Cargo Security Advisory Standards. (Walls should extend to ceiling or be at least 10 feet high and the top or roof should be resistant to forced entry.)
12. What action is taken if the high-value shipment specified in an advice message fails to arrive? (Air Freight should immediately phone the upline or downline station for an explanation.)
13. Are non-AEV items stored inside the high-value cage? (Materials other than high-value items should not be stored in high-value rooms or cages. Storage in the high-value cage or room of such items as cleaning supplies, cases of soda, and company material requires access by too many people not authorized to enter and makes inventories more difficult.)
14. Ramp transfer of high-value shipments is generally prohibited; is there any evidence of such ramp transfers? (AEV shipments should be brought back to the air freight terminal for storage if there will be any significant time between planes.)
15. Are AEV items shipped on a separate airbill or separate transfer manifest on interline shipments?

Damaged Shipments

1. What is the procedure used in handling damaged shipments tendered by shipper or forwarder? Does the procedure includc actual inspection, count, and weighing? Any exception noted?
2. What is the procedure when a trucker wants to pick up a shipment that includes damaged or broken cartons? Are the cartons signed off that way? Are items in a broken box counted and the box recoopered?
3. Is a camera available at the station during all shifts for photographing damaged shipments in case of future claims? Is a camera used regularly and routinely?
4. How are damaged goods handled? Are they locked up until they can be re-inventoried and recoopered? What records are kept? Who does the inventory and recoopering? How accessible are recoopering tools?
5. Do specific shippers consistently send poorly packaged merchandise? What, if anything, has been done about this?
6. Is there adequate protection for freight during inclement weather so as to preclude claims due to rain, snow, etc.? Are there covers on the carts? Is inside storage available?
7. Are Freight Claim and Inspection Reports (AC-18) accurately completed and immediately forwarded to designated addressees?

Law Enforcement Liaison

1. What local law enforcement agency has jurisdiction? Do officers visit the terminal or patrol the area?
2. What assistance and cooperation does the terminal receive from the FBI? From Customs? From the FAA?

Security Awareness

1. What type of supervisory and employee training is available on cargo loss prevention?
2. Are security posters of any type in evidence at the terminal? Are they in good condition? When were they last changed?

Appendix

SIGNIFICANT INCIDENTS

May 31, 1972. Lod Airport, Tel Aviv

Three Japanese passengers had debarked with 116 other passengers from an Air France Flight arriving at Tel Aviv from Paris and Rome. They picked up two valises from the conveyor belt, unzipped them, and grabbed grenades and machine guns. They began firing and lobbing the grenades indiscriminately at a crowd of about 300 in the waiting room. They killed 25 and wounded 76. Their primary target was supposed to be passengers arriving on an El Al flight about 10 minutes after they arrived. They claimed to be members of the Army of the Red Star and said the attack was in reprisal for the killing of two Arab guerrillas during an aborted hijacking of a Belgian plane at Lod Airport on May 9.

September 5, 1972. Olympic Games, Munich

In the early morning hours of September 5, 1972, Arab commandos broke into Olympic Village at Munich and began a 23-hour ordeal that resulted in the death of 17 people. The commandos stormed the Israeli compound, killing two members of the Israeli Olympic team, who stalled the commandos long enough for six of their comrades to escape. The remaining nine athletes were captured and eventually taken by helicopter to an airport 15 miles west of Munich, where a jet was going to fly them to Cairo. The nine Israelis and five of their captors, plus a West German Police Officer, were killed in the gun battle that ensued. Three of the commandos were caught and later released by the Germans in response to Arabs who hijacked a plane. The attack in Munich was made to secure the release of 200 Arab commandos imprisoned in Israel.

February 22, 1974. Samuel Joseph Byck. Delta Flight 523, Baltimore

Samuel Joseph Byck, armed with a handgun, approached a gate screening position at Baltimore where Delta Flight 523 was being processed. Without provocation, he drew his weapon and murdered the law enforcement officer at the screening checkpoint. He then boarded the aircraft, killed the copilot, and wounded the captain. He stated that he intended to crash the airliner into the White House. Byck was shot by police officers before the aircraft left the gate, and he ultimately committed suicide on the plane.

December 21, 1975. OPEC Meeting, Vienna

Six Palestinian terrorists shot their way into a meeting of the Organization of Petroleum Exporting Countries in Vienna and seized 81 hostages. They flew to Algiers, then to Tripoli, and finally surrendered in Algiers two days later. Three were killed, 7 wounded.

January 5, 1976. Japan Air Line, Manila

A Japan Air Lines DC-8 was hijacked at Manila by two Filipinos armed with guns and dynamite, who wanted to go to Tokyo but couldn't afford the price of a ticket. One of the only instances I know of where a plane was hijacked to get to a destination that was already a scheduled stop.

May 21, 1976. Philippine Airlines, Zamboanga, Philippines

A Philippine Airlines plane was hijacked by six armed men who wanted to be flown to Libya, shortly after its takeoff from Mindanao Island. The plane landed at Zamboanga where, two days later, in a battle between hijackers and ground forces, the hijackers detonated grenades aboard the plane. Thirteen were killed, 22 injured.

June 27, 1976. Air France, Entebbe, Uganda

An Air France flight from Tel Aviv to Athens and Paris was hijacked by armed terrorists out of Athens. They demanded the release of 53 terrorists in custody in five countries. They released 150 to 250 hostages. On July 3, Israeli commandos successfully carried out a spectacular rescue operation. Three of the remaining hostages were killed during the attack. Thirty individuals were killed and 42 were injured during the entire incident.

September 10, 1976. TWA Flight 355, New York. Hijacking by Croatian Nationals.

Hijacked a 727 scheduled to fly from New York to Chicago. Four men and one woman, all Croatian nationals, flew over London to Paris where they dropped leaflets. Hijackers had also planted a bomb in a locker at Grand Central Station, New York, which killed one policeman and injured three others. Passengers and crew were released unharmed.

This hijacking was interesting because it reflected some interesting facets of the "Stockholm Syndrome." The hostages were released at intervals, and their attitudes toward the hijackers varied but were not based upon the amount of time they were in custody. Those who had significant negative contact and those who had little contact with the hijackers advocated maximum penalties for them. Those who had positive contact with the hijackers and who discussed their case and their motivation for the hijacking with them became very sympathetic toward them. Some used vacation time to attend the trial, some began a defense fund for the hijackers, and others recommended defense counsel to them and refused to cooperate with law enforcement officers.

December 2, 1976. Seizure of Dutch Train by South Moluccans

This incident, which involved the hijacking of a train at Assen, the Netherlands, and the detention of the passengers for 12 days, became known as the "Dutch Train Incident." The hijacking was accomplished by seven South Moluccan youths. This was a classic example of a politically powerless group using terrorism effectively to dramatize and publicize their cause.

The South Moluccan Islands are 100 miles west of New Guinea and were once known as the Spice Islands. A group of South Moluccans who had emigrated to Holland in 1950 now wanted to return to an independent republic. Soon after the train was taken, another group of seven South Moluccans invaded the Indonesian consulate in Amsterdam, seizing more than 32 hostages, including 17 children. The final toll was three dead hostages and one wounded terrorist on the train, and one fatality at the consulate when a hostage died after jumping from a window. The hijackers were each sentenced to fourteen years.

March 9, 1977. Seizure of Buildings in Washington, D.C., by Hanafi Muslims

A dozen Hanafi Muslims seized and held 134 hostages in three buildings in Washington, D.C for 40 hours. The buildings held were the International Headquarters for B'nai B'rith, the Islamic Center, and the District Building (Washington's City Hall). The siege arose out of a dispute between the

Hanafi Muslims and the Black Muslims. One bystander was killed and a dozen hostages were injured.

March 26, 1977. Bomb at Airport, Las Palmas. Tenerife Crash

A bomb placed by the leader of the Canary Island Independence Movement exploded in a flower shop in the terminal at Gondo Airport, Las Palmas, Canary Islands. Aircraft had to be temporarily diverted to nearby airports, including Santa Cruz, Tenerife, where a Pan Am 747 and a KLM 747 collided on the ground, killing 580 persons, in the world's worst airline disaster.

May 8, 1977. Northwest Orient Flight 22, Tokyo to Honolulu

A mentally unstable American held a razor to the throat of a stewardess and demanded to be flown to Moscow. This hijacking is unusual only in that it ended when the male flight attendant hit the hijacker in the head with a fire axe.

May 23, 1977. Seizure of Train and School by South Moluccans .

Nine South Moluccan terrorists seized an express train near Groningen, the Netherlands and a school in Bovensmilde. Fifty-six hostages were taken on the train and 110 were held in the school. The terrorists demanded the release of 21 South Moluccans from Dutch prisons. Troops stormed the train and the school on June 11. Six terrorists and two hostages were killed.

October 13, 1977. Lufthansa Flight 181, Mogadishu, Somalia

During the 110-hour hijacking of Flight 181 from Majorca, the West German government refused to bow to the demands of the hijackers. In a daring raid on October 18 reminiscent of Entebbe, West German commandos, Grenzschutzgruppe 9 (G.S.G.9), freed 82 passengers and four crew members and killed three of the four hijackers. These terrorists were sadists who flaunted their cruelty. They were led by a man who called himself "Captain Mahmud" and executed the pilot, Captain Juergen Schumann. The Palestinian hijackers claimed solidarity with West German terrorists who had kidnapped industrialist Hanns-Martin Schleyer on September 5. Within two days of the rescue mission, Schleyer's body was found in the trunk of a car in France. About an hour after the news broke of the rescue, Andreas Baader, one of the founders of the Baader-Meinhof Gang (Red Army Faction) committed suicide in his jail cell, as did two of his fellow terrorists.

March 9, 1978. China Airlines, Taiwan to Hong Kong, Hijacking by Crew Member.

I include this incident only because it is illustrative of several instances where a plane has been hijacked by a crew member or airline employee. On this China Airlines flight, the flight engineer attacked the pilot and copilot with a hammer and scissors. He was killed by a cabin guard.

August 11, 1982. Explosion Aboard Pan Am Flight 830, Honolulu

An explosion in the passenger cabin of Pan Am Flight 830 occurred just a few minutes before the plane landed at Honolulu. It blew a hole in the floor of the aircraft but only caused the fuselage skin to bulge outward. One passenger was killed, 20 injured.

An eight-year investigation into this bombing ultimately implicated the Iraqi regime of Saddam Hussein. The bomb is believed to have been planted by a Palestinian named Mohammed Rashid, a top operative of the May 15 Organization. He is currently awaiting trial in Athens under the 1971 Montreal Convention which permits those charged with attacks on airliners to stand trial in the country holding them. Rashid is also a prime suspect in the April 1986 bombing of TWA Flight 840 from Rome to Athens[1]

August 25, 1982. Explosive Device Found on Pan Am Flight 441, Rio de Janeiro

A similar type bomb to that which exploded on Flight 830, and planted by the same group, was found in the passenger cabin. The bomb had been on the plane for 14 days and 40,000 miles but had not exploded because the bombers inadvertently broke off the safety pin, leaving the tip stuck in the bomb.[2] It was safely removed.

July 15, 1983. Orly Airport, Paris

A suitcase bomb exploded at a crowded Turkish Airlines counter, killing seven and injuring 56 others. A member of the Armenian Secret Army for the Liberation of Armenia (ASALA) was arrested two days later and admitted planting the bomb.

---

[1] Jay Peterzell, "The Life and Crimes of Middle East Terrorist," *Time,* 14 Jan 1991.

[2] Ibid.

June 14, 1985. TWA Flight 847, Athens to Rome.

TWA Flight 847, enroute from Athens to Rome with 145 passengers, was hijacked by two terrorists and flown to Beirut. A third terrorist had been unable to get a seat and was not allowed to board. The terrorists demanded the release of over 776 Shi'ite prisoners held in Israeli jails. At the first stop in Beirut, 19 women and children were released. Then, the plane began to shuttle from Beirut to Algiers to Beirut to Algiers and back to Beirut, all between Friday, June 14, and Sunday, June 16. The names of Captain John Testrake, First Officer Phillip Maresca, Flight Engineer Benjamin Zimmerman, Flight Attendant Uli Derickson, and murdered Navy diver Robert Stethem became familiar to all TV viewers. At the first Algiers stop, which lasted five hours, 21 more passengers were released. Then, the plane flew back to Beirut.

It was there that U.S. Navy diver Robert Stethem, age 23, was beaten, murdered and his body thrown out onto the tarmac. About a dozen additional hijackers were brought on board and six or eight passengers with Jewish sounding names were removed. Then the plane flew back to Algiers. The hijackers released three hostages on arrival and after negotiations with Algerian officials, they released 58 others. Then, the plane went to Beirut for the third time. Finally, 39 of the original 153 passengers and crew remained. Those hostages were taken to locations in Beirut while the three man crew was kept on board. Seventeen days after the hijacking began, the hostages were released.

On January 15, 1987, Mohammed Ali Hamadi, age 22, was arrested at Frankfurt airport trying to enter West Germany with liquid explosives concealed in liquor bottles. He was charged with the hijacking and has subsequently admitted his role, denying however, the killing of Stethem. On May 17, 1989, Hamadi was convicted by the Hesse State Supreme Court and sentenced to life imprisonment for the hijacking, the murder of Robert Stethem, and the possession of explosives.

June 22, 1985. Air India Explosion. Montreal to London

An Air India jetliner, traveling from Montreal to London with 329 people aboard, exploded over the Atlantic Ocean near Ireland. Less than one hour before the crash, a bomb exploded at Tokyo's International Airport in baggage from a Canadian Pacific Flight from Vancouver, B.C. Two Sikh fugitives wanted in connection with an alleged plot to assassinate Indian Prime Minister Rajiv Gandhi were suspected.

July 1, 1985. Explosion. Leonardo da Vinci Airport, Rome

A bomb exploded in a suitcase being moved between flights at Leonardo da Vinci International Airport, wounding 15 airport employees. Kashmir Liberation Organization claimed responsibility.

July 1, 1985. Explosion, British Airways Office, Madrid

A British Airways office in downtown Madrid was bombed, killing one persons and injuring 27. The Organization of the Oppressed claimed responsibility. The explosions in Madrid and Rome on July 1, followed bombings on June 19 at Frankfurt Airport and June 23 at Narita Airport, Tokyo.

October 7, 1985, Achille Lauro

The Italian cruise liner, the *Achille Lauro*, enroute from Alexandria to Port Said with 123 passengers and 315 crew aboard, was hijacked by Palestinian gunmen, apparently in retribution for the 1500-mile Israeli bombing raid on Arafat's P.L.O. headquarters near Tunis, two weeks before. During the 52-hour hijacking, American passenger Leon Klinghoffer, a stroke victim confined to a wheelchair, was shot in the forehead, and his body was thrown overboard.

The hijacking ended when President Hosni Mubarak of Egypt, in cooperation with Arafat's P.L.O., promised the hijackers safe passage out of Egypt in exchange for their surrender. The hijackers left on an EgyptAir 737, a full day after Mubarak claimed that they were no longer in his country. Fortunately, American intelligence sources were aware of what was occurring. A flight of four F-14 Tomcat fighters from the aircraft carrier *Saratoga,* intercepted the plane that was enroute to Tunis and forced it to land in Sicily, where the Italians took the four hijackers into custody. Italian Prime Minister Bettino Craxi refused a request from President Ronald Reagan to extradite the terrorists to the U.S. because the crime occurred on an Italian ship.

Much to the displeasure of the U.S., Italy did release two PLO representatives who accompanied the hijackers on the plane. One of them was Mohammed Abul Abbas Zaidan, better known as Abul Abbas, head of the Tunis-based faction of the Palestine Liberation Front (PLF).

December 27, 1985, Rome and Vienna Airports

Airport attacks at Rome and Vienna occurred within five minutes of each other on the morning of December 27, 1985. The first attack occurred at Rome's Leonardo da Vinci Airport where four men carrying 13 hand grenades and four AK-47 rifles, sprayed the terminal with gunfire, taking special aim at passengers waiting to Check in for an El Al flight and others at nearby Pan Am and TWA counters. The toll was 15 people dead, including three of the terrorists, and 74 wounded. One of the Rome terrorists was only 15 years old, and one of the dead was a 12-year-old American girl.

A few minutes later, an attack with AK-47s and grenades began at Vienna's Schwechat Airport. The toll there was three dead, including one of the terrorists, and 47 wounded. The Abu Nidal organization claimed responsibility for both massacres. The Vienna terrorists had evidently hoped to take a number of people at the airport hostage, commandeer an El Al plane, have it flown to Tel Aviv where they would meet a similar plane flown from Rome, and then blow up both planes with everyone aboard. This was to be in retaliation for Israeli's bombing of PLO headquarters in Tunis the previous October, where 70 people were killed and more than 100 were wounded.

From a security standpoint, it must be noted that three days before the airport attacks, airlines around the world were warned that groups of Libyan-backed terrorists were being sent to Europe for actions planned during the holidays.

April 2, 1986. TWA Flight 840, Rome to Athens, Explosion

TWA Flight 840, enroute from Rome to Athens, experienced an inflight explosion as it descended through 15,000 feet over Greece, preparatory to landing at Athens. The explosion tore a 9 x 4 foot hole in the side of the fuselage and killed four Americans, including an eight-month old girl. A bomb had been placed under seat 10F. Police centered their investigation on a Lebanese woman who had flown from Beirut to Cairo on March 25 and from Cairo to Athens on the same plane on the morning of April 2 and had sat in seat 10F. She had arrived late for the flight and had been driven to the plane in an airline car rather than on a bus. Airport officials said she was thoroughly searched even though she arrived late.

The Izzeddin Qassam unit of the Arab Revolutionary Cells, which in turn is linked to Palestinian renegade Abu Nidal, claimed responsibility. They said it was in retaliation for U.S. missile attacks on Libyan targets the previous month during the showdown over the right of foreign ships to use the waters of the Gulf of Sidra.

April 17, 1986. El Al, London

A woman boarding an El Al flight in London was found to have a bomb in her carry-on luggage. Her boyfriend, Nezar Hindawi, was convicted of the attempt. Evidence at the trial implicated Syrian intelligence, who provided the bomb. Britain broke off diplomatic relations with Syria over the incident.

May 3, 1986, Air Lanka, Colombo, Sri Lanka

A bomb blew the tail section away from an Air Lanka jet preparing to depart from Colombo Airport. Sixteen were killed and 41 were injured.

June 26, 1986, Barajas Airport, Madrid

A suitcase bomb partially detonated at the airport, wounding 11. The bomb was intended to explode on an El Al flight to Tel Aviv.

September 5, 1986. Pan Am Flight 73, Karachi, Pakistan

Four Arab males, dressed as security people and driving a vehicle similar to a security vehicle with a flashing amber light, were allowed to drive to a 747 loading passengers for Frankfurt to join those already enroute from Bombay. They stormed the aircraft with weapons, firing numerous shots in the process.

One of the controversies surrounding this incident was that the cockpit crew became aware what was occuring and immediately escaped, leaving the hijackers with 13 flight service personnel and over 300 passengers as hostages. Some felt the crew should have stayed. However, most authorities strongly supported their action.

Early into the hijacking, a businessman from California was shot and his body was dropped onto the tarmac. The climax of the 16-hour hijacking occurred when the planes auxiliary power unit, which powers the lights and air conditioning, failed. Not realizing what had caused the lights to go out, the hijackers apparently panicked. They began spraying the aircraft with automatic weapons fire and detonating grenades into the center area of the aircraft where all the hostages had been gathered. At least 22 were killed and more 100 were injured. Pakistani commandos did not respond until after all but the very seriously wounded had departed the aircraft. Intelligence indicated that the Abu Nidal group was responsible.

As a postscript, the two security guards who allowed the terrorists onto the field were sentenced to one year in prison for their negligence.

November 29, 1987. Korean Air Lines Flight 858

Flight 858 disappeared over the Andaman Sea off Burma, killing all 115 people on board. Two North Korean intelligence agents were arrested in Bahrain on December 1. The man committed suicide by taking a cyanide pill. The woman tried, but failed. She subsequently confessed that she and her companion had planted a bomb on board the airliner when they boarded it in Baghdad. The two later disembarked in Abu Dhabi. The woman, Kim Hyon Hui, also stated that the operation had been planned by North Korean intelligence officials as part of a plan to destabilize the South Korean government and disrupt the 1988 Olympic Games which were scheduled to be held in Seoul in September and October. Indeed, she advised that she was recruited by Kim Jong II, son of North Korea's dictator, Kim II Sung.[3] Kim Hyon Hui was sentenced to death, but her sentence was commuted. She now lives in a government house with freedom of movement. When interviewed in 1990 she said she still feels guilty and punishes herself by not buying all the dresses she wants.

December 7, 1987. PSA Flight 1771, Los Angeles to San Francisco.

David Burke was fired from USAir on November 19, 1987, after being caught on camera stealing $69 in liquor receipts. The Friday before the incident, Burke had held his girlfriend and her six-year-old daughter at gunpoint on a forced six-hour drive. Burke boarded Flight 1771 after bypassing screening as a familiar airline employee, using airline I.D., which had not been turned in when he was terminated. Over Paso Robles, California, he shot his target, Raymond Thomson, the USAir Customer Service Manager in Los Angeles. He then shot the cockpit crew, causing the aircraft to crash. All 43 people on board were killed.

April 5, 1988. Kuwait Airways Flight 422, Bangkok to Kuwait

The hijacked Kuwaiti Airways Flight 422, bound from Bangkok to Kuwait, which was forced to land at Mashhad, Iran. On April 8, it arrived in Larnaca, Cyprus, where two passengers were killed. On April 13, the plane arrived in Algiers, where the 15-day hijacking ended peacefully. It gained the distinction of being the longest uninterrupted hijacking. The seizure of TWA Flight 847 in 1985 lasted 17 days, but the hostages were taken from the aircraft and hidden in Beirut after the first few days.

The hijackers' demand was to free 17 terrorists jailed by Kuwait since 1983 for the bombings of the U.S. and French embassies in Kuwait. The hijacking was alleged to have been masterminded by Imad Mughniyen, a brother-in-law to one of the 17 hostages, and who is also believed to have organized the hijacking of TWA Flight 847. The hijackers are believed to belong to Hizballah, a radical Shi'ite group loyal to the Ayatullah Khomeini. The Algerian Government had promised the hijackers safe passage, and the Algerian Interior Minister, el-Hadi Khediri, refused to disclose details of their whereabouts.

December 21, 1988. Pan Am Flight 103, Lockerbie, Scotland

Pan Am Flight 103, which originated at Frankfurt and was enroute to New York, exploded in the air over Lockerbie, Scotland, about one hour out of London. The explosion killed 258 passengers and crew, and another 11

---

[3] *U.S. News and World Report*, (January 25, 1988):12

people were killed on the ground. A double detonator device was used; the first trigger was activated by barometric pressure and in turn activated a timing device. This type of device is believed to have been designed to thwart the airline practice of putting cargo through a pressure chamber to activate barometric triggers.

This incident aroused particular controversy because of an anonymous call received at the U.S. Embassy in Helsinki on December 5, 1988, advising that sometime within the next two weeks a bombing would be attempted against a Pan Am flight from Frankfurt to New York. The information was given to American diplomatic missions in Europe and the Middle East and to Pan Am but not to the general public. (See Chapter 4 for complete discussion).

September 19, 1989. UTA Flight 772, Niger

UTA Flight 772, on the second leg of a Brazzaville, Congo/N'Djamena, Chad/Paris flight, exploded over Niger killing all 171 people on board. Two groups have claimed responsibility: Hizballah and a previously unheard of Chadian group opposed to French support for Chadian President Hissen Habre's government.

November 27, 1989, Avianca Flight 203. Bogota

A bomb placed under a seat exploded, killing 107. A new word, "narcoterrorism," is born.

INDEX